POLYMERIZATION
PROCESS
MODELING

ADVANCES IN INTERFACIAL ENGINEERING SERIES

Microstructures constitute the building blocks of the interfacial systems upon which many vital industries depend. These systems share a fundamental knowledge base—the molecular interactions that occur at the boundary between two materials.

Where microstructures dominate, the manufacturing process becomes the product. At the Center for Interfacial Engineering, a National Science Foundation Research Center, researchers are working together to develop the control over molecular behavior needed to manufacture reproducible and reliable interfacial products.

The books in this series represent an intellectual collaboration rooted in the disciplines of modern engineering, chemistry, and physics that incorporates the expertise of industrial managers as well as engineers and scientists. They are designed to make the most recent information available to the students and professionals in the field who will be responsible for future optimization of interfacial processing technologies.

Other Titles in the Series

POLYMERIZATION PROCESS MODELING

Neil A. Dotson
Rafael Galván
Robert L. Laurence
Matthew Tirrell

VCH

Neil A. Dotson
Eastman Chemical Company
Kingsport, TN 37662

Rafael Galván
Dow Chemical Company
Midland, MI 48640

Robert L. Laurence
Department of Chemical Engineering and Polymer Science
and Engineering
University of Massachusetts
Amherst, MA 01003

Matthew Tirrell
Department of Chemical Engineering and Materials Science
University of Minnesota
Minneapolis, MN 55455

This book is printed on acid-free paper. ⊗

Library of Congress Cataloging-in-Publication Data

Polymerization process modeling / by Neil A. Dotson ... [et al.].
    p.   cm.—(Advances in interfacial engineering series)
    Includes bibliographical references and index.
    ISBN 1-56081-693-7 (alk. paper)
    1. Polymerization.   I. Dotson, Neil A., 1962–   . II. Series.
TP156.P6P6187   1995
660'.28448—dc20   95-14939
                                                                 CIP

Printed in the United States of America

ISBN 1-56081-693-7 VCH Publishers, Inc.

Printing History:
10 9 8 7 6 5 4 3 2 1

Published jointly by

VCH Publishers, Inc.	VCH Verlagsgesellschaft mbH	VCH Publishers (UK) Ltd.
220 East 23rd Street	P.O. Box 10 11 61	8 Wellington Court
New York, New York	69451 Weinheim, Germany	Cambridge CB1 1HZ
10010		United Kingdom

PREFACE

Polymerization is one of the fundamental paths—along with self-assembly, association, crystallization, and other physical routes—to building microstructured organic materials. In contrast to the other processes, polymerization involves covalent bond formation, building up large molecules which, in turn, impart a range of unique and desirable properties to the polymeric materials thus made. In conjunction with self-assembly, crystallization, and other physical processes, polymerization can be used to lend permanence and mechanical robustness to engineered microstructures. We emphasize the term "microstructure" because this book is about how to model the chemical microstructure of macromolecules produced in polymerization reactors. The details of polymeric microstructure determine the properties of the product, often in multiphase systems—systems with interfaces. Specification of the desired characteristics of a polymeric material is a more complex endeavor than for any other kind of chemical product, because of the variety and complexity of microstructure.

Patricia Layman, writing in *Chemical and Engineering News* (October 31, 1994), reports on some industrial perspectives for the future on polymer manufacturing. Prominent on the list of new processes are approaches "to tailor a polyethylene molecule more accurately." The language itself reflects the significantly different kind of challenge that a polymerization engineer faces; no one talks about "tailoring" ethylene. Its polymer, however, can be tailored, and among the means to do such are new catalysts, new comonomers, and the modification of molecular weight distribution. Fundamentally, each of these new approaches requires the engineered adjustment of the polymeric microstructure—in this case, of molecular weight distribution or copolymer attributes—to obtain better performance in processing or end-use application. Tools of both the polymer scientist and the engineer can be used to these ends; new synthetic chemistry and new reactor configurations and control schemes each play a role. Ultimately, however, for the engineer to have any power to predict the effects that these innovations will have on the complex aspects of polymer product quality, the relationship between *formation* (which involves both chemistry and reactor environment) and *structure* must be understood.

This book lays out, in a comprehensible manner, the tools available to do this for a wide range of situations, emphasizing fundamental methods and approaches. This is not an advanced, state-of-the-art compendium or review book for the practicing polymerization engineer, though some in this situation might well find the material useful. The aim, rather, is to educate and inform those who are either engineers entering the field (say as chemical or mechanical engineering students) or chemists, engaged in or collaborating on a polymerization reaction engineering project. We view this book as a bridge between the *chemistry of polymerization*, which focuses on mechanism and kinetics, and *reactor design* and *control*, where the predictive power of the engineer must be implemented. George Odian's *Principles of Polymerization* is an excellent source dedicated to polymerization kinetics and mechanism; *Control of Polymerization Reactors,* by Joe Schork et al., aims at the implementation phase of reactor design. This book, we feel, fills a need between these two. The motivation for analysis of the development of microstructure (particularly molecular weight distribution) is given in the first chapter. In subsequent chapters, we bring out the particular features of different categories of polymerizations which are industrially significant. We try to instill, in the text and in worked examples, an understanding of how the various tools of analysis work and when to employ them. More importantly, through the development and solution of models of polymerization processes, we try to bring out, in a way we have not seen brought together in a single volume, a firm understanding of how various types of polymerization work—that is, what they develop in the way of product distributions of molecular weight, composition, sequence, and so on. The book is reasonably self-standing in that not very much polymer science is assumed, though a basic polymer science course would be a very useful predecessor or prerequisite to a course derived from this book. On the other hand, it is expected that the reader is familiar with chemical kinetics and basic notions of mass and energy balances.

The material contained in this volume has been used as the basis for courses in polymerization modeling at the University of Massachusetts and the University of Minnesota (and elsewhere) numerous times over the last decade or more. The material can be covered in a ten-week, quarter-length course directed at engineering seniors and graduate students. Matt Tirrell, Rafael Galván, and Neil Dotson have each instructed such a course at Minnesota. A fuller development of this material can be accomplished in a fifteen-week, semester-length course, as has been done by Bob Laurence at Massachusetts. The text contains tried and true problems for homework or for the self-teacher.

Assistance from the Center for Interfacial Engineering at the University of Minnesota was essential in bringing this book to its final form; Hertha Schulze, Mark Swanson, and Susan Mehle were of particular help. Before the final stages, though, we received generous financial contributions from E. I. duPont de Nemours Co., Inc., EXXON Corporation, and S. C. Johnson and Sons, all of which were vital in keeping the work on this book moving. Colleagues from these companies and others also provided useful feedback on the contents of this book, as did those who peer-reviewed the book.

Matt Tirrell was the beneficiary of the Shell Oil Company Distinguished Chair in Chemical Engineering at Minnesota and the Olaf A. Hougen Visiting Professorship in Chemical Engineering at the University of Wisconsin, which provided both material and moral support, as well as time, that was applied to portions of this work. Neil Dotson acknowledges Dr. M. E. Ekart, Dr. R. L. Christiansen and especially S. C. Myers of Eastman Chemical Company for time in reviewing the final manuscript and encouragement. Colleagues at Minnesota were very helpful at various stages of the project, particularly Chris Macosko and Gus Aris, as were the many students who were taught from the various unrefined versions of this text over the years, who offered corrections and made suggestions. There are many other colleagues who offered help and encouragement during this work, for which we are grateful—and who will remain unnamed to avoid offense to those whose names escape us.

Neil A. Dotson
Kingsport, TN

Rafael Galván
Midland, MI

Robert L. Laurence
Amherst, MA

Matthew Tirrell
Minneapolis, MN

September 1995

CONTENTS

NOMENCLATURE

The tables that follow list nomenclature for variables and acronyms sufficiently common to be present in the text without local definition (i.e., definition within the section or subsection in which an item appears). The text contains more variables and acronyms than are listed here, but all others are defined locally or used only locally (e.g., in the analysis of a particular problem).

Variable	Chapters	Meaning
A	2, 5	concentration of A-type functional groups
A_p	7	total surface area of polymer particles
a_s	7	specific surface area of surfactant
A_0	2, 5	initial concentration of A-type functional groups
$(A_2)_0$	2	initial concentration of A_2 monomers
B	2	concentration of B-type functional groups
B_0	2	initial concentration of B-type functional groups
$(B_2)_0$	2	initial concentration of B_2 monomers
C_m	5, 7	transfer to monomer constant, $k_{tr,m}/k_p$
C_p	5	transfer to polymer constant, $k_{tr,p}/k_p$
C_s	5, 6	transfer to agent constant, $k_{tr,s}/k_p$
d	3, 6	parameter in free-radical polymerization, defined in Chapter 3
Da	6	Damköhler number of the first kind, being the ratio of rate of reaction to rate of exit from reactor
DP_n	1–7	number-average degree of polymerization, μ_1/μ_0
DP_w	1–6	weight-average degree of polymerization, μ_2/μ_1
DP_n^{inst}	3	instantaneous number-average degree of polymerization of dead chains
DP_w^{cumu}	3, 5	cumulative weight-average degree of polymerization of dead chains
DP_w^{inst}	3, 5	instantaneous weight-average degree of polymerization of dead chains
DP_z	1, 6	z-average degree of polymerization, μ_3/μ_2
DP_{z+1}	1, 6	(z + 1)-average degree of polymerization, μ_4/μ_3
$E(N_A^{in})$	2, 5	expected number of monomers looking into an A group
$E(N_A^{out})$	2, 5	expected number of monomers looking out of an A group
$E(W_A^{out})$	2, 5	expected weight looking out of an A group
$E(W_B^{out})$	2, 5	expected weight looking out of a B group
f	3–6	initiator efficiency factor
	5, 6	functionality of multifunctional monomer
f_i	4, 6	mole fraction of monomer i in the comonomer mixture
$f_{i,0}$	4	initial mole fraction of monomer i in the comonomer mixture
$f(\theta')d\theta'$	6	residence time distribution
F_i	4, 6	instantaneous mole fraction of monomer i incorporated into copolymer

Variable	Chapters	Meaning
$G(s)$	2–7	moment generating function; $G(1)$ equals μ_0
$G_A(s)$	2	in $A_2 + B_2$ polymerization, the moment generating function of chains with A_2 groups on both ends; $G_A(1)$ equals the concentration of such chains
$G_B(s)$	2	in $A_2 + B_2$ polymerization, the moment generating function of chains with B_2 groups on both ends; $G_B(1)$ equals the concentration of such chains
$G_M(s)$	2	in $A_2 + B_2$ polymerization, the moment generating function of chains with an A_2 group on one end and a B_2 group on the other; $G_M(1)$ equals the concentration of such chains
$H(s)$	2–6	moment generating function of intermediate chains or polymeric by-product (cyclic species); $H(1) = \lambda_0$ (or R)
I	3–7	initiator concentration
I_0	3, 4, 6	initial initiator concentration
k	2, 5, 6	polycondensation rate constant
	3, 6	propagation rate constant for ionic polymerization
k_d	3–6	initiator decomposition rate constant
k_{ij}	4, 6	the rate constant for propagation of a radical of type i with a monomer of type j (terminal model)
k_p	3, 5–7	propagation rate constant
k_t	3, 6, 7	overall termination rate constant
k_{tc}	3, 5	rate constant for termination by combination
k_{td}	3, 5	rate constant for termination by disproportionation
k_{tij}	4, 6	termination rate constant between i- and j-type radicals (similarly defined for k_{tc} and k_{td})
$k_{tr,x}$	3, 5	rate constant for transfer to agent X
M	3–6	monomer concentration
M_{Af}	2, 5	mass of monomer A_f ($f = 2$ in Chapter 2)
M_{Bg}	2, 5	mass of monomer B_g ($g = 2$ in Chapter 2)
M_i	4	concentration of monomer i
$(M_i)_0$	4	initial concentration of monomer i
M_n	1, 2, 4, 5	number-average molecular weight
M_w	1, 2, 4, 5	weight-average molecular weight
M_z	1	z-average molecular weight
M_0	3, 5, 6	initial monomer concentration
$\bar{n}$	7	average number of radicals per particle
N	7	number of polymer particles
$(N_i)_n$	4	number-average sequence length for monomer i
$(N_i)_n^{cumu}$	4	cumulative number-average sequence length for monomer i
$(N_i)_n^{inst}$	4	instantaneous number-average sequence length for monomer i
$(N_i)_w$	4	weight-average sequence length for monomer i
p	2, 5, 6	conversion of (limiting) functional groups in stepwise polymerization
	3–6	conversion of monomers in chainwise (co)polymerization
P	2	molar concentration of polymer
p_c	5	critical conversion for gelation
$P(F_i^{out})$	5	probability of a finite structure looking out of monomer i
p_i	4	conversion of monomer i in free-radical copolymerization

Variable	Chapters	Meaning
P_i	1–6	molar concentration of chains of length i
$P_{i,j}$	2, 4	molar concentration of chains comprising i A_2 units and j B_2 units (for $A_2 + B_2$ step polymerization), or dead chains comprising i M_1 units and j M_2 units (for chain copolymerization)
$P(X_{i,f})$	5	probability of an f-functional crosslinker having i arms to the network
P_0	2, 6	initial molar concentration of polymer
$P_{1,0}$	5	initial molar concentration of A_f monomers
	6	concentration of monomer in stream 0
q	3, 5–7	probability of propagation
Q	1–3, 5–7	polydispersity, DP_w/DP_n (generally)
	6	volumetric flow rate for continuous reactors
r	2, 4, 5	stoichiometric ratio of mutually reactive groups, $A_0/B_0 \leq 1$ generally; $r = A_0/(B_0 + C_0) \leq 1$ for $A_2 + B_2 + C_2$ polymerization
R	3, 6	molar concentration of intermediate (active) chains (e.g., radicals)
	1, 5, 7	ideal gas law constant
r_i	4	the ratio of homo- to cross-propagation rate constants (k_{ii}/k_{ij}) for binary copolymerization (terminal model)
R_i	3, 5	molar concentration of intermediate (e.g., active) chains of length i
R_p	3, 6, 7	rate of chainwise polymerization, or equivalently of monomer consumption (generally free radical)
s	2, 3, 5–7	dummy variable for the generating function $G(s)$, defined within the unit circle on the complex plane
S	3, 5	concentration of transfer agent
	7	surfactant concentration
$S_{i,k}$	4	sequence length distribution; number or concentration of sequences of length k of monomer type i
$S_{i,k}^{cumu}$	4	cumulative sequence length distribution; number of sequences of length k of monomer type i in accumulated polymer
t	2–7	time
T_g	1, 7	glass transition temperature
T_m	1, 7	crystalline melting temperature
V	6	reactor volume
W	2, 3	normalization constant for molecular weight distribution
w_{Af}	2, 5	weight fraction of A_f monomers ($f = 2$ in Chapter 2)
W_i	1–4	molecular weight distribution
$w_{pendant}$	5	weight fraction of material with only one arm to the network
w_{sol}	5	weight fraction of soluble material
x	5	mass conversion of monomer to polymer in long-chain branching polymerizations
α	5	branching probability
	5	elongation ratio
θ	6, 7	residence time; average residence time $= V/Q$
λ_k	3, 5	kth moment of the intermediate chain length distribution $H(s)$
μ	7	volumetric growth rate of polymer particles

Variable	Chapters	Meaning
μ_k	1–6	kth moment of the chain length distribution $G(s)$
ν	3, 7	kinetic chain length; the number of monomers a radical adds to during its lifetime
ξ	3, 5, 6	fraction of chain-ending steps which are by combination
ρ_a	7	rate of absorption of radicals into a polymer particle
ρ_i	7	rate of initiation of radicals absorbed into polymer particles
τ	2	rescaled time, $d\tau = k\,dt$ (for AB step polymerization, or AB + XB step polymerization); $d\tau = 2k\,dt$ (for A$_2$ + B$_2$ step polymerization)
	3, 5, 6	rescaled time, eigenzeit variable, $d\tau = kM\,dt$
υ_2	5	volume fraction of polymer (species 2) in a swelling experiment
ϕ_m	7	volume fraction of monomer in swollen polymer particle
χ	5, 7	Flory–Huggins polymer–solvent interaction parameter

Acronym	Chapters	Meaning
ABS	1, 6, 7	acrylonitrile–butadiene–styrene resin
cmc	7	critical micelle concentration
CSTR	7	continuous stirred tank reactor
GPC	1	gel permeation chromatography
HCSTR	6	homogeneous continuous stirred tank reactor
HIPS	1, 7	high impact polystyrene
IR	1, 7	infrared (spectroscopy)
LFTR	6	laminar flow tubular reactor
NMR	1	nuclear magnetic resonance (spectroscopy)
PFR	6, 7	plug flow reactor
PVC	1, 6	poly(vinyl chloride)
SBR	1, 7	styrene–butadiene rubber
SCSTR	6	segregated continuous stirred tank reactor

1

DISTINCTIVE FEATURES OF POLYMERS AND POLYMERIZATION REACTORS

1.1 Significance and Distinctions of Polymerization Reactions

Why should we devote a text to aspects of reactors for polymerization in particular? Certainly, polymers are industrially significant. The plastics, synthetic fibers, rubber, and coatings (such as paints and adhesives) found in the materials we use from day to day confirm this. The diversity of these polymers is known from common usage: "nylons," "vinyl," "rubber," "polyesters," and "acrylics." An idea of the scale of production of these common materials is given by the annual production figures for the United States as listed in Tables 1.1 through 1.4.[1] In this country alone, approximately 30 million metric tons of polymers were produced in 1992. With such large production figures comes the need for large employment in the area.

Scale, important as it is, does not necessarily warrant separate attention to polymerization; if it did, we should first write about sulfuric acid, which exceeds polymers in annual tonnage. Polymerization reactors are worthy of separate study, not merely because of sheer volume, but because here the engineer encounters problems that fall beyond the realm of ordinary chemical reactor design and analysis. Polymerization differs qualitatively from reactions producing small molecules in two broad ways. The first of these is best seen by contrasting the production of styrene with the production of polystyrene.

Styrene is made from ethyl benzene, which in turn is made from benzene in one of two Friedel–Crafts alkylation routes.[2] The more common of these employs an acid/alkene pair, often done with $AlCl_3$ catalysts with a small amount of HCl[3]:

1

TABLE 1.1 / 1992 Production Values for Major Plastics and Thermosets

Polymer	1992 Production ($\times 10^{-9}$ kg)
Polyethylene	9.89
High density polyethylene (HDPE)	4.45
Low density polyethylene (LDPE) and linear low density polyethylene (LLDPE)	5.44
Poly(vinyl chloride) and copolymers	4.53
Polypropylene	3.82
Polystyrene	2.29
Phenolic thermosets	1.33
Acrylonitrile–butadiene–styrene (ABS) plastics	1.18
Polyester thermoplastics	1.09
Urethane thermosets[a]	0.94
Urea thermosets	0.70
Polyester thermosets, unsaturated	0.53
Polyamides (nylon)	0.30
Epoxy thermosets	0.21
Melamine thermosets	0.11
Styrene–acrylonitrile copolymers	0.05
Other vinyl plastics	0.09

[a]1990 production value.[1b]
Source: Reference 1a.

$$\text{C}_6\text{H}_6 + \text{H}_2\text{C}=\text{CH}_2 \xrightarrow[160\text{-}180°\text{C}]{\text{AlCl}_3} \text{C}_6\text{H}_5\text{-CH}_2\text{CH}_3 \qquad (1.1.1)$$

The resulting ethyl benzene is then subjected to a dehydrogenation reaction to produce styrene:

$$\text{C}_6\text{H}_5\text{-CH}_2\text{CH}_3 \xrightarrow[580\text{-}660°\text{C}]{\text{CrO}_3 \cdot \text{Al}_2\text{O}_3} \text{C}_6\text{H}_5\text{-CH}=\text{CH}_2 + \text{H}_2 \qquad (1.1.2)$$

In the production of styrene or any other small molecule, the engineer should optimize rate, minimize hazardous wastes, and so forth, as would be the case if the product were a polymer. With regard to the product itself, though, the engineer is primarily

TABLE 1.2 / 1992 Production Values for Major Synthetic Fibers

Polymer	1992 Production ($\times 10^{-9}$ kg)
Polyester	1.62
Nylon	1.16
Olefin	0.90
Cellulosics (acetate and rayon)	0.22
Acrylic	0.20

Source: Reference 1a.

TABLE 1.3 / 1992 Production Values for Synthetic Rubber

Polymer	1992 Production ($\times 10^{-9}$ kg)
Styrene–butadiene rubber (SBR)	0.80
Polybutadiene	0.46
Ethylene–propylene rubber (EPR)	0.21
Nitrile rubber (NR)	0.07
Polychloroprene	0.07
Other	0.32

Source: Reference 1a.

concerned with questions of composition: Is the yield as high as desired? What contaminants are present, and at what concentrations? Properties of the product, insofar as purity is not concerned, are not optimized, because simple compounds such as styrene have defined physical properties (such as viscosity), which are not alterable.

This is not the case for polymers; different polystyrenes can have different properties—for example, different viscosities. The polymerization reaction engineer must not only meet specified rates, yields, and purities, but also obtain a product of certain processing characteristics and end-use properties that are, in practice, the true measures of the performance of a polymerization reactor. Differences in properties imply that the structures of the materials are not the same, and this is possible because polystyrene, unlike styrene, is not a singular product. Even if nearly identical in chemical composition, polymers may differ in a number of other ways. This is the first distinction: polymers have *numerous structural characteristics,* which determine their properties.

The definition of a *polymer* ("many units") provides us with the clearest example of such a characteristic. Styrene may be polymerized by certain chain reactions (see Chapter 3), so that the double bonds are opened and the monomers linked into chains, shown in common shorthand as follows:

$$
\left[\begin{array}{c} \underset{H}{\overset{H}{C}} - \underset{}{\overset{H}{C}} \\ \bigcirc \end{array}\right]_i
$$

(1.1.3)

TABLE 1.4 / 1992 Production Values for Paints and Coatings

Polymer	1992 Production ($\times 10^{-9}$ L)
Architectural	2.12
Product[a]	1.37
Special purpose	0.70

[a]Preliminary 1991 data.
Source: Reference 1a.

The index i indicates how many of these monomers are linked together in a given polymer chain; the nature of the end groups, although important, is neglected here. What is the value of i? (How many units are there in a chain?) Do all molecules, or polymer chains, have the same value of i? If not, what are the concentrations of chains of different lengths i? Thus, the term "polymer" immediately implies a diversity in average *chain length* and in *chain length distribution* that has no analogue in small molecules. Details of isomerism, morphology, and composition (if a "copolymer") only add to the number of structural characteristics. One can thus reasonably ask for polystyrenes of different viscosities, viscosity being largely determined by the average chain length and its distribution.

Ideally, then, the engineer should produce a polymer meeting many specified structural characteristics. Therefore, it is worthwhile to discuss both these characteristics of polymers and how to model the relation between polymer formation and structure. The engineer needs to understand why reaction engineering of polymers differs so greatly from that of small-molecule products; it is also essential to grasp the rudiments of the structural differences leading to different properties in compositionally similar materials. In practice experience often dictates what reaction conditions will lead to a polymer of acceptable properties, and the relation between structure and property may be nearly unknown. This is not the most satisfactory situation, however, and an understanding of fundamentals can assist in either rectifying resulting inadequacies or identifying gaps in knowledge.

The second distinction is that the polymeric nature of the product affects the physics underlying the *reaction kinetics* in ways unknown to small-molecule systems. The chain length i is often large, on the order of 10^2 to 10^4, meaning that molecular weights of tens or hundreds of thousands are encountered. Over the course of the polymerization, when either polymer chains are growing to these lengths or the concentration of such chains is increasing, the viscosity rises greatly, often by six orders of magnitude or more. The polymerization process becomes more difficult: mixing is impeded, as is heat removal. Polymerizations are often quite exothermic, with heats of polymerization $-\Delta H$ typically in the range of 60 to 90 kJ/mol.[4-6] Thus if no heat were removed, the temperature might, in theory, rise hundreds of degrees kelvin. Poorness of mixing only makes this problem worse. At the end of the process, the high viscosity may also impede the removal of volatiles, such as by-product or unreacted monomer. Even if these problems were solved, the characteristics of the polymer, such as the chain length distribution, might be drastically affected by goodness of mixing. These problems make scale-up a more serious challenge for polymerization reactors than for reactors producing small molecules.

The kinetics of polymerization may be complex as well, for two reasons. The first is simply the number of different reactions that occur, whether undesirable side reactions (as also occur in small-molecule reactions) or intentionally concurrent reactions (as will usually be the case for the polymerization of styrene). The kinetics of these reactions, though, are strongly influenced by the extreme physical changes in the system, such as the increase in viscosity. The combination of vast increases in viscosity and the involvement of long-chain molecules in the reactions leads to diffusional control of some reactions. Diffusion can then determine the rate of the reaction, and a direct link between the physics of polymers and the kinetics of polymerization is established.

Many other physical problems can plague polymerizations, such as the large change in volume upon reaction. We do not delve into these matters here, but rely on the few examples just cited to suggest the extent to which polymer physics affects polymerization. We will give specific examples of this second distinction at the appropriate points in the text. We must return soon to the first distinction, that of the numerous structural characteristics of polymers, inasmuch as their importance in determining properties defines the focus of this book. Before that, though, we briefly introduce the physical state of polymer solutions and melts.

1.2 Polymers in Bulk and in Solution

So far, we have left the boundary between small molecules and polymers ill-defined. In this text, we will use two operating definitions of "polymer": one regarding the structure–property relations, the other, the formation–structure relations. Certain physical properties, such as viscosity, vary over the whole range of molecular weights; but others, such as the melting point and other thermal transitions, rapidly approach a constant value with increasing chain length. It is convenient to define a polymer as a molecule with a chain length sufficient for these properties to be fairly constant; generally this corresponds to molecules of mass exceeding a few thousand grams per mole. The useful physical properties of polymers derive from this large size, in terms of both mass and spatial extent, as we shall soon see. This definition, still somewhat loose, is appropriate for the relations between structure and physical properties to be discussed in this section. When examining the relation between formation and structure, we will find it conceptually and mathematically easier to define all reactive species, including monomer and species of intermediate length, *oligomers,* as polymer. For the present, we will maintain the first definition (> ~ 1000 g/mol); for final products the difference between the two will often be unimportant.

The simplest polymers are covalently bonded strings of atoms and have *large spatial extent.* The *connectivity* granted by the bonds allows interactions between distant regions tens to thousands of ångstroms away. This range of interaction, orders of magnitude greater than that for small molecules, can propagate effects (e.g., stress and structure) from one region to another, affecting thermodynamic phase behavior, fluid dynamics, and the ability of the polymers to associate with other molecules in solution, among other things. Given the large spatial extent and inability of polymer chains to pass through one another (uncrossability), the polymers will certainly be intertwined, and will become *entangled* as the molecular weight and concentration in solution increase. Thus the dynamic properties, related to diffusion and to fluid flow (*rheological* properties) of sufficiently dense fluids containing long molecules exhibit unusual features, which we will touch on in later sections.[7,8]

The architecture or configuration of polymers may be more intricate than the simple linear structure indicated for polystyrene; variations can be built into macromolecules through the degree and nature of branching, for example. Nonlinearity in the backbone structure of a polymer affects its solution properties in several ways, mostly because branched polymers are denser and more compact than their linear counterparts of equal molecular weight. Sufficient increase in the degree of branching and interconnectedness gives one a *polymer gel,* a virtually infinite structure.[7,9] Such a *network* can be made in several ways. Chemical crosslinks can be introduced between different chain backbones, either by polymerization of multifunctional monomers (nonlinear polymerization) or by chemical reactions between preformed polymers. Physical crosslinks can arise between polymers owing to attractive interactions between certain regions of different macromolecules, producing a network by "physical gelation." Gels may have low polymer concentration, but they differ from solutions in that the three-dimensional connectivity confers on a gel the properties of a solid (albeit in many instances a soft, pliant one), whereas a true solution is a liquid. Even as solids often have yield stresses, deforming perfectly elastically at small strains, gels generally exhibit more elasticity than polymer solutions, even if the polymer solids content of the gel is as low as that of a dilute solution.

The large number of interactions experienced by the same chain also grants polymers special properties. A small interaction between individual segments and other molecules can be multiplied into a large intermolecular interaction by virtue of the number of times the interaction is duplicated within the same molecule. Two specific manifestations of this phenomenon may be noted: (1) styrene and cyclohexane are miscible in all proportions at most temperatures of interest, while high molecular weight polystyrene and cyclohexane have a critical temperature in the neighborhood of 35°C below which phase separation occurs,[10] and (2) with rare exceptions, two different polymers exhibit only limited mutual miscibility.[7]

1.2.1 Conformation of Polymer Chains

Given the unique properties that spatial extent and connectivity impart, it is worthwhile to look at the question of the conformation* of a polymer chain. The primary model of the liquid state

---

*For the terms "conformation" and "configuration," we use the notation of Odian[6] rather than that of Flory.[9] "Conformation" refers to the arrangement of the atoms of a chain in space, while "configuration" refers to the chain topology, those aspects of the chain structure unchanged by bond rotations and only changed by the breaking of chemical bonds, or unlikely events such as rotation about double bonds.

Figure 1.1.
A short random walk in
two dimensions.

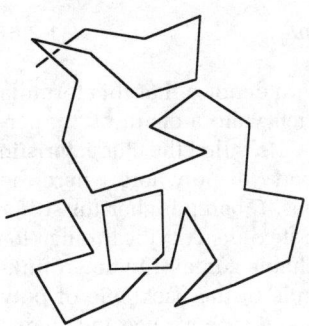

conformation of a linear polymer is that of the random walk. Figure 1.1 shows such a short random walk executed in two dimensions; we will consider a longer walk of N steps of length a, in three dimensions. The length of the walk, r, given as the distance between the starting and ending points, varies from walk to walk and so is described by a distribution, which for large N is approximated very well by the Gaussian distribution[11]

$$W(r,N)dr = \left(\frac{\beta}{\pi^{1/2}}\right)^3 \exp(-\beta^2 r^2)\, 4\pi r^2 dr \qquad (1.2.1)$$

where

$$\beta = \left(\frac{3}{2Na^2}\right)^{1/2}$$

From this, the mean square length of the walk can be found to be:

$$\langle r^2 \rangle = Na^2 \qquad (1.2.3)$$

which is true even for short random walks (N small). The characteristic length of the walk thus varies (or "scales") as the square root of the number of steps. This key result is the actual basis for much of the observed behavior of polymers.

In applying the random-walk results to a polymer comprising n identical units or segments of length l, we might be tempted to identify n with N, and a with l. (For polymers such as polystyrene, l would be the length of the —(—C — C—)— repeat unit, approximately 2.5 Å). The random walk applied to polymer chains in this way gives the so-called freely jointed chain. Real polymer chains, however, are not freely jointed: familiar from the organic chemistry of small molecules are fixed bond angles and steric hindrances to free rotation about the backbone. Therefore correlations exist between the orientations of neighboring segments that are absent in the random walk. Introducing a fixed bond angle θ between the bonds of a formerly freely jointed chain causes the average dimensions to increase[12]:

$$\langle r^2 \rangle_{\text{freely rotating}} = nl^2 \frac{1 + \cos\theta}{1 - \cos\theta} \qquad (1.2.4)$$

that is, by a factor of about 2 for the tetrahedral angle ($\theta = 109.5°$). The freedom of rotation may be restricted as well, since a typical flexible polymer will exhibit a potential energy surface with respect to rotations around the backbone segments. Averaging the rotational angles with respect to a Boltzmann weighting with a rotational potential energy gives the excellent approximation:[13,14]

$$\langle r^2 \rangle_{\text{hindred rotation}} = nl^2 \frac{1 + \cos\theta}{1 - \cos\theta}\, \frac{1 + \langle\cos\Phi\rangle}{1 - \langle\cos\Phi\rangle} \qquad (1.2.5)$$

where $\langle\cos\Phi\rangle$ is the average cosine of the angle of rotation.

The interactions arising from bond angles and rotational hindrance are local or short-ranged in their origin, and they do not destroy the random-walk scaling:

$$\langle r^2 \rangle_0 = Cnl^2 \qquad (1.2.6)$$

so that $\langle r^2 \rangle_0^{1/2} \sim n^{1/2}$. The subscript zero denotes that this formula is limited to local interactions, which yield a chain of "unperturbed" dimensions. The prefactor, C, is called the characteristic ratio and is a conformational property of polymers where the information on local stiffness resides, C being higher for stiffer polymers and lower for those more flexible. A typical range for values of C is about 4 to 12, with chains containing heterolinkages, such as the flexible ether bonds in the backbone of poly(ethylene oxide), occupying the low end of this range and chains with bulky side groups (e.g., polystyrene) or rigid backbone links (e.g., polyimides) at the high end. Values of the characteristic ratio have been tabulated for many polymers.[14] Since the effects of these local interactions can be lumped into the parameter C, it is possible to recover the simple random-walk model by redefining the segments of the random walk, that is, by defining an equivalent freely jointed chain with N segments of length a to obey equation (1.2.3). N and a refer to the number and length of so-called Kuhn statistical segments.[15] Since the contour length of the chain must be preserved, $nl = Na$, and $N = n/C$ and $a = Cl$. The chain can be treated as a random walk, except on a small scale along the chain ($<$ about C monomers). The chain is stiff over these small scales but on larger scales executes a random walk.

A further difference between real chains and random walks remains, though. A random walk, executed in time, can cross itself, whereas the polymer chain, existing in space and taking up volume, cannot. Some volume is excluded; moreover, solvation of the chain in a good solvent encourages different segments to maximize their solvation by avoiding proximity to one another. Such interactions are termed "long-range" because they may involve segments distant along the contour of the chain, even if near in space. This self-avoidance characteristic, which is strong in good solvents, further increases the end-to-end distance, "perturbing" it so as to destroy the validity of the random-walk scaling. The basic result, obtained by the simple, nonrigorous approach developed by Flory,[7,9] shows that the scaling is altered in the following way:

$$\langle r^2 \rangle^{1/2} \sim n^{3/5} \qquad (1.2.7)$$

Because excluded volume exists between the segments irrespective of the solvent in which they reside, we may wonder whether the random-walk model is useful in any situation. It is, in two cases. The first is that of polymers in a so-called *theta solvent,* a weak solvent in which the tendency to swell due to self-avoidance is exactly compensated by a slight collapsing tendency induced in the polymer chain by the weak solvent. Theta solvents are important conceptually in polymer science and are useful for characterization, since in them chains adopt ideal dimensions. More important for us is the second situation, the pure polymer fluid or "melt." Here, the swelling tendency is negated by the intervention of segments coming from other polymers, screening the effects of self-avoidance. Chains in melts have ideal random-walk properties.

Having established the validity of the random-walk model in the case of most importance to us (melts), let us compute one other average conformational property useful at this point. The *radius of gyration*, $\langle s^2 \rangle_0^{1/2}$ is defined as the root-mean-square distance of any segment (i.e., statistical segment) from the center of mass of the chain. This is given as follows[9]:

$$\langle s^2 \rangle_0 = \frac{Na^2}{6} = \frac{\langle r^2 \rangle_0}{6} \qquad (1.2.8)$$

The proportional relation between mean radius of gyration and end-to-end distance implies that the two quantities can be used interchangeably to describe the size of a linear Gaussian chain. The radius of gyration, though, gives a better measure of the volume occupied by an individual polymer chain. [Proportionality also exists for chains with long-range interactions, since $\langle s^2 \rangle$ for a perturbed chain scales with n as in equation (1.2.7).]

With this rough measure of the volume in which a polymer coil resides, we are in a position to estimate the density of an individual polymer coil as follows:

$$\text{density of segments within a polymer coil} = \frac{n}{(4/3)\pi \langle s^2 \rangle_0^{3/2}} \qquad (1.2.9)$$

This shows that the density of segments decreases with chain length, either as $n^{-1/2}$ or $n^{-4/5}$ depending on whether the chain is a random walk. For long chains, the segment density is thus very low, and the chain is a very sparse object. This relationship is the basis for the large spatial extent of polymer chains. The remaining space within this sparse object does not remain unfilled, of course; in a dilute solution the volume of a polymer coil is filled with solvent, and in a bulk polymer the remaining space is filled with other polymer chains. This is the basis for the large number of interactions, either with solvent, or with segments of other chains, intertwined or entangled. The conformations assumed by polymer chains ensure these features, and so are the source of the properties of polymers. Although we will have only a few occasions to return explicitly to the Gaussian distribution, its importance should be borne in mind, especially since bulk or near-bulk concentrations will be our focus.

1.2.2 Physical State of Bulk Polymer

The most common analogy made to help visualize the intertwined physical state of the melt or bulk polymer as just described is that of a plate of spaghetti. While the analogy is fine for static properties, it is not adequate for dynamic properties, since the chains in bulk polymer may be mobile. This is what we assumed for the melt: that rotation about the backbone bonds allow the chain to writhe and continually change its conformation owing to Brownian motion, unlike a plate of motionless spaghetti. Rotation may not be permitted, however, because the bulk polymer may be either *glassy* or *semicrystalline*.

The *glassy state* is that in which the random-walk conformation holds true, but rotation of the backbone bonds is severely restricted. The individual chains are thus frozen in their particular random conformations. The restriction of segmental motion comes from insufficient thermal energy to overcome the barrier to rotation, which in bulk polymer is a highly cooperative process. This implies two things.

1. Sufficient thermal energy may exist for noncooperative thermal motions, such as rotation of pendant methyl groups. Thus, it is not that all motion ceases, but only that the segmental motion required to change the conformation of the chain ceases (or at least becomes very slow).

2. With increasing temperature, sufficient thermal energy should be attained so that a more fluid state is reached. Thus, there exists a *glass transition temperature* T_g above which the polymer is a melt and below which the polymer is glassy.

The glass transition is marked by a decrease in the modulus by a factor of approximately 3 orders of magnitude (a "softening"), and in practice by changes in the heat capacity and in the thermal expansion coefficient as in Figure 1.2. Thus, it resembles a second-order transition. Nonetheless, various kinetic effects are manifested. For example, T_g, measured during a temperature decrease, depends on the rate of cooling, as does the magnitude of the change in coefficient of thermal expansion. It is not our point here to determine exactly the extent to which this transition

Figure 1.2.
Specific volume versus temperature for two polymers of identical glass transitions, one amorphous and one semicrystalline. (After Odian,[6] used by permission of John Wiley & Sons.)

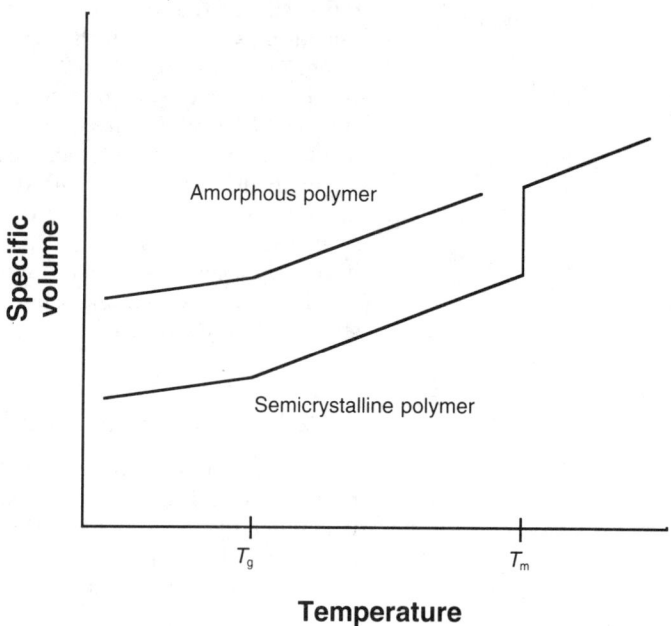

is dynamic or thermodynamic; the point rather is that in practice, the glassy state has a drastic effect on properties, imparting a brittleness and a higher modulus, and thus T_g is an important property of polymers.

The glass transition temperature is determined largely by chain stiffness, although to some degree by intermolecular forces as well. Thus T_g exhibits quite a range: from well below room temperature for polymers such as poly(dimethylsiloxane) ($T_g = -127°C$), polyethylene ($T_g = -125°C$), and natural rubber ($T_g = -73°C$), to well above room temperature for polymers such as polystyrene ($T_g = 100°C$), poly(methyl methacrylate) ($T_g = 105°C$), and polytetrafluoroethylene ($T_g = 117°C$).[6] The relevant stiffness here is the dynamic stiffness, not the static stiffness expressed in the characteristic ratio C. Thus, one cannot as a rule claim that a polymer of higher C will necessarily have a higher T_g. The degree of correlation between the dynamic and static stiffnesses depends on the relative effects of bulky side groups on $\langle \cos \Phi \rangle$ and the height of the energetic barrier to rotation. A number of other effects are of course possible for chains of more complicated backbones, such as those with aromatic rings in the backbone or chains with two backbones, which have a topology like that of a ladder. There also exist other thermal transitions below T_g which correspond to the cessation of other motions, such as torsional oscillations of chain segments, pendant methyl group rotation, and motion of larger pendant groups or parts thereof.

Segmental motion may be facilitated by physical agents as well as by increased temperature. These agents, which decrease T_g, may be part of the polymer or not. The presence of chain ends promotes rotation of backbone bonds, and thus T_g depends on the chain length of the polymer. This dependence is often correlated as a decrease with the inverse of the average chain length from a value of T_g appropriate for infinitely long chains. For polymers such as polystyrene and poly(methyl methacrylate), a decrease in T_g of more than 10°C can be expected for polymers of less than 20,000 g/mol. Likewise, the random incorporation of a comonomer characterized by a lower T_g will also serve to decrease the T_g of the copolymer, according to its weight fraction. Agents external to the polymer, known as *plasticizers,* are essentially solvents compatible with the polymer, of low molecular weight yet nonvolatile. They find extensive use in polymers such as poly(vinyl chloride) (PVC). Perhaps a more familiar plasticization process, although temporary, is the ironing of garments. The water serves to decrease T_g and thus the necessary ironing temperature; what maintains the integrity of these fibers is a crystalline structure that persists above T_g.[16]

If the polymer chains can pack into an ordered structure, the polymer may exist in a *semicrystalline state* and so exhibit a *melting temperature,* $T_m > T_g$. This is a true first-order transition, as attested to by the discontinuity in density at T_m (see Figure 1.2) and by a lack of dependence of T_m on the rate of temperature increase, disregarding superheating. This does not mean that crystallization does not exhibit kinetic effects. The intertwined or entangled state of a polymer melt ensures that crystallization is a relatively slow process and that the resulting crystals will not be perfect. Thus, thermal history during crystallization and subsequent annealing (or perfecting of the crystals) determines T_m, which is below the ideal melting point for perfect crystals. Likewise, crystallinity is rarely, if ever, complete in polymers; there always persist amorphous regions that are either glassy or meltlike, depending on the temperature. Thus we denote these polymers as *semi*crystalline.

Fitting the polymer into a crystal demands a high degree of regularity on the part of the polymer chains. Irregularities caused by branch points, the presence of comonomers, chain ends, and stereochemistry, will impede crystallization. (We will deal with the stereochemical factors in greater depth shortly.) In many cases, the irregular chains or parts of chains are excluded to the amorphous regions; in other cases the amorphous regions consist of chains that are like those in the crystalline regions but have not crystallized for kinetic reasons. In the crystals, the polymer chains no longer adopt random walk configurations but rather are highly extended, in planar zigzag form as in polyethylene, or in helical forms as is the case with polymers of substituted ethylenes, such as propylene or tetrafluoroethylene.

Crystalline morphologies are varied and depend on the conditions during crystallization, but often they take the form of lamellae. The axis of a chain, surprisingly, lies close to the perpendicular of the plane of the lamella. Since this shorter dimension of the crystal is often on the order of 10 nm, only about 40 monomer units can be accommodated within the crystal. Upon emerging from the crystal face, the chains either reenter the same crystal or wander into the surrounding amorphous material, perhaps to participate in another crystal. Such tethers between lamellae are expected for polymer crystallized in the melt, in which large arrangements of lamellae, called spherulites, are formed. Spherulites are often sufficiently large to scatter visible light, giving the polymer a cloudy appearance, as in the plastic milk bottles made of high-density polyethylene.

As in the case of a glass, segmental motion within the crystal is hindered by the ordered structure. This occurs even for temperatures exceeding T_g; the polymers give up segmental mobility to achieve a lower energy state. Also as was the case for glasses, there may be other transitions below T_m which correspond to the hindering of other molecular motions. In addition to these, though, there may be significant crystal–crystal transitions. For example, polytetrafluoroethylene undergoes a transition at 19°C from a crystal based on a 13/1 helix to one based on a 15/1 helix. Polybutylene at room temperature slowly transforms from its melt-crystallized 11/3 helix to a 3/1 form. Such changes are of course accompanied by a density change, hence may be very important in some applications.[17]

In a single polymer crystal, the modulus measured in the plane of the crystal is orders of magnitude lower than that in the direction perpendicular. Polycrystallinity, the different orientations of the crystals within the spherulites, and twists in the lamellae themselves usually prevent the bulk plastic from behaving anisotropically. This will not be the case for polymers in which orientation occurs during the processing, such as in the production of fibers. The mechanical properties of semicrystalline polymers depend on the fraction of polymer that is crystalline and the morphology, among other things (such as whether $T > T_g$). Materials with very low crystallinity may behave as weakly reinforced composites, while those of greater crystallinity will behave as physical gels. Materials of high crystallinity may resemble metals bearing a large number of defects.

1.3 Structural Characterization of the Polymer Backbone

While the long-chain nature of polymers plays the dominant role in determining properties, it does not act alone. For example, crystallinity is determined by the regularity of the polymer on the smallest scales. Polymer properties are governed by characteristics on all length scales, from that of covalent bonds (Å) to that of larger morphologic features, such as spherulites, observable by visible light microscopy. We proceed to review this hierarchy of structural characteristics, to roughly assess (through anecdote) the influences of those characteristics on polymer properties, and to discuss the methods applied to characterize them. The hierarchical divisions are admittedly arbitrary, but we use the following, listed in order of increasing length scale: polymer backbone, chain length, chain architecture, morphology, and bulk characteristics. We shall deal with each of these in turn. On the smallest scale, that of the polymer backbone, we find details of composition, sequence, and stereochemistry.

1.3.1 Composition

Overall composition should be as important for a polymer as for a small molecule. Composition is a trivial question for a true homopolymer but not for copolymers, in which monomers of different types are connected together. Copolymer composition influences a number of properties; for example, the glass transition temperature of a copolymer can be increased by the addition of a comonomer of higher T_g. The best example of this is poly(styrene-co-cutadiene). The glass transition temperature of polybutadiene is $-90°C$ while that for polystyrene is $100°C$, so that T_g of a copolymer of the two is expected to fall roughly between these values.[16] Styrene–butadiene rubber (SBR), which is used in automobile tires, has approximately 25 mol % styrene, whereas the copolymers used in paint latices are made with much higher amounts of styrene. The main difference is the higher glass transition temperature of the latter (-30 to $20°C$, vs. $-60°C$ for SBR). In this case, the bulky aromatic side group of the styrene monomer has increased T_g above the value for the homopolymer of butadiene. One may do the opposite, of course, as in the addition of acrylate monomers to methacrylates. In other cases, such as copolymerization with acrylonitrile to produce nitrile rubber, T_g may be increased because of the polarity of the monomer.

A comonomer may be added for reasons other than altering T_g.[18] It may be desired to change the degree of crystallinity or T_m, or to make the polymer more compatible with a plasticizer or a dye. The presence of a comonomer may serve to increase or decrease swelling; in the case of nitrile rubber, swelling in oil or gasoline is decreased by the presence of the polar monomer. Both poly(vinyl acetate) and poly(vinylidene chloride) often contain comonomers to enhance stability. Poly(vinylidene chloride) discolors (yellows) because of the formation of long sequences of conjugated double bonds accompanied by the evolution of HCl. Since, however, the mechanism requires long sequences of vinylidene chloride monomers, if these are broken up by an inactive monomer, such as ethyl acrylate, yellowing may be hindered. In most instances, copolymerization serves to address several demands at once; copolymers of vinyl chloride (87%) and vinyl acetate (13%), which are used for phonograph records, have to have such properties as low melt viscosity and resistance to shrinkage upon cooling.[19] Multiple demands may require polymerizations of three or more monomers.

The composition of the polymer is thus very important, but because the monomers are usually unequally reactive, the initial monomer composition will not be the same as the polymer composition at all extents of reaction. Moreover, one may desire a fairly uniform composition (i.e., all chains having similar composition), and attempts to achieve this goal are complicated by the unequal reactivity of the monomers as well. In the case of SBR, the butadiene is more reactive than styrene, so the feed must be rich in styrene to result in the desired polymer at the beginning of the reaction. Because of the discrepancy between monomer and polymer composition, one needs to have methods of characterizing the composition of the polymer. These include: nuclear magnetic resonance spectroscopy (NMR), infrared (IR) and ultraviolet (UV) spectroscopies (both of which are probably the easiest to implement on line), mass spectrometry, and elemental analysis, if the monomers are different enough in elemental composition to allow distinguishing.

1.3.2 Sequence

The placement of the monomers along the chain is important as well. This was implicit in the example of the stabilization of poly(vinylidene chloride); if the ethyl acrylate were not distributed fairly randomly, breaking up the long vinylidene chloride sequences, the comonomer would be of no use in preventing yellowing. There must be a difference between a polymer with a blocky structure:

$$\sim\sim\sim AAAAAAAAAABBBB\sim\sim\sim \qquad (1.3.1)$$

and one with a more nearly random structure:

$$\sim\sim\sim AABAAABAAABABA\sim\sim\sim \qquad (1.3.2)$$

In the former case, the ethyl acrylate (B) monomer would be

useless in prohibiting yellowing, although it might very well grant the polymer other attractive properties.

The arrangement of the monomers on this smallest scale can have drastic implications for bulk properties other than stabilization. If we return to the case of a styrene–butadiene copolymer, it matters a great deal whether those two monomers are "randomly" copolymerized or exist in blocks. Both the copolymer destined for tires and that for latices were "random," but we can also make a block copolymer of the two monomers. What we have stated as usually true of different polymers is also true of the different blocks—they are thermodynamically incompatible, and hence will phase-separate. In this case, though, the covalent bonds linking the two blocks forbid macroscopic phase separation but allow microphase separation into glassy polystyrene regions and meltlike polybutadiene regions. The glassy regions act as large crosslinks, and thus the polymer is a physical gel. It is referred to as a *thermoplastic elastomer:* "elastomer" because of the elastic properties imparted by the network structure, "thermoplastic" because when the material is heated above the T_g of polystyrene, the polymer will flow. The NMR spectrum of the block copolymer looks like a mixture of the two homopolymers; absent are the splittings and chemical shift changes indicative of a significant fraction of the two monomers being cross-polymerized. There are other tests for blockiness as well. The clearest of these is to use thermal or mechanical analysis to see whether there are two transitions (indicative of two different glass transition temperatures), corresponding to the homopolymeric glass transition temperature of each phase. If the copolymerization is "random," the system will be homogeneous and only one transition will be seen.

Just as the glass transition is affected by blockiness, so is the melting behavior. For example, the random copolymer of ethylene and propylene will be amorphous, whereas the corresponding homopolymers and thus the block copolymeric counterpart are highly crystalline. This is because the random copolymer does not have sufficient regularity to allow for crystallation. In both this case and the case of the SBR block copolymer, characterization by scattering techniques is possible because the system comprises different phases.

Considering only the two limiting cases of blocky and random copolymers is an obvious simplification; there is a range of behavior between truly random and completely blocky. The best way to characterize this is by the sequence length, which is the number of monomers in an uninterrupted sequence of monomers of the same type. Sequence length and sequence distribution are discussed in Chapter 4. This subject is tied, for example, with the amount of discoloration one can expect in the vinylidene chloride–ethyl acrylate copolymers, but also with questions of multiphase structure—for even short sequences, if very incompatible with the major polymer, may phase-separate, and this may or may not be desirable for the end-use properties. Sequence length is best characterized by NMR, specifically ^{13}C NMR, which shows great sensitivity toward the different monomer sequences that can be formed. Other analytical techniques such as IR may be used to correlate with sequence length.

1.3.3 Chain Configuration

The term *configuration* denotes more permanent, topological features than the different conformations allowed by bond rotation in Section 1.2.1. Here we imply features of polymer chain structure that are altered only by the breaking of bonds. Odian[6] gives a fairly extensive list of the different kinds of configurational features that may arise in polymers; we deal only with the two most common, and simplest to understand.

The first is the stereochemistry of a polymer formed from an α,α-disubstituted ethylene, which has the following form:

$$H_2C = C \overset{R_1}{\underset{R_2}{\diagup}} \qquad (1.3.3)$$

The most common singly unsaturated monomers (e.g., styrene) have this form. If R_1 is alkyl and R_2 is a hydrogen, we term this an α-*olefin*. If we polymerize this monomer in some way, we may examine the particular configuration in which the chain was formed. Since this configuration is unaltered upon conformational changes, we are free to examine the configuration of a chain in the most convenient conformation available: planar zigzag form. This is obtained by stretching the chain to its maximum extension such that all the backbone carbons lie in a plane and inscribe a zigzag in that plane.* The substituents of the chain, the R_1 and R_2 on the α-carbon and the two protons on the β-carbon, lie on either side of the plane. This can be shown with the following familiar projection:

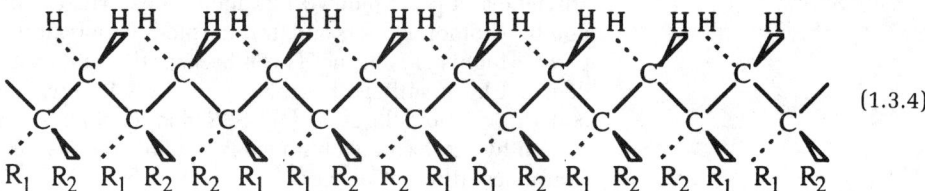

$$(1.3.4)$$

The broken lines represent bonds falling behind the plane of the page, the wedges those in front. The α-carbons form the lower line of backbone carbons, the β-carbons the upper. If $R_1 \neq R_2$, then along the length of the chain, different arrangements of each of these substituents are possible with respect to the side of the plane on which they lie. Two extreme cases of regularity are denoted as *isotactic* and *syndiotactic*. An *isotactic* chain is one in which the substituents always lie on the same side of the plane:

The fact that the bulkiness of side groups may not allow this conformation in reality is of no importance.

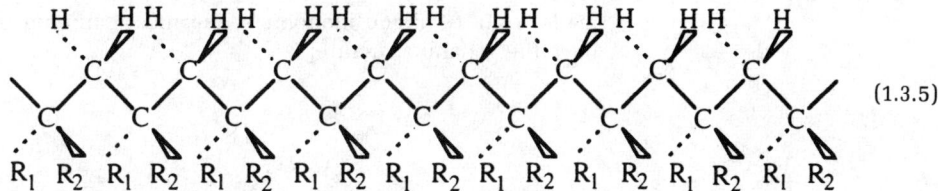

(1.3.5)

A *syndiotactic* chain is one in which the substituents alternate from one side to the other on successive α-carbons:

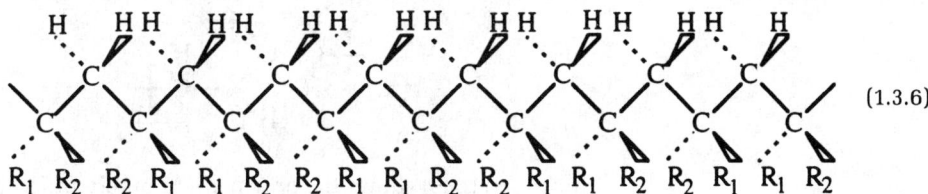

(1.3.6)

These are extreme cases, and a range of behavior between these two limits may be expected; returning to our first chain, we see that there is a mixture of dyads (successive pairs of α-carbons) which are either isotactic (meso or *m* dyads) or syndiotactic (racemic or *r* dyads):

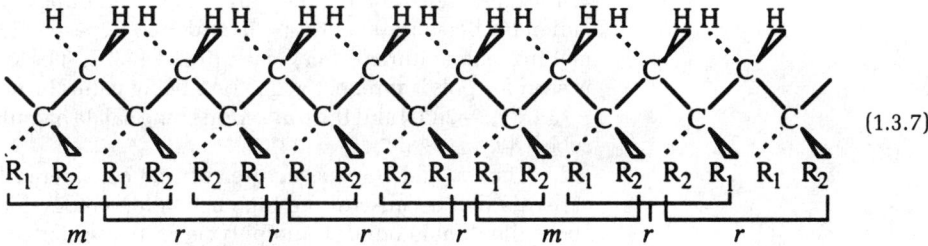

(1.3.7)

Such a mixed chain, which is termed *atactic,* may loosely be considered as a copolymer of meso and racemic dyads.

Tacticity is a measure of a particular kind of regularity, a stereochemical regularity, which must therefore exert a great influence on the tendency to crystallize. Thus while isotactic polystyrene is semicrystalline, the usual commercial polystyrene, which is atactic, is amorphous, even though it is often called "crystal" polystyrene because of its optical clarity. NMR (usually ^{13}C NMR) is the general basis for directly characterizing tacticity, although tacticity can of course be correlated to other properties.

Another kind of isomerism is encountered in polymers

made from 1,3-diene monomers, the most common of which have the general structure:

$$H_2C = C \diagup^R \atop \diagdown_H C = CH_2 \tag{1.3.8}$$

The more common dienes are butadiene (R = H), isoprene (R = CH_3), and chloroprene (R = Cl). The cis–*trans* form of isomerism arises when the polymerization proceeds to form a 1,4-polymer:

$$\tag{1.3.9}$$

Equation (1.3.9) shows the polymer in a purely *trans* configuration. The polymer may be formed in a purely *cis* configuration,

$$\tag{1.3.10}$$

or *in* some mixture of the two analogous to atacticity. These different geometric configurations affect properties drastically. While glass transition temperatures of *cis*- and *trans*-1,4-polybutadiene differ only slightly (−95 and −83°C, respectively), the melting temperatures differ greatly (6 and 145°C). The same trend is seen for polyisoprene, the *cis* form being natural rubber ($T_g = -73°C$, $T_m = 28°C$) and the *trans* form being balata or gutta-percha rubber ($T_g = -73°C$, $T_m = 74°C$).[16]

The *cis* and *trans* structures can be considered topologically distinct because of the high activation energy for rotation about the double bond. If the polymerization of butadiene proceeds not by 1,4-polymerization but by 1,2-polymerization, then the issue of tacticity arises just as discussed before, the only difference being that the pendant group is unsaturated. If R ≠ H, then the number of different configurations increases, because 1,2- and 3,4-polymerizations yield different products. We will not deal with such complications[6] here, but leave the topic by saying that these configurational distinctions, like tacticity, are best directly measured by NMR, although they too can be correlated with bulk properties.

1.4 Characterization of Chain Length

As implied in the preceding discussion on polystyrene, we generally expect that not all chains will have the same length *i*. Indeed, we expect a distribution of chain lengths, which we denote by P_i, the molar concentration of species of length i. P_i $(i = 1, \infty)$ designates the *chain length distribution*. This distribution may be described in part by *moments* of that distribution, μ_i, of which the most often encountered are the first three:

$$\mu_0 = \sum_{i=1}^{\infty} P_i \qquad (1.4.1)$$

$$\mu_1 = \sum_{i=1}^{\infty} iP_i \qquad (1.4.2)$$

$$\mu_2 = \sum_{i=1}^{\infty} i^2 P_i \qquad (1.4.3)$$

Depending on the type of the distribution, these moments may contain all the information on the distribution; in general, though, higher moments, similarly defined, are needed.

We may further define average *degrees of polymerization* by the ratios of successive moments. The number-average degree of polymerization, DP_n, is the ratio of the number of monomers to the number of molecules, or the ratio of the first moment to the zeroth:

$$DP_n = \frac{\mu_1}{\mu_0} \qquad (1.4.4)$$

One can define other average chain lengths by the ratio of any two consecutive moments. The next most useful, and often the better indicator of the properties of the polymer, is the weight-average degree of polymerization:

$$DP_w = \frac{\mu_2}{\mu_1} \qquad (1.4.5)$$

This is the average number of monomers on a molecule chosen by randomly selecting a monomer. The z-average degree of polymerization is:

$$DP_z = \frac{\mu_3}{\mu_2} \qquad (1.4.6)$$

and higher degrees of polymerization, DP_{z+1}, $DP_{z+2} \ldots$, may be similarly defined. Notice that in the definitions of the moments [equations (1.4.1)–(1.4.3)], species of all lengths have been included, and thus we have adopted the second convention for the definition of polymer, the formation–structure convention (see Section 1.2). It might seem that degrees of polymerization so defined would not correlate well with polymer properties, but in the cases of interest the species of small size will be so scarce that the difference between the two conventions will be negligible.

If all the chains were of one length i' so that $P_i = 0$ for $i \neq i'$, then all these different average degrees of polymerization would be equal. Synthetic polymers, though, are never *monodisperse,* but rather are *polydisperse,* and thus $DP_n < DP_w < DP_z < DP_{z+1} < \cdots$. The most common descriptor of the breadth of the distribution is referred to as the *polydispersity,* which we call Q, although other notations for this abound. The polydispersity is defined as follows:

$$Q = \frac{DP_w}{DP_n} \qquad (1.4.7)$$

An equivalent alternative to the chain length distribution is the *molecular weight distribution* W_i, which is expressed in mass per volume:

$$W_i = P_i m_i \qquad (1.4.8)$$

where m_i is the molar mass of a polymer of length i. Often a good approximation consists of saying that this mass is proportional to i through the mass of the monomer (i.e., $m_i = im_{monomer}$); departures from this arise from chain ends and branch points. We can define the number- and weight-average molecular weights, M_n and M_w, in a fashion similar to that for DP_n and DP_w:

$$M_n = \frac{\sum_{i=1}^{\infty} W_i}{\sum_{i=1}^{\infty} W_i/m_i} \qquad (1.4.9)$$

$$M_w = \frac{\sum_{i=1}^{\infty} m_i W_i}{\sum_{i=1}^{\infty} W_i} \qquad (1.4.10)$$

If $m_i = im_{monomer}$, M_n is proportional (through $m_{monomer}$) to the ratio of the zeroth moment to first negative moment, and M_w is proportional to the ratio of the first moment to zeroth moment of the molecular weight distribution (not the chain length distribution). It then follows that

$$M_n = m_{monomer} DP_n \qquad (1.4.11)$$

$$M_w = m_{monomer} DP_w \qquad (1.4.12)$$

Thus the polydispersity will be the same, regardless of whether it is based on degree of polymerization or on molecular weight.

1.4.1 Significance of Molecular Weight and Its Distribution

We know that the molecular weight of polymers is important because this property distinguishes polymers from small molecules. In addition to the effects (e.g., on T_g) mentioned earlier, which are weak for high polymers, much stronger effects of mo-

Figure 1.3.
Viscosity of a low-density polyethylene melt at different temperatures. (From ref. 21, used by permission of Carl Hanser Verlag.)

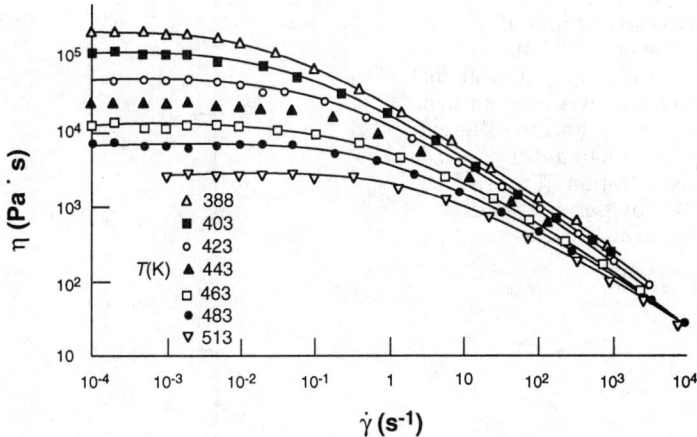

lecular weight are seen in both the flow (or *rheological*) properties and in the related mechanical properties important for end use. Graessley states that the "molecular weight distribution and long chain branching influence all flow properties, profoundly in some cases, and their effects are especially important in the entanglement regime."[20]

The most familiar flow property is viscosity, the ratio of stress τ to rate of strain $\dot{\gamma}$. It is important both in the processing of the polymer and in some end uses. For the familiar small-molecule liquids, such as water or glycerin, the viscosity is a constant, μ, independent of the rate of strain, and the fluid is "Newtonian." For more complex fluids such as suspensions or polymeric solutions, the viscosity is a function, η, of $\dot{\gamma}$, and the fluid is "non-Newtonian." The usual behavior is that the viscosity decreases with strain rate, or "shear-thins" as shown in Figure 1.3.[21] The viscosity is constant at both sufficiently low and high $\dot{\gamma}$, η_0 and η_∞, respectively; at intermediate $\dot{\gamma}$ the viscosity exhibits a power-law dependence on $\dot{\gamma}$. The intermediate power law regime thus appears linear on a plot of $\log(\eta)$ versus $\log(\dot{\gamma})$, the slope typically being in the range -0.4 to -0.9. The infinite shear viscosity η_∞ is generally not attainable for polymer melts because of shear degradation. For a melt, the zero-shear viscosity depends on the weight-average molecular weight very strongly:

$$\eta_0 \sim M_w^1 \qquad M_w < M_c \qquad (1.4.13)$$
$$\sim M_w^{3.4} \qquad M_w > M_c$$

as shown for several polymers in Figure 1.4.[22] Here M_c is a critical molecular weight for entanglement, a stronger concept than the mere intertwining we have spoken of before. The value of M_c depends on the polymer in question but generally corresponds to degrees of polymerization in the range of 100–300.[20] The molecular weight distribution itself does not appear to have much of an effect on η_0 (except possibly for highly polydisperse samples), but it does have an effect on $\eta(\dot{\gamma})$. While the shear rate at which shear-thinning becomes significant increases with both concentration and molecular weight, it decreases as the distribu-

Figure 1.4.
Viscosity of nine different polymer melts versus molecular weight. Note that each axis has been shifted arbitrarily for the different polymers to aid in visualization. (From ref. 22, used by permission of Springer-Verlag.)

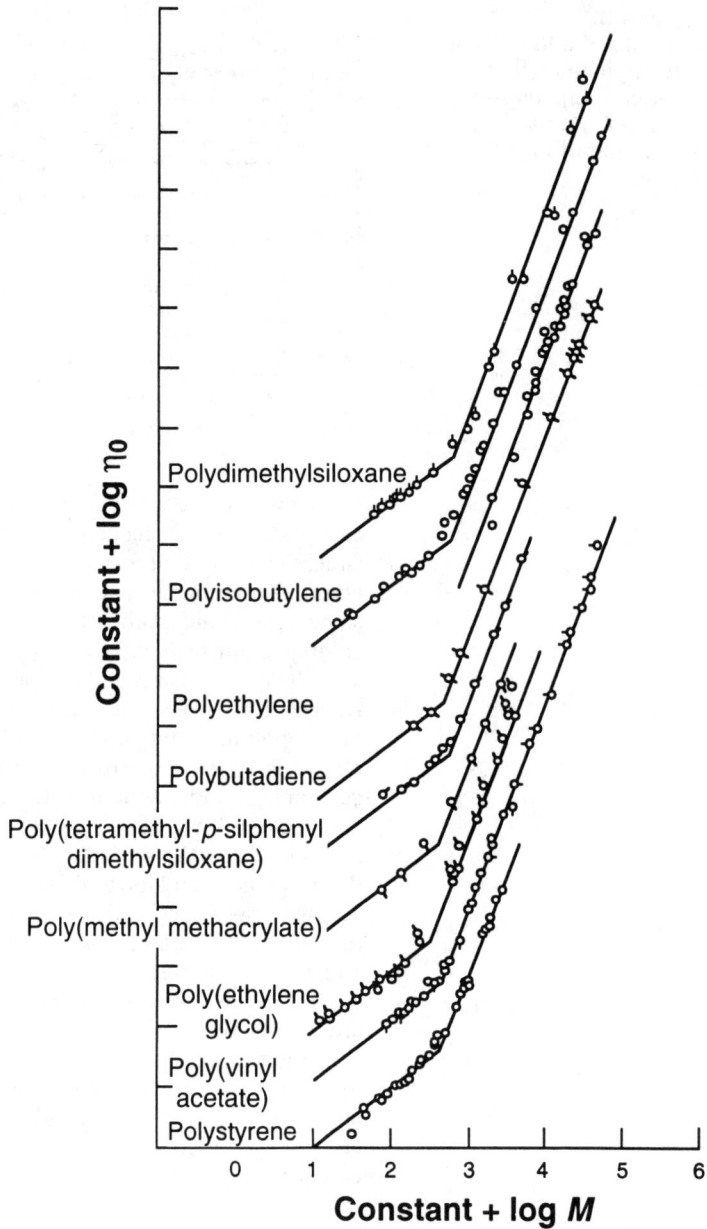

tion becomes broader. Increased breadth of the distribution enlarges the range over which power-law behavior is seen[8] and also decreases the power-law index.[23]

Besides exhibiting non-Newtonian viscosity, polymer solutions have other unusual rheological properties that are even more highly dependent on molecular weight. One of these is the first normal stress difference.[8] Because this is a less familiar concept than viscosity, we introduce it in a particular idealized flow: laminar flow between two parallel planar surfaces (see Figure 1.5). Here the fluid flows in the x-direction and a velocity gradient exists in the y-direction. For a Newtonian fluid the stress

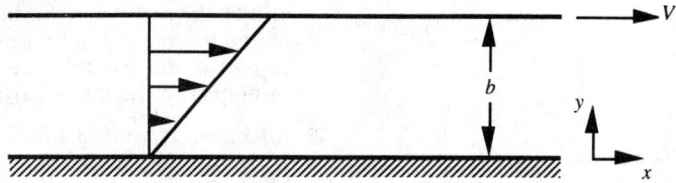

Figure 1.5.
Steady simple shear flow between two parallel plates. The shear rate $\dot{\gamma} = V/b$, where V is the velocity of the upper plate and b the distance between the two.

solely comprises τ_{yx}, a (shear) stress on a surface of constant y exerted in the x-direction. In a polymeric fluid, though, there exist stresses τ_{xx}, τ_{yy}, and τ_{zz} arising from stretching, along the streamlines, of the chains. These are somewhat analogous to pressures in that they constitute stresses perpendicular to the surface on which they act; they are not, however, isotropic. Because of this we define the (nonzero) first and second normal stress differences, $(\tau_{xx} - \tau_{yy})$ and $(\tau_{yy} - \tau_{zz})$, respectively. The shear stress τ_{yx} can be related to the material property η as follows:

$$\tau_{yx} = -\eta(\dot{\gamma})\,\dot{\gamma} \tag{1.4.14}$$

Therefore these stress differences can be related to two more material functions, the first and second normal stress coefficients, as follows[8]:

$$(\tau_{xx} - \tau_{yy}) = -\Psi_1(\dot{\gamma})\,\dot{\gamma}^2 \tag{1.4.15}$$

$$(\tau_{yy} - \tau_{zz}) = -\Psi_2(\dot{\gamma})\,\dot{\gamma}^2 \tag{1.4.16}$$

where Ψ_1 is the more important material property of the two, Ψ_2 generally being much smaller and of opposite sign (negative).

The practical consequences of nonzero normal stress differences are many and often rather striking. In the idealized flow of Figure 1.5, the main outcome is that a pressure is exerted on the two plates, making it necessary to apply a force to keep the plates at a constant separation. This constraint is due to the preference of the polymer to be in a random-walk conformation rather than the conformation stretched along the streamlines. This effect indicates some elastic character; the polymer chains, if given the chance, will snap back into their random-walk conformations. Less idealized situations are characterized by rheological effects due at least in part to the normal stress differences.[8] Such effects include:

1. *Extrudate (or "die") swell* (due mostly to the first normal stress difference but also in part to the second). This effect can be understood very easily in terms of the ideal flow of Figure 1.5. In the channel of a cylindrical die, as in the channel of Figure 1.5, there is force exerted on the wall because of the polymer, and upon exiting the die, the polymer can relax appropriately, a behavior which swells the extrudate radially, often by a factor of 3 or more. Because many polymers go through an extrusion step at least once during processing, this swelling can be significant to the process.

2. *Rod climbing.* If one stirs a Newtonian fluid with a rod, a vortex is formed; a stirred polymeric solution, however,

climbs the rod. As in the case of extrudate swell, this effect is due mostly to the first normal stress difference but also in part to the second. The implications for mixing with impellers in the chemical reactor are obvious.

3. *Effects on secondary flows.* Secondary flows are associated with inertial and elastic effects, the influence of the latter often running counter to that of the former. Thus, just as with the unexpected reversal from Newtonian solutions seen in rod climbing, secondary flows in polymeric solutions are often reversed in direction from those in Newtonian fluids.

The first normal stress coefficient depends on $\dot{\gamma}$ much like the viscosity η. In entangled melts, the zero-shear first normal stress difference depends strongly on M_w:

$$\Psi_{1,0} \sim M_w^7 \qquad (1.4.17)$$

which gives an indication of how strongly the effects just enumerated will be influenced by molecular weight.

A related experiment, which illustrates the effect of the molecular weight distribution, is the creep recovery experiment. Here the polymer sample has been subjected to a constant and small shear stress τ for some time, hence is in steady shearing flow. The stress is removed at a certain time, and the polymer recoils or "recovers" somewhat. The strain γ is observed as a function of time, and the final value of the strain γ_∞ defines a recoverable or steady state compliance $J_e^0 (= \gamma_\infty/\tau)$. This is found to be:

$$J_e^0 = \frac{\Psi_{1,0}}{2\eta_0^2} \qquad (1.4.18)$$

and thus for entangled melts is independent of M_w or a weakly increasing function of it.[24] Compliance is, however, strongly dependent on the breadth of the distribution through fairly high moments[25]:

$$J_e^0 \sim \frac{DP_z DP_{z+1}}{DP_w^2} \qquad (1.4.19)$$

The exceptions to this rule generally point to an even stronger dependence on the distribution, and the theoretical dependence bears this out[25]:

$$J_e^0 \sim \frac{DP_{z+1} DP_{z+2}}{DP_w DP_z} \qquad (1.4.20)$$

The recoverable compliance is closely related to the melt elasticity $((\tau_{xx} - \tau_{yy})/\tau_{yx})$, and both are measures of the elastic energy stored during steady-state flow. The high melt elasticity resulting from a broadly distributed polymer results in long elastic memory, high melt tension during fiber spinning, and good bubble stability during film blowing.[26]

Figure 1.6.
Typical behavior of the modulus $G(t)$ for a noncrystalline polymer.

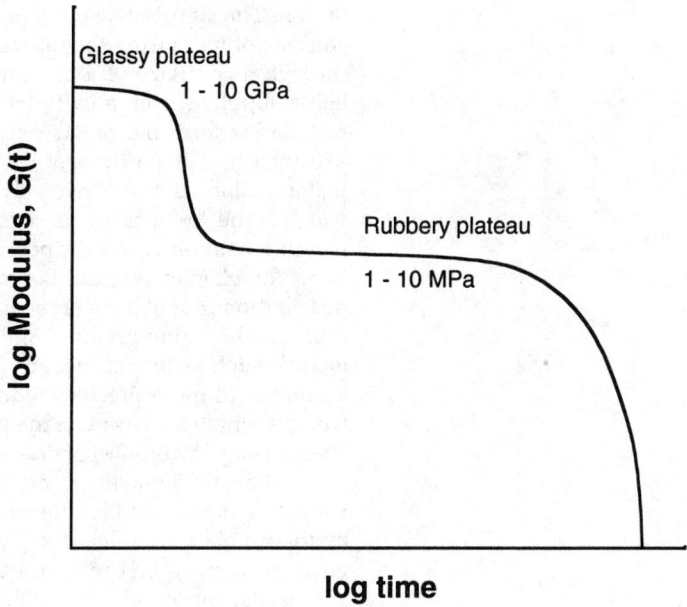

Glassy plateau
1 - 10 GPa

Rubbery plateau
1 - 10 MPa

log Modulus, G(t)

log time

Polymer solutions and melts, even though fluids, thus have some elastic character and so are generally referred to as *viscoelastic.* They have both viscous (fluid) and elastic (rubbery) attributes. Non–steady-state flow conditions especially should manifest this. An example is the experiment in which a (small) strain γ is imposed on a polymer melt and the decay of the force or stress with time observed. This defines a modulus $G(t)$ ($= \tau/\gamma$). The failure of the stress to decay instantaneously is indicative of the elasticity or memory of the polymer; that $G(t)$ does decay to zero (if the polymer is not crosslinked) indicates that the polymer is a fluid. The modulus of a polymer melt above T_g has the form shown in Figure 1.6. Two plateaus are noted, the first at very short times indicative of temporal glassy behavior (sufficient time has not elapsed for conformational changes to have occurred), the second indicative of rubbery behavior caused by entanglements. The rubbery plateau modulus is independent of the molecular weight as long as the molecular weight is above M_c (a plateau exists). The length of the plateau increases with increasing molecular weight, and increasing the breadth of the molecular weight distribution tilts the plateau so that it is no longer flat.

Even though a number of the foregoing experiments are far removed from practical situations, they do substantiate the strong effects of molecular weight and its distribution on the processing characteristics and on the end properties.

1.4.2 Measurement of Molecular Weight

It is clear that rheological properties, such as $G(t)$ (or its Fourier transform), could be used as a correlative tool to measure average molecular weights and the breadth of the molecular weight distribution. The simplest example of this is the *melt index*. This test consists of measuring the rate of extrusion of a molten polymer through a capillary of a certain length and diameter, the melt being supplied from a cylinder by the weight of a piston. The melt index for a particular polymer is the amount of polymer extruded in a given time, at specified temperature and mass of piston,[17] and is thus inversely related to average molecular weight. The test has in its favor simplicity, a relatively short measurement time, and the possibility of being implemented on-line. The characterization is not unique (polymers of different distributions might have the same melt index), but that complication may be unimportant if one can be certain that misleading factors, such as branching, are prohibited. There are, however, a number of more precise techniques for measuring molecular weights which are based on the properties of dilute solutions.[27,28] We give only the briefest of descriptions of these techniques here.

The number-average molecular weight is the molecular weight as usually defined for small molecules: the mass divided by the number of molecules. Two approaches may be taken to measure M_n. The first relies on the polymer possessing identifiable units, the number of which is independent of the chain length or mass. For example, a linear chain has two ends. If these ends are distinguishable by a spectroscopic technique such as NMR, and if the ends are not terribly dilute (i.e., if the chains are relatively short), then M_n may be measured. Similarly, M_n may be found by titration if the end group is appropriately reactive. Both techniques rely on the assumptions of linearity (two ends per molecule) and identity of the end groups. Side reactions that introduce branch points or alter the end groups invalidate such an approach. (It is possible to measure M_n of branched polymers in this way only if all molecules have the same, known number of branches.)

To overcome these limitations, we may measure M_n directly through any of a number of colligative techniques, such as freezing point depression, or osmometry. This approach is applicable regardless of the chain architecture or the identity of end groups. The colligative techniques are thus the more popular methods of determining M_n, with osmometry being the most widely used. Data on osmotic pressure and its concentration dependence have been obtained traditionally by direct measurement of the pressure resulting after equilibration of a polymer solution with pure solvent across a membrane impermeable to polymer and perfectly permeable to solvent. For sufficiently low mass concentrations c of polymer, the osmotic pressure Π follows an ideal-gas-like equation:

$$\lim_{c \to 0} \left(\frac{\Pi}{RT} \right) = \frac{c}{M_n} \qquad (1.4.21)$$

Extrapolation of Π/cRT to infinite dilution thus yields the inverse of M_n. Typical plots of Π/cRT versus c, however, show curvature coming from the neglected higher order terms in concentration (a virial expansion), and this affects the accuracy of the extrapolation. A plot of the square root of Π/cRT versus c is usually linear,[9] the intercept being $(1/M_n)^{1/2}$. There are also many technical points to consider: for example, adequate time must be allowed for equilibration; membranes must be leak-free, truly impermeable to polymer, and nonadsorptive.[29] The method is therefore unsuitable for very low molecular weights, for which impermeability is hard to achieve, as well as for high molecular weights ($> \sim$50,000), for which the osmotic pressure is too low to measure.

Although useful, M_n does not indicate the breadth of the distribution, and, more to the point, properties such as viscosity are better related to M_w than to M_n. To measure M_w, we need a technique sensitive not to the number of chains but to their mass. Light scattering is one such technique. The experimental observable in a scattering experiment is the total, time-averaged intensity of light scattered at a certain angle θ, proportional to R_θ. We can relate R_θ to M_w as follows:

$$\lim_{c,\theta \to 0} \left(\frac{Kc}{R_\theta} \right) = \frac{1}{M_w} \qquad (1.4.22)$$

where the constant K contains the refractive index increment (actually its square) of the polymer in the solvent, which provides the contrast necessary for scattering. The technique is generally limited to polymers of molecular weight between 10^5 and 10^8.

Equation (1.4.22) indicates a double extrapolation to zero concentration and zero angle. As for osmotic pressure, the concentration dependence is given as a virial expansion. The virial expansion appears because the fluctuations in refractive index giving rise to scattering are related to the concentration dependence of the osmotic pressure, since it is against the osmotic pressure that the molecules must move to bring about a local increase in the concentration of the solution above the mean. The mathematics behind this is explained in detail by Yamakawa.[27] Suffice it to say here that the concentration dependence allows the measurement of the second virial coefficient, A_2. (Often, for the purposes of achieving a linear extrapolation, a square root extrapolation as suggested for osmotic pressure is performed for light-scattering data as well.) The angular dependence of Kc/R_θ results from destructive interference within single polymer coils and so is related to the spatial extent of the polymer. The angular dependence allows one to measure the z-averaged square radius of gyration, which is defined as follows:

$$\langle s^2 \rangle_z = \frac{\displaystyle\sum_{i=1}^{\infty} \langle s_i^2 \rangle i^2 P_i}{\displaystyle\sum_{i=1}^{\infty} i^2 P_i} \qquad (1.4.23)$$

A number of useful techniques are sensitive mainly to the hydrodynamic volume pervaded by the polymer in solution. The first of these, measurement of intrinsic viscosity, provides another measure of average molecular weight. The analysis is built directly on the Einstein theory for the viscosity of dilute suspensions of hard spheres,[30] the result of which is:

$$\frac{\eta - \eta_s}{\eta_s} = 2.5\phi \tag{1.4.24}$$

where ϕ is the volume fraction of the spheres, η the viscosity of the suspension, and η_s that of the pure solvent. At first glance, this relation appears to be of no help toward measuring the hydrodynamic volume of the spheres, since only their volume fraction, and not their size, enters in. Most polymers, however, exist in solution not as hard spheres but as quite sparse coils, and the higher the molecular weight, the more sparse [see equation (1.2.9)]. Thus the equivalent volume fraction occupied by these hydrodynamic spheres depends on molecular weight, even at the same mass concentration. Appropriate alterations of (1.4.24) give the Flory–Fox equation[31] for the intrinsic viscosity of a polymer solution, $[\eta]$, again based on an extrapolation to zero concentration:

$$[\eta] \equiv \lim_{c \to 0} \left(\frac{\eta - \eta_s}{c\eta_s} \right) = \frac{\Phi \langle s^2 \rangle^{3/2}}{M} \tag{1.4.25}$$

The constant Φ is experimentally found to be nearly constant for many polymers with a value between 2.0 and 2.6 $\times$ 10^{23} mol^{-1}; the best theory predicts Φ to be asymptotically constant at 2.25 $\times$ 10^{23} mol^{-1}.[27] The intrinsic viscosity is thus a measure of the pervaded volume of the polymer per unit mass, which is an increasing function of molecular weight (at least for linear chains). The experimental molecular weight dependence of a polymer sample is usually expressed in the empirical form known as the Mark–Houwink–Sakurada equation:

$$[\eta] = KM^a \tag{1.4.26}$$

Values of the empirical constants K and a are extensively tabulated.[5] Significantly, the exponent a is found to vary between 0.5 (in theta solvents) and 0.8 (in good solvents). The molecular weight measured in this way is the so-called viscosity-average molecular weight M_v, which is based on fractional moments:

$$M_v = \left(\frac{\sum\limits_{i=1}^{\infty} m_i^a W_i}{\sum\limits_{i=1}^{\infty} W_i} \right)^{1/a} \tag{1.4.27}$$

The viscosity-average molecular weight is often nearly the weight-average molecular weight.

Thus far, all the techniques give a measure only of some average molecular weight. A number of methods sensitive to hydrodynamic volume allow one to obtain, if not the complete distribution, more than one average. The most important of these techniques, gel permeation chromatography (GPC) or (better) size exclusion chromatography (SEC), performs separation on the ba-

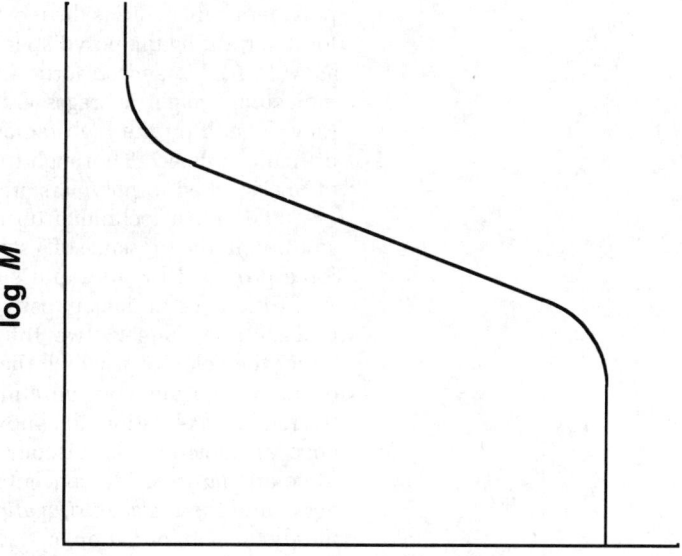

Figure 1.7.
Typical calibration curve
for size exclusion
chromatography
experiment.

log _M_

Time or elution volume

sis of hydrodynamic volume. Consider a polymer solution flowing through a porous medium,* the pores of which are "polydisperse" but comparable in size to the polymers in solution. The
larger chains will be effectively excluded from pores too small
to contain them, and the larger the chain, the more pores from
which it will be excluded. Hence, the larger chains will pass
through the medium faster because they are allowed less volume
to explore. Thus a pulse of polymer solution injected into the
stream at the entrance to the column will emerge from the column
not as a pulse, but distributed over time, based on the hydrodynamic volume of the chains.

The elution time distribution of a polydisperse sample is
thus related to the molecular weight distribution of the polymer
(at least after correction for instrument broadening), but this
relation needs to be known. Generally, $\log(M)$ is proportional to
the elution time or volume, as shown in Figure 1.7, but only
in a certain range, because polymer sufficiently large will be
excluded from all pores and polymer sufficiently small will be
permitted to explore all. The proportionality and range of validity
is peculiar to the polymer and the column, and so some calibration needs to be performed. This is most commonly done with
narrowly distributed standards of the polymer in question (if
available), although it can also be done by "universal calibration"
using intrinsic viscosity measurements.[32]

The molecular weight distribution, rather than the chain
length distribution, is measured because the elution time is usually detected by a technique such as IR or differential refrac-
tometry, both being sensitive to the mass, not the number, of

*The porous solid is sometimes a gel of crosslinked polymer (hence the
term GPC). More frequently used today, however, is a microporous silica
that has received a surface treatment to minimize polymer adsorption.

polymer chains. SEC is the most common and convenient means for determining the polydispersity of a polymer sample, as well as M_n, M_w, M_z, and so forth, although the successively higher molecular weight averages are increasingly uncertain because they depend on the high molecular weight tail, which may be difficult to detect. Thorough treatments of the practical aspects of SEC applied to polymers are available.[33]

The third technique turns on the behavior of a polymer solution in the presence of a strong force, generally a centrifugal force provided by the rapid spinning of a tube about its axis. Any difference in density between the solvent and the polymer will act to separate the two. Information can then be gained from either the velocity at which the sedimentation boundary moves or the equilibrium concentration profile. These are governed by the radial acceleration, the buoyant force arising from the difference in densities, the viscous drag, and the diffusive driving force arising from the concentration gradient. The equilibrium version of the *ultracentrifugation* experiment allows most commonly the determination of M_w and M_z, although higher averages are possible. The sedimentation velocity and the diffusion coefficient can both in principle be measured from the dynamic experiment. This route, however, provides a less certain way of measuring the molecular weight, since the relationship between diffusion coefficient and molecular weight must be known or assumed (as is also the case in the dynamic light-scattering experiment in which the diffusion coefficient is measured directly).

1.5 Characterization of Chain Architecture

Some details of chain architecture or topology, such as block versus random copolymers, have been considered. In those cases the chains were linear; but many polymers occur in branched form. Branching strongly affects the polymeric properties and end uses in several ways. First, branches often constitute a deviation from regularity, and as such strongly decrease the degree of crystallinity. This is exactly the difference between low- and high-density polyethylene—that is, the low-density polyethylene is made by a process that allows for the presence of short branches and thus gives a polymer of moderate (40–60%) crystallinity and low density (0.91–0.93 g/mL), while the high-density polyethylene is made by a process that for the most part forbids short-chain branching and thus yields a highly crystalline (70–90%) and dense (0.94–0.96 g/mL) product (see Chapters 5 and 7 for more about each of these). The more crystalline polymer has many advantages over the less, such as increased strength, better chemical resistance, and upper use temperature, although in many applications (e.g., when a more flexible product is desired), the less crystalline polymer is superior.

In noncrystalline polymer solutions or melts, branches directly affect rheological properties in two ways. The first is that branched chains have a higher segment density than linear chains of the same molecular weight. If we compare radii of gyration of the two (end-to-end distance no longer being a useful concept), we can define their ratio, g, as follows:

$$g \equiv \frac{\langle s^2 \rangle_{0,b}}{\langle s^2 \rangle_{0,l}}$$ (1.4.28)

where the subscripts b and l refer to branched and linear chains of the same total degree of polymerization. The branching ratio is less than unity, in general, because branched polymers are more compact than their linear counterparts of the same molecular weight. The more numerous the branches, the greater the decrease in g. Calculations of g can be made for regular and randomly branched polymers in the unperturbed state.[27,34] An idealized regularly branched polymer is monodisperse both in the number of branches and in their length; the branches or "arms" radiate from a central point and are f in number. For these f-armed "stars," $g = ((3f - 2)/f^2)$,[34,35] and so reductions in radius of gyration of more than 20% for a three-armed star and almost 40% for a four-armed star are seen. The hydrodynamic radius is likewise affected, so that both intrinsic viscosity and solution viscosity are reduced. This reduction in effective size will also be seen for randomly branched (polydisperse) systems; but note that in this case interpretation of data from techniques such as SEC becomes more difficult because two different structures may share the same hydrodynamic radius.

No reduction in viscosity with branching is observed in more concentrated solutions or melts for polymers with branches long enough to themselves be entangled. This is because the branches eliminate certain diffusive modes available to linear chains or those with only short, unentangled branches.[36] As a consequence, stress relaxes much more slowly and the material properties are profoundly affected. For example, the viscosity increases exponentially with the molecular weight rather than as a power law, and is independent of the number of branches.

If the material is sufficiently branched to be a gel or network, the gel itself, although susceptible to swelling, cannot be dissolved in any solvent that is not degradative. Thus, the most obvious test for the existence of gel is to determine whether the system contains an insoluble fraction. There also exist clear rheological tests, since the gel has solidlike character and should be able to withstand a strain without relaxing. In other words, the polymer should exhibit a nonzero modulus, or, in terms of the experiment for $G(t)$ described in Section 1.4.1, $G(\infty)$ will exceed zero. Other tests exist as well, and the structure of such a system can be investigated by analysis of the soluble part of the system, and by elasticity and swelling measurements, all of which are discussed in greater length in Chapter 5.

The presence of branches may be confirmed spectroscopically, as has been done for polyethylene, especially if they are numerous and short. Longer branches with correspondingly fewer branch points are more difficult to determine this way, however, and correlations with the aforementioned rheological changes would be better in such cases.

1.6 Morphology

Morphologic features reside at large length scales, visible by light or at least electron microscopy, yet are determined by polymer characteristics on the smallest scales: crystallinity is possible only for chains of sufficient stereoregularity; microphase separation becomes thermodynamically necessary only when copolymers are sufficiently blocky. Heterogeneous or anisotropic structure greatly contributes to, or dominates, the behavior of the material. Increasing crystallinity takes a material from the form of a very flexible polymer to something more like a rigid gel, one with large physical crosslinks; something similar may happen in block copolymers microphase-separating into rubbery and glassy phases. The effects may be desirable or not: the rubbery dispersed phase in high impact polystyrene (HIPS) may increase impact strength; a copolymer with too much blocky content may appear hazy and unattractive. Morphology may also arise naturally during the polymerization, as when the polymer is not soluble in its monomer. This is the case with poly(vinyl chloride), which very early in the reaction precipitates from its monomer. For PVC, the resulting morphology is as important, if not more so, than the details of chain length distribution, and also affects the course of the polymerization. Cataloguing all the effects of various morphologies on all the end properties of the polymer is quite beyond us here (see, e.g., Ferry[24] for the effects of crystallinity on mechanical properties).

The kinds of morphology, also too numerous to mention, are determined not only by the physical cause but also by, for example, thermal or shear history. This is particularly true of semicrystalline polymers, for which the perfection and thickness of the lamellae, as well as the size of the spherulite, depend on thermal history, and these morphologic features are manifested in the value of T_m. The kind of crystalline morphology and the anisotropy are greatly affected by the drawing process used to make fibers. Well-controlled multiblock copolymers, such as di- or triblock polymers, may have regular morphologies (e.g., lamellar or cylindrical). For polymers such as PVC, which precipitate from their monomer, or networks that deswell during their formation, porosity becomes important, as do pore size and its distribution (obviously so in the porous gels used for GPC). Scattering techniques and microscopy provide the most direct tools to probe these structures, but thermal, mechanical, and spectroscopic techniques are also important.

1.7 Particle Characteristics

A number of polymers are produced as particles, either because of the convenience of the process or because the product is desired in that form. Paints and adhesives such as white glue are familiar examples of polymeric latices. The porous gels used for GPC or for ion-exchange resins are other examples of product that is desired in the form of particles. More about the processes that give these products will be found in Chapter 7.

Two general characteristics are introduced if the polymer is particulate: particle size (both average particle size and particle size distribution) and surface characteristics. Both are crucial in final properties, such as the stability of a polymeric latex and the properties of films drawn from such emulsions, and in the rheology of the material during production. While discussion of surface characteristics is postponed until Chapter 7 (in which context desirable surface characteristics will become apparent), particle size is simple enough. Techniques for measuring particle size fall into two rough categories: those that chromatographically separate the particles and those that do not.[37] The various light-scattering techniques and microscopy are good examples of the latter. Separation may be effected by sedimentation and by size exclusion chromatography as for dissolved polymer, but also by a number of related techniques based on hydrodynamic separation. The standard for these measurements, though, remains microscopy, often electron microscopy because of the small size of the particles ($\sim$0.1 μm). That microscopy is the basic method is due in part to the ability to determine whether the particles are spherical, as the other techniques assume, and whether aggregation has occurred (so that even spherical precursors may yield nonspherical products). Even microscopy is not without its faults in this respect, for if the sample preparation deforms the particles, particle shape will not be represented accurately. Reviews of particle size measurement are available (e.g., ref. 38).

1.8 Additives

Additives are not *necessarily* correlated to the polymer at any particular length scale, and thus are somewhat outside the hierarchy of length scales. Additives are not peculiar to polymers. Styrene, for example, is generally not sold pure because of its tendency to thermally polymerize; rather, inhibitors in $\sim$10 ppm amounts are added to prevent premature polymerization and the accompanying safety hazards. Likewise, additives may be present in the polymer product as well, not only to stabilize the product, but also for plasticization, reinforcement, or pigmentation.

As important as inhibition is for monomers in storage, prevention of degradation is important for the polymeric products.[16,39] "Degradation" is a rather broad term referring to any mechanism that causes undesirable changes in the properties of the polymer. The agents of such change are many: heat, ultraviolet radiation, chemical agents (e.g., atmospheric O_2, ozone, water, salt spray), mechanical strain, and biological attack. A single agent may act in a variety of ways; water may plasticize, crosslink, hydrolyze, serve as a solvent for another reactive species, or act at an interface to effect delamination. The most common additives used to hinder degradation are those that stabilize the polymer against attack by radiation or chemical agents. These are antioxidants, antiozonants, and ultraviolet absorbers (which more efficiently absorb this radiation and release the energy in less destructive forms). A host of other concerns must be addressed when designing an appropriate system for stabilization.

The introduction of nonvolatile small molecules, plasticizers, to decrease the T_g of a polymer, makes the polymer less brittle and more flexible. External plasticizers are often used for PVC. They generally are petroleum-based oils, fatty acids, or esters. Choice of a plasticizer depends on the obvious issues of T_g, modulus, toxicity, odor, color, loss characteristics (by diffusion, dissolution in a neighboring solvent, evaporation), and so forth. The addition of the plasticizer may be profoundly affected by the morphological and particulate characteristics of the polymer. Many materials, rather than being too brittle, have insufficient modulus or strength. Thus polymers such as rubbers and coatings such as adhesives may be compounded with reinforcing fillers, which may be either particulate or fibril.[40,41] In many cases these fillers may dominate the mechanical behavior. The most familiar is the carbon black present in styrene–butadiene rubber for tires; silica may also be used. At times, other fillers are used, not to change the final properties of the polymer, but to improve its processing characteristics. Again, there are a number of design concerns similar to those for plasticizers.

1.9 Classification of Mechanisms of Polymerization

Inasmuch as this text deals for the most part with the formation–structure relation, it is convenient to divide polymerization mechanisms into different classes within which general formative features are shared.[6,16,42,43] We make the common distinction between stepwise and chainwise mechanisms (see Table 1.5). The overriding difference between the two in this text is taken to be whether monomers only are allowed to react with active groups, or whether species of all sizes react with one another. This distinction is chosen because it dictates the evolution of structural characteristics, and thus the mathematical form of the equations governing the evolution of P_i. Since this text concentrates on such mathematics, it is more convenient to group reactions in this way.

The reader should know that other distinctions are used by other authors; Odian,[6] for example, takes the rate of buildup of molecular weight as the distinguishing factor and thus classifies living polymerizations as stepwise. This approach is not incorrect, but it would be confusing in our context. It should also be noted that the division between these two mechanisms in particular cases may not always be as hard and fast as Table 1.5 implies. Chainwise and stepwise reactions may proceed concurrently or in competition; a particular mechanism for linking

TABLE 1.5 / Distinctions Between Stepwise and Chainwise Polymerization

Characteristic	Stepwise Polymerization	Chainwise Polymerization "Living"	Chainwise Polymerization Usual
Number and kind of reactions	Only one necessary reaction, that between two functional groups, usually dissimilar	Only two reactions: Initiation Propagation	At least three reactions: Initiation Propagation Termination May also have transfer and inhibition
Reacting species	Two dissimilar functional groups; species of any size may react with one another	Active species (anion or cation) with a monomer; active species of any size may react with monomer only in propagation step	Active species (e.g., free radical) with a monomer; active species of any size may react with monomer only in propagation step
Convention as to what is considered polymer	All species considered to be polymer	Unreacted monomer is considered to be distinct from polymer	Unreacted monomer is considered to be distinct from polymer
Polymer concentration with conversion, p	P_0 decreasing curve, P vs p	P_0 constant horizontal line, P vs p	P increasing linear line vs p
Degree of polymerization with conversion, p	DP_n increasing curve from 1, vs p	DP_n increasing linear line from 1, vs p	DP_n constant horizontal line, vs p

monomers may have some chainwise and some stepwise character. Nonetheless, the distinctions are directly useful in most cases and, in the intermediate cases, serve to clarify the situation. In Chapters 2 and 3, then, we introduce in turn stepwise and chainwise polymerizations.

Problems

1.1. How can you determine the amount of atactic polypropylene in a sample of that polymer?

1.2. How can you measure the composition of a styrene–acrylonitrile (SAN) copolymer? For a styrene–butadiene rubber? How might you be able to know whether these were random or block? How could you determine, if it were nominally random, whether a significant number of long acrylonitrile sequences existed in the SAN copolymer?

1.3. How could you distinguish between a terpolymer of acrylonitrile, butadiene, and styrene and a blend of polybutadiene with poly(acrylonitrile-*co*-butadiene), which is what ABS resins are in practice?

1.4. Suppose you have two samples of polystyrene, each with a narrow distribution of molecular weights so that they may be considered to be monodisperse. The first has a molecular weight of 10^5 and the second 10^6. Calculate M_n, M_w, and M_z for:
(a) a mixture of 0.5 mol of the first and 0.5 mol of the second
(b) a mixture of 0.5 g of the first and 0.5 g of the second

1.5. If you were to actually make up the mixtures suggested in Problem 1.4, how could you check the answers experimentally? In other words, what different techniques could be used to measure M_n, M_w, and M_z, and would they work well? What would be the best method(s)?

1.6. How could you tell whether a sample of polyethylene had been crosslinked by radiation? How could you distinguish between polyethylene with long and short branches (low-density polyethylene) and that with just short branches (linear low-density polyethylene)? Suppose the latter to be a random copolymer of ethylene and propylene; how would it be distinct from a block copolymer of the two?

1.7. What sort of particle size distribution is desirable to obtain a higher percentage of solids in a polymeric emulsion?

References

1. (a) *Chem. Eng. News,* **71(26)**, 44–45 (1993). (b) B. F. Greek, *Chem. Eng. News,* **69(23)**, 39 (1991).

2. T. W. G. Solomons, *Organic Chemistry,* 5th ed. Wiley, New York (1992).

3. L. F. Albright, *Processes for Major Addition-Type Plastics and Their Monomers,* 2nd ed. Krieger, Malabar, FL (1985).

4. H. Sawada, *Thermodynamics of Polymerization.* Dekker, New York (1976).

5. J. Brandrup and E. H. Immergut, Eds., *Polymer Handbook* 3rd ed. Wiley, New York (1989).

6. G. Odian, *Principles of Polymerization,* 3rd ed. Wiley, New York (1991).

7. P.-G. de Gennes, *Scaling Concepts in Polymer Physics.* Cornell University Press, Ithaca, NY (1979).

8. R. B. Bird, R. C. Armstrong, and O. Hassager, *Dynamics of Polymer Liquids, Vol. 1: Fluid Mechanics,* 2nd ed. Wiley, New York (1987).

9. P. J. Flory, *Principles of Polymer Chemistry*. Cornell University Press, Ithaca, NY (1953).

10. A. R. Schultz and P. J. Flory, *J. Am. Chem. Soc.*, **74**, 4760 (1952).

11. S. Chandrasekhar, *Rev. Mod. Phys.*, **15**, 1 (1943).

12. F. T. Wall, *J. Chem. Phys.*, **11**, 67 (1943).

13. M. H. Benoit, *J. Chim. Phys.*, **44**, 18 (1947).

14. P. J. Flory, *Statistical Mechanics of Chain Molecules*. Wiley, New York (1969).

15. M. Doi and S. F. Edwards, *The Theory of Polymer Dynamics*. Oxford University Press, Oxford (1986).

16. F. Rodriguez, *Principles of Polymer Systems,* 2nd ed. McGraw-Hill, New York (1982).

17. N. G. McCrum, C. P. Buckley, and C. B. Bucknall, *Principles of Polymer Engineering*. Oxford University Press, Oxford (1988).

18. T. Alfrey, Jr., J. J. Bohrer, and H. Mark, *High Polymers, Vol. 8: Copolymerization*. Wiley-Interscience, New York (1952).

19. J. F. Gabbett and W. M. Smith, in *High Polymers, Vol. 18: Copolymerization,* G. E. Ham, Ed., pp. 587–637. Wiley-Interscience, New York (1964).

20. W. W. Graessley, *Adv. Polym. Sci.*, **16**, 1 (1974).

21. J. Meissner, *Kunstoffe,* **61**, 576 (1971).

22. G. C. Berry and T. G. Fox, *Adv. Polym. Sci.*, **5**, 261 (1968).

23. S. Middleman, *The Flow of High Polymers*. Wiley-Interscience, New York (1968).

24. J. D. Ferry, *Viscoelastic Properties of Polymers,* 3rd ed. Wiley, New York (1980).

25. R. B. Bird, C. F. Curtiss, R. C. Armstrong, and O. Hassager, *Dynamics of Polymer Liquids, Volume 2: Kinetic Theory,* 2nd ed. Wiley, New York (1987).

26. D. W. van Krevelen, *Properties of Polymers: Correlations with Chemical Structure*. Elsevier, New York (1972).

27. H. Yamakawa, *Modern Theory of Polymer Solutions*. Harper & Row, New York (1971).

28. H. Fujita, *Polymer Solutions*. Elsevier, New York (1990).

29. R. U. Bonnar, M. Dimbat, and F. H. Stross, *Number Average Molecular Weights*. Wiley-Interscience, New York (1958).

30. (a) A. Einstein, *Ann. Phys.*, **19**, 289 (1906); *ibid.*, **34**, 591 (1911). (b) A. Einstein, *Investigations on the Theory of Brownian Movement*. Dover, New York (1956).

31. P. J. Flory and T. G. Fox, *J. Am. Chem. Soc.*, **73**, 1904 (1951).

32. Z. Grubisic, P. Rempp, and H. Benoit, *J. Polym. Sci., Polym. Lett.*, **B5**, 753 (1967).

33. J. Janca, Ed., *Steric Exclusion Liquid Chromatography of Polymers*. Dekker, New York (1984).

34. W. Burchard, *Adv. Polym. Sci.*, **48**, 1 (1983).

35. B. H. Zimm and W. H. Stockmayer, *J. Chem. Phys.*, **17**, 1301 (1949).

36. D. S. Pearson, *Rubber Chem. Technol.*, **60**, 439 (1987).

37. C. A. Silebi, course notes from "Advances in Emulsion Polymerization and Latex Technology" short course, Lehigh University (1993).

38. H.G. Barth, *Modern Methods of Particle Size Analysis.* Wiley, New York (1984).

39. C. E. Schildknecht, in *High Polymers, Vol. 10: Polymer Processes,* C. E. Schildknecht, Ed., pp. 525–550. Wiley-Interscience, New York (1956).

40. H. L. Gerhart and E. W. Moffett, in *High Polymers, Vol. 10: Polymer Processes,* C. E. Schildknecht, Ed., pp. 761–836. Wiley-Interscience, New York (1956).

41. G. S. Garvin, in *High Polymers, Vol. 10: Polymer Processes,* C. E. Schildknecht, Ed., pp. 679–760. Wiley-Interscience, New York (1956).

42. H. Gerrens, *ChemTech,* **12**, 380, 434 (1982).

43. P. C. Hiemenz, *Polymer Chemistry.* Dekker, New York (1984).

2

STEP GROWTH
POLYMERIZATION

2.1 Introduction

Stepwise polymerization occurs by the reaction of functional
groups on molecules of any size to form combined molecules of
larger size capable of further reaction. The functional groups are
often dissimilar, as in the reaction of a hydroxyl and a carboxylic
acid to form an ester linkage; similar groups may also react, however, as in the formation of an ester linkage by ester exchange.
The more common chemical mechanisms for stepwise polymerization, which involve the production of a condensate of low
molecular weight, are also known as "condensation polymerizations."[1] While the absence or presence of a condensate does have
important ramifications for the reaction engineer, the kinetic
schemes in either case have more in common with each other
than with chainwise systems. Hence, we will not make the formation of a condensate a primary distinction. Books solely focused
on stepwise polymerizations are available.[2,3]

   The more common fibers of Table 1.2, polyesters and nylons, are good examples of stepwise polymers, both of which are
also finding increasing use as plastics (Table 1.1). Numerous
chemical reactions can be used for stepwise polymerization; the
more common of these will be discussed in Section 2.3. A recurring feature bears mentioning now: with only a few exceptions,
stepwise mechanisms produce chains with heteroatomic backbones. Chainwise mechanisms, however, can also be used to
make chains with heteroatomic backbones, and some of these
polymers can be made by either route.

   Neither a condensate product nor a heteroatomic backbone
is thus the true distinctive feature of stepwise polymerization;
rather, the mutual reaction of functional groups on species of all
sizes is. Moreover, if this reaction is strictly random, mathematical description becomes simple, and probability theory can be
used to predict the chain length distribution, for example. The
idea of randomness in such polymerizations has historically
played a large role in the understanding of these polymerizations;
the main credit for this work belongs to Flory.[4] The aspect of
randomness is essential to any probabilistic treatment and is
founded on the equal reactivity assumption[5] (see Section 2.7).

We begin with an idealized case, or paradigm, which indicates some of the general features of stepwise polymerizations. It also provides an introduction to nearly all the mathematical approaches we may take to describe any polymerization. These approaches will be both deterministic (derived from kinetic rate equations familiar to readers from standard reactor courses) and probabilistic. The brief survey in Section 2.3 of the more common commercial stepwise systems indicates the complications covered in successive sections: stoichiometric imbalance between dissimilar groups (Section 2.4), the presence of monofunctional agents or impurities (Section 2.5), reversibility (Section 2.6), violations of the equal reactivity assumption (Section 2.7), and cyclization (Section 2.8).

2.2 Paradigm: Linear AB Step Polymerization

To introduce the basic features of step polymerization, we take as an example the self-condensation of an amino acid to form a polyamide or nylon:

$$i H_2N{-}(CH_2)_{10}{-}\overset{\overset{\displaystyle O}{\|}}{C}{-}OH \rightarrow H{\left[\overset{\overset{\displaystyle H}{|}}{N}{-}(CH_2)_{10}{-}\overset{\overset{\displaystyle O}{\|}}{C}\right]}_i OH + (i-1)\,H_2O \qquad (2.2.1)$$

11-aminoundecanoic acid   nylon 11

This particular nylon is a specialty polymer of excellent moisture resistance, dimensional stability, and electrical properties.[6] The reaction can be written:

$$(AB)_n + (AB)_m \rightarrow (AB)_{n+m} \qquad (2.2.2)$$

where A corresponds to a carboxylic acid group and B to an amine. We will take this example of batchwise AB step polymerization as our paradigm.*

2.2.1 Kinetic Solution

We first define the rate constant for the condensation reaction by the following equation:

$$\frac{dA}{dt} = \frac{dB}{dt} = -kAB \qquad (2.2.3)$$

---

*An equivalent paradigm would be A_2 homopolymerization, where the A groups react with one another. Although this example is more industrially important [for poly(ethylene terephthalate) in particular], AB step polymerization offers some pedagogical advantages. (See Appendix 2A for more about A_2 homopolymerization.)

where A and B are the functional group concentrations. The chemical system chosen forces the initial functional group concentrations, A_0 and B_0, to be equal, while the chemistry expressed in equation (2.2.3) forces these two to disappear at equal rates, so that at all times A is equal to B, and moreover equal to the total polymer concentration P, under the convention that all species are treated as polymer. Thus:

$$\frac{dP}{dt} = -kP^2 \qquad (2.2.4)$$

which, with the initial condition that at $t = 0$, $P = P_0$, has the solution:

$$P = \frac{P_0}{1 + ktP_0} \qquad (2.2.5)$$

Consistent with the foregoing definition of the rate constant, we can write the following equation for the disappearance of the monomer, P_1:

$$\frac{dP_1}{dt} = -2kP_1P \qquad (2.2.6)$$

where the factor of 2 appears because of the two distinguishable reactions (see Appendix 2A). To find the entire distribution, we must write down the differential equations describing the evolution of the concentration of each species, which has the general form:

$$\frac{dP_i}{dt} = k\sum_{j=1}^{i-1} P_j P_{i-j} - 2kP_iP \qquad i > 1 \qquad (2.2.7)$$

where the convolution term is the creation term, which was absent in equation (2.2.6) for P_1 (again, see Appendix 2A for a detailed explanation of the form of this equation). Each equation successively depends on the solution to all previous equations; physically, such dependence is the result of making the larger species from the smaller. Thus we have an infinite set of differential equations to solve.

2.2.1.1 Direct Sequential Solution

The most obvious route toward solution of the infinite set of equations is successive solution. This route, certainly not the most efficient, is possible because the evolution equation for species i depends only on knowledge of the previous $i - 1$ solutions. If the solutions of higher equations were necessary, as they would be if smaller species were being made from larger (e.g., for reversible polymerizations or polymer degradation), this would not be possible. If we desire only the concentrations of a limited number of species, either analytical or numerical solution is possible in a finite amount of time. If we are interested in the entire distribution, however, an analytical solution is necessary.

For the monomer P_1, we have from equations (2.2.6) and (2.2.5):

$$\frac{dP_1}{dt} = -2kP_1P_0 \frac{1}{1 + ktP_0} \tag{2.2.8}$$

with the initial condition $P_1 = P_0$ at $t = 0$, since we have only monomer at the beginning. This equation is a separable differential equation, easily solved to yield:

$$P_1 = P_0 \left(\frac{1}{1 + ktP_0} \right)^2 \tag{2.2.9}$$

The evolution equation for the dimer, P_2, is given as follows:

$$\frac{dP_2}{dt} = kP_0^2 \left(\frac{1}{1 + ktP_0} \right)^4 - 2kP_2P_0 \left(\frac{1}{1 + ktP_0} \right) \tag{2.2.10}$$

subject to the initial condition that $P_2 = 0$. Solving by the method of variations of parameters yields:

$$P_2 = P_0 \left(\frac{1}{1 + ktP_0} \right)^2 \left(\frac{ktP_0}{1 + ktP_0} \right) \tag{2.2.11}$$

The corresponding equation for the trimer, P_3, is:

$$\frac{dP_3}{dt} = 2kP_0^2 \left(\frac{1}{1 + ktP_0} \right)^4 \left(\frac{ktP_0}{1 + ktP_0} \right) - 2kP_3P_0 \left(\frac{1}{1 + ktP_0} \right) \tag{2.2.12}$$

which, under the initial condition that $P_3 = 0$, has the solution:

$$P_3 = P_0 \left(\frac{1}{1 + ktP_0} \right)^2 \left(\frac{ktP_0}{1 + ktP_0} \right)^2 \tag{2.2.13}$$

From these solutions, we can suggest that the general form might be:

$$P_i \overset{?}{=} P_0 \left(\frac{1}{1 + ktP_0} \right)^2 \left(\frac{ktP_0}{1 + ktP_0} \right)^{i-1} \tag{2.2.14}$$

We can prove this by induction. Assuming this form for P_{i-1}, the kinetic equation for P_i becomes:

$$\frac{dP_i}{dt} = (i - 1)kP_0^2 \left(\frac{1}{1 + ktP_0} \right)^4 \left(\frac{ktP_0}{1 + ktP_0} \right)^{i-2} - 2kP_iP_0 \left(\frac{1}{1 + ktP_0} \right) \tag{2.2.15}$$

Because the solution to the homogeneous equation is always the same, we can write:

$$P_i = P_0 \left(\frac{1}{1 + ktP_0} \right)^2 \int_0^t (i - 1) \left(\frac{ktP_0}{1 + ktP_0} \right)^{i-2} \left(\frac{1}{1 + ktP_0} \right)^2 kP_0 dt \tag{2.2.16}$$

This equation is easily integrated to give:

$$P_i = P_0 \left(\frac{1}{1 + ktP_0} \right)^2 \left(\frac{ktP_0}{1 + ktP_0} \right)^{i-1} \qquad (2.2.17)$$

Since we know that the suggested form holds for $i = 3$, and since we have proven that if it holds for $(i - 1)$ it must hold for i, we have proven the distribution of equation (2.2.17) through inductive reasoning.

From equation (2.2.5), the conversion of functional groups, p, is:

$$p = \frac{A_0 - A}{A_0} = \frac{B_0 - B}{B_0} = \frac{P_0 - P}{P_0} = \frac{ktP_0}{1 + ktP_0} \qquad (2.2.18)$$

so that equation (2.2.17) can be rewritten:

$$P_i = P_0(1 - p)^2 p^{i-1} \qquad (2.2.19)$$

This distribution is called the geometric distribution, the most probable distribution, or the Flory–Schulz distribution; we will use the first term. The properties of this important distribution, which reappears fairly regularly throughout the text, are discussed in Section 2.2.1.3.

This chain length distribution can be transformed into a molecular weight distribution as in Chapter 1 (see Section 1.4), but the relations are complicated for condensation polymerizations because the condensate is not lost from the end groups.[7] The geometric chain length distribution and the molecular weight distribution are shown in Figures 2.1 and 2.2; the mass of condensation products has been ignored in the latter. The chain length distribution is a monotonically decreasing function, while the molecular weight distribution exhibits a maximum.

Figure 2.1.
Geometric chain length distribution shown at different conversions.

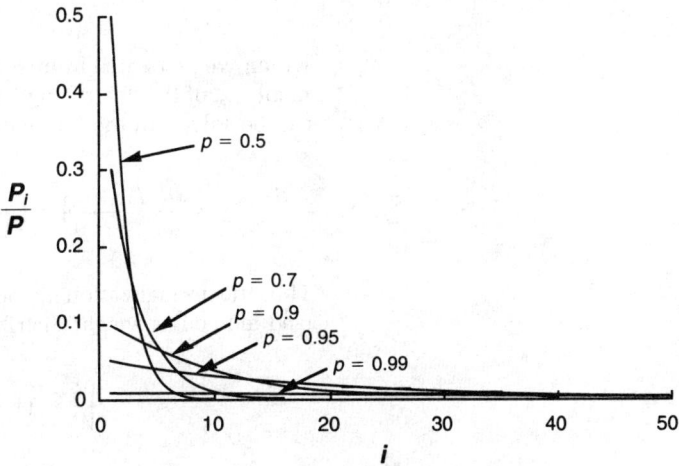

Figure 2.2.
Geometric molecular
weight distribution shown
at different conversions.

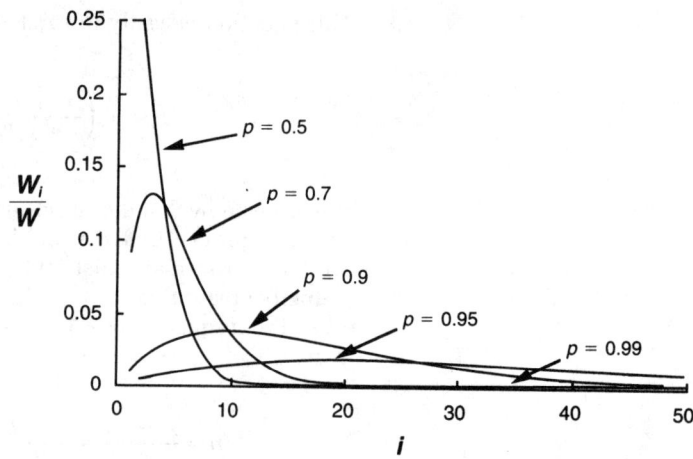

Example 2.1

Derive the molecular weight distribution for AB step polymerization, normalized like P_i/P so that the distribution sums to unity. Neglect the mass of the condensation products.

Solution:

The relation between the chain length distribution and the molecular weight distribution is shown in equation (1.4.8). Neglect of the mass of the condensation products gives m_i exactly proportional to i. The proportionality constant, equal to the mass of a monomer unit, is unimportant in this case because it will cancel out in the normalization. Thus, we let

$$W_i = iP_0(1 - p)^2 \, p^{i-1} \tag{2.2.20}$$

The normalization constant, W, is given as:

$$W = \sum_{i=1}^{\infty} W_i = \sum_{i=1}^{\infty} i \, P_0(1 - p)^2 p^{i-1} = P_0(1 - p)^2 \sum_{i=1}^{\infty} i \, p^{i-1} \tag{2.2.21}$$

which we recognize from equation (1.4.2) as the first moment, μ_1, of the chain length distribution. The summation can be solved in the following way:

$$\sum_{i=1}^{\infty} i \, p^{i-1} = \frac{d}{dp} \sum_{i=1}^{\infty} p^i = \frac{d}{dp} \left(\frac{1}{1 - p} \right) = \left(\frac{1}{1 - p} \right)^2 \tag{2.2.22}$$

Thus the normalization factor is merely P_0, and the normalized molecular weight distribution is:

$$\frac{W_i}{W} = i(1 - p)^2 \, p^{i-1} \tag{2.2.23}$$

2.2.1.2 Discrete Transformation Methods

Although sequential solution with induction was successful in the case of AB step polymerization, the technique seems somewhat uncertain or fragile because it relies on the discernment of an emerging pattern. The pattern was clear enough in this case but it may not always be so. Thus, it would be better to have a more systematic technique. A discrete transform method is such a technique, and it will often provide a concise representation of the rate equations, if not the solution.

Howe[8] apparently was first to make use of discrete transform methods in the analysis of polymerization (specifically free-radical) kinetics, introducing the generating function $G(s)$ to obtain the "characteristic function" of a chain length distribution. This procedure has broad application in the analysis of any stochastic processes.[9,10] An equivalent technique introduced by Scanlan,[11] and used in sample data control systems,[12] is the z-transform, which has been applied to the analysis of step growth polymerization.[13] The significant advantage of both the generating function $G(s)$ and the equivalent z-transform in the analysis of polymer kinetics is the large body of information available on the properties of the transform. Before returning to the specific case of AB step polymerization,* we briefly define the generating function and give its more important properties.

Let $P_i(t)$ be a function of a discrete variable i, $i = 1, 2, 3, 4, \ldots$ and of a continuous variable t. The generating function of $P_i(t)$ is defined as follows:

$$G(s,t) = \sum_{i=1}^{\infty} s^i P(t) \tag{2.2.24}$$

where s is a complex number in the unit circle.† The time dependence will often be left implicit. The corresponding operator we may call $\mathbf{G}$:

$$\mathbf{G} \equiv \sum_{i=1}^{\infty} s^i \tag{2.2.25}$$

Four properties of $G(s)$ are particularly useful. The first is the *shifting property*:

$$\mathbf{G}(P_{i+1}) = s^{-1}\,\mathbf{G}(P_i) - P_1 \tag{2.2.26}$$

The second property is the *convolution property*. If P_i and R_i are both functions of the discrete variable i, then:

$$\mathbf{G}\left(\sum_{j=1}^{i-1} P_{i-j}R_j\right) = \mathbf{G}(P_i)\,\mathbf{G}(R_i) \tag{2.2.27}$$

---

*Source material for the definitions and the properties may be found in the texts on sample data control systems[14,15] and theory of complex variables.[16]

†The z-transform replaces s with z^{-1}, so that G(z) converges without rather than within the unit circle.

$G(s)$ is called a generating function because of its property of *moment generation*. The kth moment of the distribution P_i is derived from $G(s)$ as follows:

$$\mu_k = \lim_{s \to 1} \left(\frac{\partial^k G(s)}{\partial (\ln s)^k} \right) \qquad (2.2.28)$$

The final property is that of *inversion*. It should be possible to express $G(s)$ as a power series in s. Identifying each term in the series comprises inversion. A more elegant, though not necessarily more useful, method of inversion is available using integration in the complex plane:

$$P_i = \frac{-1}{2\pi(-1)^{1/2}} \oint G(s)s^{-(1+i)}ds \qquad (2.2.29)$$

Table 2.1 lists other useful properties and discrete transforms of common distributions.

Returning to AB step polymerization, we transform equations (2.2.6) and (2.2.7) to:

$$\frac{\partial G(s)}{\partial \tau} = G^2(s) - 2\,G(s)\,G(1) \qquad (2.2.30)$$

TABLE 2.1 / Properties of the Generating Function

Function u_i	Generating Function $U(s) = \sum_{i=1}^{\infty} s^i u_i$
1. $aP_i + bR_i$	$aG(s) + bH(s)$
2. P_{i+j}	$s^{-j}G(s) + \sum_{k=1}^{j} u_k s^{k-j}$
3. $c^{ai}P_i$	$G(c^a s)$
4. c	$cs(1 - s)^{-1}$
5. $\delta(j - i)$	s^j
6. c^i	$cs(1 - cs)^{-1}$
7. ic^i	$cs(1 - cs)^{-2}$
8. $i^2 c^i$	$cs(1 + cs)(1 - cs)^{-3}$
9. iP_i	$s\dfrac{\partial G(s)}{\partial s}$
10. $\dbinom{i + k}{i} c^i$	$(1 + cs)^{k+1} - 1$
11. $\dfrac{c^i}{i}$	$-\ln(1 - cs)$
12. $\dfrac{c^i}{i!}$	$e^{cs} - 1$
13. $\dfrac{c^{i-1}}{(i - 1)!}$	se^{cs}
14. $P_i = 0 \quad i \le k$ $\quad\ = R_{i-k} \quad i > k$	$s^k H(s)$
15. $\sum_{k=1}^{i-1} P_{i-k} R_k$	$G(s)H(s)$

where $d\tau = k\,dt$. Since the feed is pure monomer with concentration P_0, the initial condition is $G(s,0) = P_0 s$. The evolution of the total concentration of polymer, $G(1)$, is found by letting $s = 1$:

$$\frac{dG(1)}{d\tau} = -G(1)^2 \qquad (2.2.31)$$

which is equivalent to equation (2.2.4), as $G(1) = P$, and thus has the same solution, equation (2.2.5). Equation (2.2.30) can be easily solved using generating functions normalized by $G(1)$[17]:

$$y = \frac{G(s)}{G(1)} \qquad (2.2.32)$$

Equation (2.2.30) can be rewritten as follows:

$$\frac{\partial y}{\partial G(1)} = \frac{y(1-y)}{G(1)} \qquad (2.2.33)$$

which leads to

$$G(s) = \frac{P_0 \left(\dfrac{G(1)}{P_0}\right)^2 s}{1 - \left(1 - \dfrac{G(1)}{P_0}\right) s} \qquad (2.2.34)$$

Proofs of these last two equations are left as an exercise to the reader. To invert equation (2.2.34) back to the discrete i-space, we make use of identity 6 in Table 2.1, so that

$$G(s) = P_0 \left(\frac{G(1)}{P_0}\right)^2 \sum_{i=1}^{\infty} \left(1 - \frac{G(1)}{P_0}\right)^{i-1} s^i \qquad (2.2.35)$$

The ith term of this series corresponds to P_i:

$$P_i = P_0 \left(\frac{G(1)}{P_0}\right)^2 \left(1 - \frac{G(1)}{P_0}\right)^{i-1} \qquad (2.2.36)$$

which is again the geometric distribution given in equations (2.2.17) and (2.2.19).

2.2.1.3 Direct Solution for Moments

The entire distribution, or its unique transform $G(s)$, may be more than we want to know; specific moments of the distribution may suffice. Because the geometric distribution will recur throughout this text, it is worthwhile examining its first three moments, which allow calculation of DP_n, DP_w, and the polydispersity Q. The first three moments are:

$$\mu_0 = \sum_{i=1}^{\infty} P_i = P_0 \sum_{i=1}^{\infty} (1-p)^2\, p^{i-1} = P_0\,(1-p)^2 \sum_{j=0}^{\infty} 2p^j \qquad (2.2.37)$$

$$\mu_1 = \sum_{i=1}^{\infty} iP_i = P_0 \, (1 - p)^2 \sum_{i=1}^{\infty} i \, p^{i-1} = P_0 \, (1 - p)^2 \sum_{j=0}^{\infty} (j + 1)p^j \qquad (2.2.38)$$

$$\mu_2 = \sum_{i=1}^{\infty} i^2 P_i = P_0 \, (1 - p)^2 \sum_{i=1}^{\infty} i^2 \, p^{i-1} = P_0 \, (1 - p)^2 \sum_{j=0}^{\infty} (j + 1)^2 p^j \qquad (2.2.39)$$

Since p is less than 1, the infinite sums converge[18] and the moments are given as:

$$\mu_0 = P_0 \, (1 - p) \qquad (2.2.40)$$

$$\mu_1 = P_0 \qquad (2.2.41)$$

$$\mu_2 = P_0 \frac{1 + p}{1 - p} \qquad (2.2.42)$$

Proofs of these equations are left as an exercise for the reader [approaches like that shown in equation (2.2.22) are helpful]. Equations (2.2.36) and (2.2.37) are in retrospect obvious; μ_1 is the total concentration of mers, always equal to the initial concentration, and μ_0 is the total polymer concentration equal to P and $G(1)$.

The average degrees of polymerization are therefore given by:

$$DP_n = \frac{1}{1 - p} \qquad (2.2.43)$$

$$DP_w = \frac{1 + p}{1 - p} \qquad (2.2.44)$$

and the polydispersity Q by

$$Q = \frac{\mu_2 \mu_0}{\mu_1^2} = 1 + p \qquad (2.2.45)$$

The remarkable characteristic of linear stepwise polymerizations, which was indicated in Table 1.5, is the slow growth in chain length in terms of conversion. Figure 2.3 shows the increase of DP_n and DP_w with conversion, which must be very near unity before chains of appreciable length are produced. Thus, whereas for small-molecule reactions a conversion of 95% might be acceptable, for an AB step polymerization this corresponds to a DP_n of only 20, insufficient for many end uses. Thus, the demands on conversion are greater for step polymerizations because this affects not only the economics through the yield, but also the kind of product obtained. The polydispersity Q is seen to increase steadily with conversion, approaching 2, just as the chain length and molecular weight distributions broaden with conversion as seen in Figures 2.1 and 2.2.

2.2.1.4 Moment Integration
If the moments of the distribution were all we wanted, it would be worthwhile to develop ways that avoid first having to find

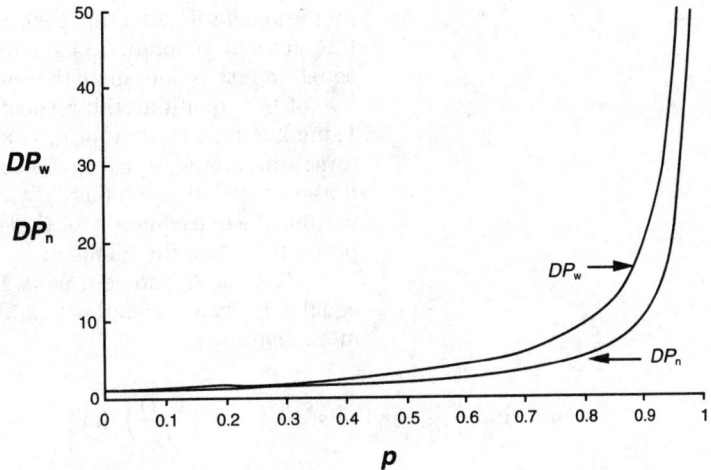

Figure 2.3.
DP_n and DP_w versus conversion for AB step polymerization (geometric distribution).

the distribution. We would prefer to solve directly for the moments. This can be done by applying the operator M_k to the evolution equations (2.2.6) and (2.2.7):

$$M_k \equiv \sum_{i=1}^{\infty} i^k \qquad (2.2.46)$$

Doing this gives us equations for the evolution of the first three moments:

$$\frac{d\mu_0}{d\tau} = -\mu_0^2 \qquad (2.2.47)$$

$$\frac{d\mu_1}{d\tau} = 0 \qquad (2.2.48)$$

$$\frac{d\mu_2}{d\tau} = 2\mu_1^2 \qquad (2.2.49)$$

with initial conditions given by $\mu_k(\tau = 0) = P_0$, for all k. Equation (2.2.47) is of course equivalent to equations (2.2.4) and (2.2.31). Solving these equations yields the following:

$$\mu_0 = \frac{P_0}{1 + ktP_0} \qquad (2.2.50)$$

$$\mu_1 = P_0 \qquad (2.2.51)$$

$$\mu_2 = P_0 (1 + 2P_0kt) \qquad (2.2.52)$$

These three equations are equivalent to equations (2.2.40)–(2.2.42), through the relationship between time and conversion given in equation (2.2.18).

2.2.1.5 From Generating Function
An equivalent route to obtaining the evolution equations for the moments of the distribution makes use of the kinetic equation

for the generating function $G(s)$, equation (2.2.30). One can apply the moment generating characteristic of $G(s)$ to the evolution equation just as one can to the solution; by taking the kth derivative of the equation with respect to $\ln(s)$ and evaluating at $s = 1$, the kinetic equation for μ_k is obtained. This, however, leaves three differential equations to solve. There is no reason in the present case (although there may be in others; see Chapter 5) not to solve the one equation for $G(s)$ and use the moment generation property to find the moments.

The zeroth moment μ_0 is simply equal to $G(1)$, which is equal to P given in equation (2.2.5). The first moment, μ_1, requires more math:

$$\mu_1 = \lim_{s \to 1} \frac{\partial G(s)}{\partial (\ln s)} = \lim_{s \to 1} s \frac{\partial G(s)}{\partial s} = P_0 \left(\frac{G(1)}{P_0} \right)^2 \lim_{s \to 1} \frac{s}{\left[1 - \left(1 - \frac{G(1)}{P_0} \right) s \right]^2} = P_0 \quad (2.2.53)$$

The second moment, μ_2, is found similarly:

$$\mu_2 = \lim_{s \to 1} \frac{\partial^2 G(s)}{\partial (\ln s)^2} = \lim_{s \to 1} \left[s \frac{\partial}{\partial s} \left(s \frac{\partial G(s)}{\partial s} \right) \right]$$

$$= P_0 \left(\frac{G(1)}{P_0} \right)^2 \lim_{s \to 1} \frac{\left[1 + \left(1 - \frac{G(1)}{P_0} \right) s \right] s}{\left[1 - \left(1 - \frac{G(1)}{P_0} \right) s \right]^3} \quad (2.2.54)$$

$$= 2P_0 \frac{P_0}{G(1)} - P_0 = P_0(1 + 2ktP_0)$$

The average degrees of polymerization and the polydispersity then follow immediately, and all results are consistent with those found earlier.

2.2.2 Statistical Solution

The feature unique to polymers (or systems with aggregated species in general) is that there is an infinite (though countable) number of species to be treated. This is the simplest aspect of the structural diversity discussed in Section 1.1, and the problems it entails have no analogue in small-molecule systems. Nonetheless, we have established techniques that allow solution within the formalism familiar from common reactor engineering. We can also use a less familiar formalism: statistical or probabilistic reasoning. Historically, statistical approaches predate the use of techniques based on kinetic equations, such as those just presented. In many cases they offer a more straightforward route to desired properties, either the entire distribution or simply the average degrees of polymerization. Although the probabilistic approaches are not as rigorously systematic as the kinetic approaches, and thus rely on the user to identify the proper statistical variables and understand the implicit assumptions, they cannot be neglected for two reasons. First, the student will no doubt encounter the statistical methods and should be familiar with them. Second, some problems are more simply solved by statistical methods, and certain problems can be solved only by statistical reasoning (see Chapter 5). The shortcomings of statistical approaches will be discussed in more detail in Section 2.2.3.

The statistical approaches are based on conversion, p, rather than time. The conversion of functional groups then becomes the probability that any particular functional group is reacted. In our example of the polyamide, we have the probability of any given carboxylic acid group or amine group having been reacted. The overall conversion can be applied to each individual group as a probability only because all groups are assumed to be equally reactive; this implicit assumption is made more explicit in the kinetic equations by independence of the rate constant k from length i.

Numerous statistical methods are used to treat polymerization problems; we will deal with three: the combinatorial approach, the formal Markov chain theory approach, and the recursive approach. The other most significant method applies a discrete transform to derive a probability generating function.[19] These techniques differ in language and power but will yield equivalent results if based on the same model of the polymerization. (They are also emphatically *techniques* rather than *models*, as they are often wrongly called.) We will delay discussion of the formal Markov chain approach until it is appropriate to present an example that shows its power (see Section 2.4).

2.2.2.1 Combinatorial Method

The mole fraction of polymer of length i is equivalent to the probability of a polymer molecule being of length i. To find this, we can visualize standing on the unreacted amine group of a polymer. For this molecule to be i units long, it must have $i - 1$ amide linkages and one unreacted carboxylic acid group. The probability of $i - 1$ amide linkages is the product of $i - 1$ separate probabilities that a carboxylic acid group is reacted, that is, p^{i-1}. The probability that the final carboxylic group is unreacted is $1 - p$, so the probability of finding a molecule with i mers is

$$\frac{P_i}{P} = (1 - p)\, p^{i-1} \qquad (2.2.55)$$

We need not include a factor of $1 - p$ to account for the probability of finding the unreacted amine group at the "beginning" of the molecule. That was requisite in having a molecule and being positioned at its end. Introducing the additional factor of $1 - p$ would have given us the molar concentration rather than the mole fraction, and since the total concentration of mers is P_0, we can recover the distribution P_i of equation (2.2.19):

$$P_i = P_0\, (1 - p)^2\, p^{i-1} \qquad (2.2.19)$$

The combinatorial reasoning leads us to the same result as the kinetic derivation, with much less math. More than that, it gives some physical insight as to why the geometric distribution appears. However, it should be borne in mind that the entire analysis is based on an implicit assumption of equal reactivity implicit in the use of conversion, p, for all functional groups. The only missing piece is the relationship between time and conversion, that is, the kinetics of the reaction. This information, necessary to the design of a polymerization reactor, must be obtained from a kinetic analysis, but one that is no more intricate than for the small-molecule reaction A + B → C. The uniquely polymeric part of the problem resides entirely in the statistical analysis.

2.2.2.2 Recursive Method

A second approach is somewhat akin to the moment approaches in the kinetic solution, although here rather than directly solving for moments one is solving for average degrees of polymerization or molecular weights (or ratios of moments of the chain length distribution). An approach such as this has taken a variety of forms in the literature; here we illustrate that first derived by Macosko and Miller.[20-22] It is termed the "recursive" approach because it relies on the statistical equivalence of all functional groups of the same kind, which is guaranteed by equal reactivity. Thus if one takes an imaginary walk along a polymer chain, one has the recurring experience of finding that each A group arrived at is statistically no different from the A groups encountered previously.

The polymer resulting from the AB polymerization is:

$$
\begin{array}{c}
\overset{\text{in}}{\overset{\leftarrow}{|}} \qquad \overset{\text{out in}}{\overset{\leftarrow \; \rightarrow}{|}} \qquad \overset{\text{in out}}{\overset{\leftarrow \; \rightarrow}{|}} \qquad \overset{\text{in}}{\overset{\rightarrow}{|}} \\
(\text{———}AB\text{———} AB\text{———} AB\text{———} \mathbf{AB} \text{———} AB \text{———})
\end{array}
\qquad (2.2.56)
$$

The original AB monomers in this drawing are indicated as B ——— A; "in" and "out" are directions in which we can walk. Picking the A group in bold type, we define the "in" direction as that toward the B group from the same original monomer unit as the A; "out" is the opposite direction, toward the B group with which the A reacted, if it did, or toward nothing if the A is unreacted. Similar definitions apply to the directions associated with B groups.

The expected (or average) mass of the polymer chain attached to a randomly chosen A group looking "out" is denoted as $E(W_A^{out})$. It is equal to the expected weight attached to an A looking "out," conditional on A having reacted with B, times the probability of A reacting with B (p), plus the expected weight attached to an A looking "out" conditional on A not having reacted with B, times the probability of A not reacting $1 - p$. In equation form this reads:

$$E(W_A^{out}) = E(W_A^{out} \mid \text{A r B})p + E(W_A^{out} \mid \text{A dnr B})(1 - p) \qquad (2.2.57)$$

where r stands for "reacts with" and dnr for "does not react with."

If we neglect the mass of the unliberated condensate at the chain end, the second term of equation (2.2.57) is zero; thus:

$$E(W_A^{out}) = E(W_A^{out} | A \text{ r } B)p \tag{2.2.58}$$

The expected weight attached to A looking "out" given that A has reacted with a B is equal to the expected weight attached to a B group looking "in":

$$E(W_A^{out} | A \text{ r } B) = E(W_B^{in}) \tag{2.2.59}$$

This, in turn, is equal to the weight of an AB-mer (after liberation of the condensate product) plus the expected weight attached to an A looking "out":

$$E(W_B^{in}) = M_{AB} + E(W_A^{out}) \tag{2.2.60}$$

Equations (2.2.58)–(2.2.60) may be solved to give the three unknown expected weights. A parallel set of three equations can be derived for the B group:

$$E(W_B^{out}) = E(W_B^{out} | A \text{ r } B)p + E(W_B^{out} | A \text{ dnr } B)(1 - p) \tag{2.2.61}$$

$$E(W_B^{out} | A \text{ r } B) = E(W_A^{in}) \tag{2.2.62}$$

$$E(W_A^{in}) = M_{AB} + E(W_B^{out}) \tag{2.2.63}$$

These two sets of recursions close simply and algebraically because walking along the chain has the mathematical result of bringing one to another position, statistically equivalent to the starting point.

The weight-average molecular weight is given by the sum of the weight of an AB-mer plus the expected weights attached to each arm looking "out":

$$M_w = M_{AB} + E(W_A^{out}) + E(W_B^{out}) \tag{2.2.64}$$

or analogously for weight-average degree of polymerization:

$$DP_w = 1 + E(N_A^{out}) + E(N_B^{out}) \tag{2.2.65}$$

where the $E(N_A^{out})$ and $E(N_B^{out})$ are the expected *numbers* of units attached to A and B. These are derived from $E(W_A^{out})$ and $E(W_B^{out})$ by replacing M_{AB} with 1. Solving equations (2.2.58)–(2.2.63), and the analogous set for the $E(N)$, gives:

$$E(W_A^{out}) = E(W_B^{out}) = M_{AB}E(N_A^{out}) = M_{AB}E(N_B^{out}) = M_{AB}\frac{p}{1-p} \tag{2.2.66}$$

$$E(W_A^{in}) = E(W_B^{in}) = M_{AB}E(N_A^{in}) = M_{AB}E(N_B^{in}) = M_{AB}\frac{1}{1-p} \tag{2.2.67}$$

By substituting into equations (2.2.64) and (2.2.65), we can write:

$$M_w = M_{AB} + M_{AB}\frac{p}{1-p} + M_{AB}\frac{p}{1-p} = M_{AB}\frac{1+p}{1-p} \tag{2.2.68}$$

and

$$DP_w = \frac{1 + p}{1 - p} \qquad (2.2.69)$$

where the unit weight M_{AB} does not include the weight of the released water. The error is in not accounting for the added weight present at the ends, for at each end there is essentially half a water.

In the preceding calculations, we have picked monomers at random; this is a "weight averaging" process for molecules—the larger the molecule, the proportionately greater chance it has of being chosen. If, instead, we choose molecules at random, by picking chain ends at random, and ask for the expected weight attached to the end group looking "in," we obtain a "number-averaged" quantity. If we pick end groups, we must statistically weight the $E(W^{in})$ by the mole fraction of each type of end group, that is, the mole fractions of unreacted A and B, (merely equal to the overall mole fractions x_A and x_B, in this case); thus:

$$M_n = x_A E(W_A^{in}) + x_B E(W_B^{in}) \qquad (2.2.70)$$

$$DP_n = x_A E(N_A^{in}) + x_B E(N_B^{in}) \qquad (2.2.71)$$

Since there are equal numbers of A and B ends in this simple case $x_A = x_B = 1/2$, we get from equations (2.2.66) and (2.2.67):

$$M_n = M_{AB} \frac{1}{2} \frac{1}{1 - p} + M_{AB} \frac{1}{2} \frac{1}{1 - p} = M_{AB} \frac{1}{1 - p} \quad (2.2.72)$$

and

$$DP_n = \frac{1}{1 - p} \qquad (2.2.73)$$

We have gone into some detail in this case because the reasoning is most likely not familiar to the student and because it is a straightforward case that illustrates the procedure simply.

2.2.3 Kinetic Versus Statistical Treatments

As we have mentioned, polymerizations were first treated with statistical approaches[4] and only later by kinetic equations. There are cases for which one or the other may be more better suited. The polymer reaction engineer should understand and be able to use both, knowing their respective difficulties and pitfalls.

The kinetic approach has many advantages. It is nearly always applicable. The number and complexity of the resulting equations may become prohibitive as the number of components increases, as differences in reactivity appear, as intricate reactor configurations are introduced, and so forth, but one *can* write the appropriate equations. In addition, the kinetic approach is less intuitive than the statistical approach and so more foolproof. The former, however, will often be more laborious than the latter.

This will become apparent especially in Chapter 5 for nonlinear polymerizations. In fact, we will find that after gelation, the kinetic approach is unable to describe the internal structure of the gel, whereupon the difficulty becomes an impossibility.

The simplicity of the statistical approach is its biggest advantage, especially compared to involved kinetic approaches. This is manifested particularly in nonlinear polymerizations, and specifically after the gel point, when only statistical methods will serve for probing the internal structure of the gel. This simplicity is purchased at a price, however. A trivial disadvantage is that the answers are not in terms of time, but conversion. Generally, the complete distribution can be obtained only from the combinatorial method, which rapidly becomes too difficult to apply as the polymerization becomes more complicated. A more serious disadvantage is the need for the user to recognize the appropriate statistical variables and to know what assumptions are implicit in the statistical formulation. The responsibility is put on the user to properly formulate the statistical problem or know if a framework (such as Markov chain theory, to be dealt with in Section 2.4.3) is really appropriate for the problem. (Statistical approaches do not constitute models, and in any event, not all models can be properly analyzed by a given technique.) While examples of flawed intuition with regard to statistical arguments are common in the literature, a number of them are tied to the last, and greatest disadvantage: that the simple statistical approach in a number of cases is made impossible by some physical attribute of the system (e.g., nonbatch reactor configuration, substitution effects, cyclization, chain growth without termination) that leads to inability to reach a statistically equivalent point. We hasten to add that this stricture applies only to simple application of statistical approaches; more sophisticated statistical derivations often are possible, but these are beyond the scope of this text. When cases are treated for which the simple statistical approach cannot be used, we will identify the source of the problem so as to deepen the reader's intuition on that subject.

From the foregoing discussion, it should be clear that both kinetic and statistical ways of thinking about polymerization problems should be at the reaction engineer's disposal. One may even wish to use them simultaneously on the same problem.[17] Even if no such synthesis is being attempted, there are cases for which one approach might be better. The paradigm may not provide the clearest example of the greater simplicity of the statistical approach, but the case of $A_2 + B_2$ polymerization, which allows the possibility of unbalanced stoichiometry, will (see Section 2.4).

2.3 Stepwise Chemistries

A number of different reactions, which should be known to the reader through organic chemistry, can produce polymers by stepwise mechanisms. Now we review those that are the most common industrially; more complete treatments of chemistry (including the many side reactions that occur) can be found in the books by Odian[6,23] and Lenz.[24]

2.3.1 Polyamides

The formation of an amide linkage can occur by the reaction of a carboxylic acid with an amine, liberating the condensate water:

$$R_{(1)}\text{—}\overset{\displaystyle O}{\overset{\|}{C}}\text{—OH} + H_2N\text{—}R_{(2)} \rightarrow R_{(1)}\text{—}\overset{\displaystyle O}{\overset{\|}{C}}\text{—}\overset{\displaystyle H}{\overset{|}{N}}\text{—}R_{(2)} + H_2O \qquad (2.3.1)$$

Polyamides are thus stepwise polymers formed by multiple amide linkages. Proteins and enzymes, those familiar biological macromolecules, also are composed of amide linkages; the synthetic polymers are much simpler, however, comprising not specific sequences of different amino acids but, in the simplest case, merely a repetition of one monomer. The polarity of the amide linkage causes semicrystalline polyamides to have high melting points (200–300°C), which was the property that made them desirable as synthetic fibers.

2.3.1.1. Synthesis from Diacids and Diamines

The most common route to the production is one of the type $A_2 + B_2$ (i.e., from diacids and diamines). The most important polyamide, nylon 6/6 ($T_m = 265°C$), is made in this way:

$$i H_2N\text{—}(CH_2)_6\text{—}NH_2 + i HO\text{—}\overset{\displaystyle O}{\overset{\|}{C}}\text{—}(CH_2)_4\text{—}\overset{\displaystyle O}{\overset{\|}{C}}\text{—OH} \rightarrow$$

1,6-hexane diamine   adipic acid

$$H\left[\overset{\displaystyle H}{\overset{|}{N}}\text{—}(CH_2)_6\text{—}\overset{\displaystyle H}{\overset{|}{N}}\text{—}\overset{\displaystyle O}{\overset{\|}{C}}\text{—}(CH_2)_4\text{—}\overset{\displaystyle O}{\overset{\|}{C}}\right]_i OH + (i-1) H_2O \qquad (2.3.2)$$

poly(hexamethylene adipamide), or nylon 6/6

The first number in the common name refers to the number of carbons in the diamine, the second to the number in the diacid. [If a nylon is made from AB polymerization, i.e., from an amino acid, it has only one index, as in equation (2.2.1), and is called a monadic nylon.]

2.3.1.2 Synthesis from Diacyl Chlorides and Diamines

Amide linkages can also be formed by the reaction of an acyl chloride with an amine:

$$R_{(1)}\text{—}\overset{\displaystyle O}{\overset{\|}{C}}\text{—Cl} + H_2N\text{—}R_{(2)} \rightarrow R_{(1)}\text{—}\overset{\displaystyle O}{\overset{\|}{C}}\text{—}\overset{\displaystyle H}{\overset{|}{N}}\text{—}R_{(2)} + HCl \qquad (2.3.3)$$

Aliphatic nylons can be made in this way, the most well-known being nylon 6/10, which is carried out interfacially between phases, one containing 1,6-hexanediamine and the other sebacoyl chloride. This reaction, popular as the "nylon rope trick," is not used commercially because the acyl chlorides are more expensive than the corresponding acid. For the production of aromatic polyamides (polyaramids), however, the reaction of the aromatic carboxylic acid is too slow, and so acyl chlorides are preferred. A common example goes by the trade name of Kevlar:

$$i H_2N - \bigcirc - NH_2 + iCl-\overset{\overset{O}{\|}}{C} - \bigcirc - \overset{\overset{O}{\|}}{C}-Cl \rightarrow$$

$$H \left[\overset{\overset{H}{|}}{N} - \bigcirc - \overset{\overset{H}{|}}{N}-\overset{\overset{O}{\|}}{C}- \bigcirc - \overset{\overset{O}{\|}}{C} \right]_i OH + (i - 1) H_2O \quad (2.3.4)$$

poly(iminocarbonyl-1,4,-phenylene)

The wholly aromatic backbone gives even higher T_m, often greater than 500°C, and so these polyaramids are used in applications requiring high heat and flame resistance.

2.3.1.3 Anhydride Synthesis
Polyamides can be made without condensate formation if an anhydride replaces the carboxylic acid:

$$R_{(1)} \overset{\overset{O}{\|}}{\underset{\overset{\|}{O}}{\overset{C}{\underset{C}{\Big\langle}}}} O + H_2N-R_{(2)} \longrightarrow HO-\overset{\overset{O}{\|}}{C}-R_{(1)}-\overset{\overset{O}{\|}}{\underset{\overset{|}{H}}{C}}-\overset{\overset{H}{|}}{N}-R_{(2)} \quad (2.3.5)$$

An example of this is the following:

$$i \, O \overset{O}{\underset{O}{\bigcirc}} O + i \, H_2N-\bigcirc-NH_2 \longrightarrow$$

pyromellitic anhydride p-phenylenediamine   (2.3.6)

$$\left[\overset{\overset{O}{\|}}{HO-C} \bigcirc \overset{\overset{O}{\|}}{C-OH} \right.$$
$$\left. HO \left[\underset{\overset{\|}{O}}{C} \bigcirc \underset{\overset{\|}{O}}{C} - \overset{\overset{H}{|}}{N} - \bigcirc - \overset{\overset{H}{|}}{N} \right] H \right.$$

In practice this polyamide is not the final product, however. It is turned into a polyimide by the reaction between the neighboring amine and carboxylic acid groups, liberating water:

$$(2.3.7)$$

Such stiff chains have high temperature resistance.

2.3.1.4 Other Routes

None of the common polyamides are formed in an AB-type polymerization, which was our paradigm. The monadic nylons, such as nylon 6, could be made from the appropriate amino acid, but generally the rate of cyclization (reaction within the monomer) is too high, so these are generally produced by a chainwise ring-opening polymerization (see Chapter 3). Exceptions include the formation of a different Kevlar by reaction of p-aminobenzoic acid,[23] a stiff monomer which will not cyclize, and possibly the polymerization of 11-aminoundecanoic acid shown in equation (2.2.1).[6]

2.3.2 Polyesters

Polyesters are the most common stepwise polymers. In contrast to polyamides, they are almost always partially aromatic. This is largely because both are intended for similar purposes, such as fibers, which require high T_g and T_m. While in nylons the polarity of the amide linkage assists in achieving these properties, the stiffness of the aromatic group is required in polyesters. The reaction to form ester linkages is similar to that forming amide linkages:

$$R_{(1)}-\overset{O}{\overset{\|}{C}}-OH + HO-R_{(2)} \xrightarrow{acid} R_{(1)}-\overset{O}{\overset{\|}{C}}-O-R_{(2)} + H_2O \qquad (2.3.8)$$

the hydroxy group replacing the amine. Again, water is liberated. The presence of an acid catalyst (not necessarily for the more reactive and favorable amidation reaction) is indicated. Polyesters are rarely, however, made by the route indicated in equation (2.3.8).

2.3.2.1 Synthesis from Diesters by Ester Interchange

The most common polyester is poly(ethylene terephthalate) (PET), which finds application in the common polyester fiber, in film (Mylar), and in food containers such as beverage bottles. It has a T_m of 270°C, comparable to that of nylon 6/6. It is made by ester interchange reaction of a single monomer, bishydroxy-ethyl terephthalate:

$$i\ \text{HO—CH}_2\text{CH}_2\text{—}\overset{\displaystyle O}{\overset{\|}{C}}\text{—}\langle\bigcirc\rangle\text{—}\overset{\displaystyle O}{\overset{\|}{C}}\text{—CH}_2\text{CH}_2\text{—OH} \rightarrow$$

bishydroxyethyl terephthalate (2.3.9)

$$\text{H}\left[\text{O—CH}_2\text{CH}_2\text{—}\overset{\displaystyle O}{\overset{\|}{C}}\text{—}\langle\bigcirc\rangle\text{—}\overset{\displaystyle O}{\overset{\|}{C}}\right]_i\text{OCH}_2\text{CH}_2\text{—OH} + (i-1)\ \text{HO—CH}_2\text{CH}_2\text{—OH}$$

This monomer is produced by esterification of terephthalic acid, or by ester exchange of the dimethyl ester thereof. The polymerization reaction proper, then, ideally comprises an A_2 type of polymerization.

2.3.2.2 Synthesis from Diacid Chloride and Diol

The most common use for this route is in the production of polycarbonates, which are polyesters of carbonic acid. The most common of these is that based on bisphenol A:

$$i\ \text{HO}\langle\bigcirc\rangle\text{—}\overset{\overset{\displaystyle CH_3}{|}}{\underset{\underset{\displaystyle CH_3}{|}}{C}}\text{—}\langle\bigcirc\rangle\text{—OH} + i\ \text{Cl—}\overset{\displaystyle O}{\overset{\|}{C}}\text{—Cl} \rightarrow$$

bisphenol A   phosgene

$$\text{H}\left[\text{O—}\langle\bigcirc\rangle\text{—}\overset{\overset{\displaystyle CH_3}{|}}{\underset{\underset{\displaystyle CH_3}{|}}{C}}\text{—}\langle\bigcirc\rangle\text{—O—}\overset{\displaystyle O}{\overset{\|}{C}}\right]_i\text{Cl} + (i-1)\ \text{HCl} \qquad (2.3.10)$$

The product, used as an engineering plastic, is usually referred to simply as polycarbonate and has a T_m similar to that of PET (270°C). The polymerization is an interfacial one; for details, see Chapter 6.

2.3.2.3 Anhydride Synthesis

Polyesters as well as polyamides can be made from anhydrides. A common one is based on the reaction between ethylene glycol and maleic anhydride:

$$i\ \ \overset{\displaystyle O=C}{\underset{\displaystyle HC}{}}\overset{O}{\underset{\displaystyle =}{}}\overset{\displaystyle C=O}{\underset{\displaystyle CH}{}} + i\ \text{HO—CH}_2\text{CH}_2\text{—OH} \rightarrow$$

$$\text{H}\left[\text{O—(CH}_2)_2\text{—O—}\overset{\displaystyle O}{\overset{\|}{C}}\text{—}\overset{\displaystyle H}{\overset{|}{C}}\text{=}\overset{\displaystyle H}{\overset{|}{C}}\text{—}\overset{\displaystyle O}{\overset{\|}{C}}\right]_i\text{OH} + (i-1)\ \text{H}_2\text{O} \qquad (2.3.11)$$

In this case, i is usually not terribly large (often ≤ 10). This unsaturated polyester resin is used as a prepolymer for further free-radical polymerization forming networks (see Chapter 5).

2.3.3 Polyethers, Polysulfones, and Polysulfides

The more common aliphatic polyethers, such as poly(ethylene oxide), are produced by chainwise mechanisms to be discussed in Chapter 3. Aromatic polyethers such as poly(phenylene oxide) are produced by oxidative coupling, a mechanism having both stepwise and chainwise attirbutes.* Other aromatic polyethers, however, are produced by a mechanism more nearly stepwise, aromatic nucleophilic substitution:

$$\phi\text{-ONa} + X\text{-}\phi\text{-Y-} \longrightarrow \phi\text{-O-}\phi\text{-Y} + \text{NaX} \qquad (2.3.12)$$

If Y is a carbonyl group, then we have a polyetherketone; if Y is SO_2, then the polymer formed is a polyethersulfone. This is usually accomplished by an $(A_2 + B_2)$-type polymerization—that is, between aromatic dihalides and bisphenolate salts. Two common examples are polyetheretherketone (PEEK) and polyethersulfone (PES):

PEEK:
$$\left[O\text{-}\phi\text{-}\overset{\overset{\displaystyle O}{\|}}{C}\text{-}\phi\text{-O-}\phi \right]_i \qquad (2.3.13)$$

PES:
$$\left[O\text{-}\phi\text{-}\underset{\underset{\displaystyle O}{\|}}{\overset{\overset{\displaystyle O}{\|}}{S}}\text{-}\phi \right]_i \qquad (2.3.14)$$

The latter can be made, not only from an $(A_2 + B_2)$-type polymerization, but also from the self-polymerization of an AB-type monomer. The polyetherketones are semicrystalline materials of high T_g (143°C for PEEK) and T_m in excess of 300°C (334°C for PEEK), while the polysulfones are amorphous but with high T_g (around 200°C). Both are used as plastics where thermal stability is required, the polyetherketones being able to withstand higher temperatures.

---

*The mechanism is stepwise in that species of all sizes react with one another, but chainwise in that chains grow (or diminish) one unit at a time and that an increase in the average chain length comes only with the reactions of a monomeric unit. See Odian[23] for a discussion of the mechanism. We will deal with this case neither here nor in Chapter 3.

Related is the formation of aromatic polysulfides, although the reaction probably involves more than the simple nucleophilic substitution:

$$i\ Cl-\langle\bigcirc\rangle-Cl + i\ Na_2S \rightarrow H-\left[-\langle\bigcirc\rangle-S-\right]_i + (i-1)\ NaCl \qquad (2.3.15)$$

p-dichlorobenzene   sodium sulfide   poly(p-phenylenesulfide)

This highly crystalline polymer has a T_g of 85°C and a T_m of 285°C.

2.3.4 Polyurethanes and Polyureas

The formation of urethane linkages proceeds by the reaction of an isocyanate with a hydroxy group without the release of a condensate product:

$$R_{(1)}-N=C=O + HO-R_{(2)} \rightarrow R_{(1)}-\overset{H}{\underset{}{N}}-\overset{O}{\underset{}{C}}-O-R_{(2)} \qquad (2.3.16)$$

while replacement of the hydroxy with an amine leads to urea linkages:

$$R_{(1)}-N=C=O + H_2N-R_{(2)} \rightarrow R_{(1)}-\overset{H}{\underset{}{N}}-\overset{O}{\underset{}{C}}-\overset{H}{\underset{}{N}}-R_{(2)} \qquad (2.3.17)$$

The usual route is $A_2 + B_2$ polymerization, but often as copolymers. An example is the reaction of toluene diisocyanate with short diols such as 1,4-butanediol and long flexible polyols (e.g., based on polypropylene oxide) to yield a microphase-separated morphology. Two of the most visible uses for polyurethanes are for foams for upholstery and in injection-molded items such as automobile fenders.

2.3.5 Epoxies

Epoxy resins are oligomeric polymers with epoxide end groups, typical resins being made by the reaction of bisphenol A with epichlorohydrin:

$$(i+1)\ HO-\langle\bigcirc\rangle-\overset{CH_3}{\underset{CH_3}{C}}-\langle\bigcirc\rangle-OH + (i+2)\ CH_2\text{-}CHCH_2Cl \rightarrow (i+2)\ HCl +$$

bisphenol A   epichlorohydrin   (2.3.18)

$$CH_2CHCH_1-\left[-O-\langle\bigcirc\rangle-\overset{CH_3}{\underset{CH_3}{C}}-\langle\bigcirc\rangle-OCH_2\overset{OH}{\underset{}{CHCH_2}}-\right]_i-O-\langle\bigcirc\rangle-\overset{CH_3}{\underset{CH_3}{C}}-\langle\bigcirc\rangle-OCH_2CHCH_2$$

Typical values of i range between 1 and 30. These resins can be cured in several ways (e.g., with amines or by ring-opening polymerization) to form crosslinked networks (see Chapter 5). They are used for coatings, adhesives, and structural composites.

2.3.6 Polysiloxanes

So far, we have dealt only with organic polymers, but inorganic polymers are also made. Polysiloxanes (silicones) represent a good example of this, the most common of these being polydimethylsiloxane (PDMS), made from dichlorodimethylsilane:

$$i\mathrm{Cl} - \underset{\underset{\mathrm{CH_3}}{|}}{\overset{\overset{\mathrm{CH_3}}{|}}{\mathrm{Si}}} - \mathrm{Cl} + (i+1)\,\mathrm{H_2O} \rightarrow \mathrm{H} \left[\mathrm{O} - \underset{\underset{\mathrm{CH_3}}{|}}{\overset{\overset{\mathrm{CH_3}}{|}}{\mathrm{Si}}} - \mathrm{N} \right]_i \mathrm{OH} + 2i\,\mathrm{HCl} \qquad (2.3.19)$$

an oversimplification because many cyclic species are formed. This chemistry is used to form low molecular weight oligomers used as oils or additives. Higher molecular weight PDMS is made from a ring-opening polymerization of the cyclic tetramer of PDMS (see Chapter 3).

2.4 A$_2$ + B$_2$ Step Polymerization: Stoichiometry

Few of the common step polymerizations given in Section 2.3 involve monomers of the AB (or A$_2$) type. Both the majority of the examples, and the most significant ones, are of the more complicated scheme in which at least two bifunctional monomers bearing dissimilar functional groups react together (e.g., A$_2$ + B$_2$). Both the most common polyamide, nylon 6/6, and the most common polyester, poly(ethylene terephthalate) fit into this scheme. Although the reaction of bishydroxyethyl terephthalate monomer to form the polymer is formally an A$_2$ polymerization, the production of that monomer from a reaction of ethylene glycol with either terephthalic acid or the dimethyl ester thereof is an A$_2$ + B$_2$ polymerization, which determines the products going into the formally A$_2$ reaction stage. The complication in A$_2$ + B$_2$ polymerizations comes from the possibility of an imbalance in stoichiometry, which was inherently impossible for the polymerization of AB monomers.

The most important result of unbalanced stoichiometry is a severe limit on the molecular weight attainable. Often this is undesirable, and so one can either ensure a balanced stoichiometry (as for nylon 6/6, by production of an amino-acid salt from the two monomers prior to polymerization) or change the polymerization into an A$_2$ type [roughly what occurs in the production of poly(ethylene terephthalate)]. However, one may not be completely successful in these measures, or one may actually desire an unbalanced stoichiometry. Thus it is important to determine the results of having a limiting reagent, which will by convention be the A functionality.* The stoichiometric imbalance r is thus defined:

$$r = \frac{A}{B} \leq 1 \qquad (2.4.1)$$

For $r < 1$, at the maximum conversion possible there will be unreacted B_2 groups, or free ends, so that infinite molecular weight cannot be obtained as for AB polymerization [see equation (2.2.43)]. Instead the molecular weight has a limiting value, which must be determined by r because it is the only parameter left in the system at $p = 1$ [where p is the conversion of the limiting reagent (A)]. To determine the relation between r and the ultimate chain length achievable, we must first attempt to describe this more difficult case mathematically.

2.4.1 Kinetic Description

For polymerization of AB monomers, a chain of length n (≥ 2) can be represented as follows:

$$AB \text{---} (\text{---} AB \text{---})_{n-2} AB \qquad (2.4.2)$$

These chains always have an A group on one end and a B on the other. For $A_2 + B_2$ polymerization, specifying the chain length n is insufficient because the number of A_2 groups (i) and B_2 groups (j) may differ, giving different end groups. Indeed, there are three types of molecules containing i A_2 groups:

type A:	$(AA\text{---}BB\text{---})_{i-1}AA$	$i \geq 1$	(2.4.3a)
type B:	$BB\text{---}(\text{---}AA\text{---}BB)_i$	$i \geq 0$	(2.4.3b)
type M:	$(AA\text{---}BB\text{---})_{i-1}AA\text{---}BB$	$i \geq 1$	(2.4.3c)

These species correspond to "odd-A" n-mers, "odd-B" n-mers, and "even" n-mers, in Flory's notation.[25] All molecules containing an even number of monomer units are of type M (of "mixed" end groups). Those containing an odd number of units are either type A (A-terminated) or type B (B-terminated). An imbalance of stoichiometry (more B_2 monomers than A_2) obviously will give more chains of type B than of type A and more odd n-mers overall. For stoichiometric balance ($r = 1$), there

---

This allows for relative lack of either amines or acids, for example; the one being the limiting reagent is assigned as A. If one insists on assigning a particular group, such as the amine, as A, then twice as many equations must be written—one set for amines in excess, another for amines in lack. Assignment of a limiting reagent simplifies this. It is easy to see that if one takes B to be the limiting reagent, certain equations, such as those for molecular weight, will take a form different from those in this text, even though they will be consistent. This should be kept in mind when returning to the original references.

will be as many odd as even polymer molecules, and the odd molecules will be equally divided between types A and type B. In this case it is intuitive that the distribution of polymer chain lengths will be geometric.

To have complete knowledge of the distribution, we must calculate, not P_n, but $P_{i,j}$, the concentration of chains comprised of i A_2 groups and j B_2 groups. Since the molecules shown in equations (2.4.3) give the allowed values of j ($j = i - 1$, i, $i + 1$), we will designate a chain containing i A_2 units as $P_{i,j-1}$ (2.4.3a), $P_{i,i+1}$ (2.4.3b), or $P_{i,i}$ (2.4.3c). The system will consist of these species [which include both A_2 monomers ($P_{1,0}$) and B_2 monomers ($P_{0,1}$)], all of which will be considered "polymer." Chains of type A can be formed only by reaction of a chain of type M with one of type A and may disappear only by reaction with chains of type B or type M. Similar restrictions exist for the chains of other types. Thus, we can write the kinetic equations:

$$\frac{dP_{i,i-1}}{dt} = 2k \sum_{k=1}^{i} P_{k,k-1} P_{i-k,i-k} - 2kP_{i,i-1} \left\{ 2 \sum_{k=0}^{\infty} P_{k,k+1} + \sum_{k=1}^{\infty} P_{k,k} \right\} \qquad i \geq 1 \qquad (2.4.4a)$$

$$\frac{dP_{i,i+1}}{dt} = 2k \sum_{k=0}^{i} P_{k,k+1} P_{i-k,i-k} - 2kP_{i,i+1} \left\{ 2 \sum_{k=1}^{\infty} P_{k,k-1} + \sum_{k=1}^{\infty} P_{k,k} \right\} \qquad i \geq 0 \qquad (2.4.4b)$$

$$\frac{dP_{i,i}}{dt} = 4k \sum_{k=1}^{i} P_{k,k-1} P_{i-k,i-k+1} + 2k \sum_{k=1}^{i} P_{k,k} P_{k-i,k-i}$$

$$- 2kP_{i,i} \left\{ \sum_{k=1}^{\infty} P_{k,k-1} + \sum_{k=0}^{\infty} P_{k,k+1} + \sum_{k=1}^{\infty} P_{k,k} \right\} \qquad i \geq 1 \qquad (2.4.4c)$$

(The production terms are ignored for the kinetic equations for the two monomers, $P_{1,0}$ and $P_{0,1}$.) Three separate generating functions can be defined:

$$G_A(s) \equiv \sum_{i=1}^{\infty} s^i P_{i,i-1} \qquad (2.4.5a)$$

$$G_B(s) \equiv \sum_{i=0}^{\infty} s^i P_{i,i+1} \qquad (2.4.5b)$$

$$G_M(s) \equiv \sum_{i=1}^{\infty} s^i P_{i,i} \qquad (2.4.5c)$$

Equations (2.4.4) can be written in terms of these generating functions:

$$\frac{\partial G_A(s)}{\partial \tau} = G_A(s)G_M(s) - G_A(s)[2G_B(1) + G_M(1)] \qquad (2.4.6a)$$

$$\frac{\partial G_B(s)}{\partial \tau} = G_B(s)G_M(s) - G_B(s)[2G_A(1) + G_M(1)] \qquad (2.4.6b)$$

$$\frac{\partial G_M(s)}{\partial \tau} = 2G_A(s)G_B(s) + \frac{1}{2}G_M^2(s) - G_M(s)[G_A(1) \qquad (2.4.6c)$$

$$+ G_B(1) + G_M(1)]$$

where $d\tau = 2k\,dt$. We could have defined a single generating function with two dummy variables s_A and s_B, corresponding to the double indexing of the concentration P, as follows:

$$G(s_A,s_B) \equiv \sum_{i=0}^{\infty} \sum_{j=0}^{\infty} s_A^i s_B^j P_{i,j} \qquad (2.4.7)$$

This allows reduction to a single equation, but vectorial generating functions are obtained. We take the simpler approach to be able to follow (although with changes in notation) earlier derivations.[17,26]

The initial conditions in the s-domain are:

$$G_A(s,0) = s(A_2)_0 = sr(B_2)_0 \qquad (2.4.8a)$$

$$G_B(s,0) = s(B_2)_0 \qquad (2.4.8b)$$

$$G_M(s,0) = 0 \qquad (2.4.8c)$$

Expressions for the total amount of each of the molecular types can be readily obtained by letting $s = 1$ in equations (2.4.6) and integrating. Thus

$$G_A(1) = r(B_2)_0 \frac{(1-r)\exp[-2(B_2)_0(1-r)\tau]}{1 - r\exp[-2(B_2)_0(1-r)\tau]} \qquad (2.4.9a)$$

$$G_B(1) = (B_2)_0 \frac{1-r}{1 - r\exp[-2(B_2)_0(1-r)\tau]} \qquad (2.4.9b)$$

$$G_M(1) = 2r(B_2)_0(1-r)\frac{\exp[1 - (B_2)_0(1-r)\tau] - \exp[-2(B_2)_0(1-r)\tau]}{\{1 - r\exp[-(B_2)_0(1-r)\tau]\}\{1 - r\exp-2(B_2)_0(1-r)\tau]\}} \qquad (2.4.9c)$$

At the completion of the polymerization (as $\tau \to \infty$), chains of types A and M vanish, with the result that all remaining chains are end-capped with the more abundant B_2 groups. The concentration of these chains is $G_B(1,\infty) = (B_2)_0(1-r)$. Figures 2.4 through 2.6 show the trends for these three kinds of species for highly unbalanced systems.

Figure 2.4.
Total concentration of A_2-capped polymers, or odd-A n-mers for stoichiometrically imbalanced $A_2 + B_2$ polymerization.

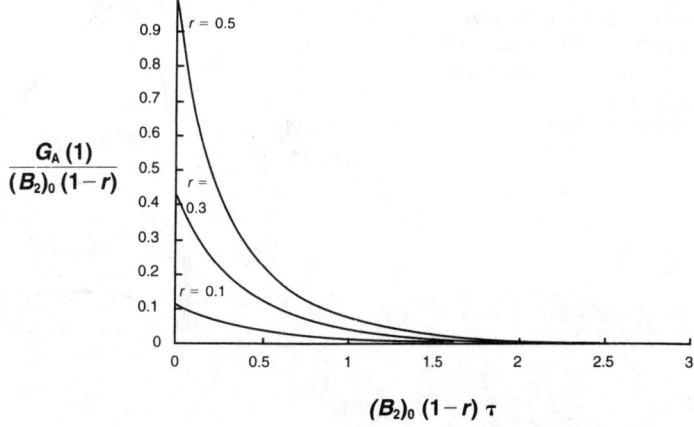

Figure 2.5.
Total concentration of B_2-capped polymers, or odd-B n-mers for stoichiometrically imbalanced $A_2 + B_2$ polymerization.

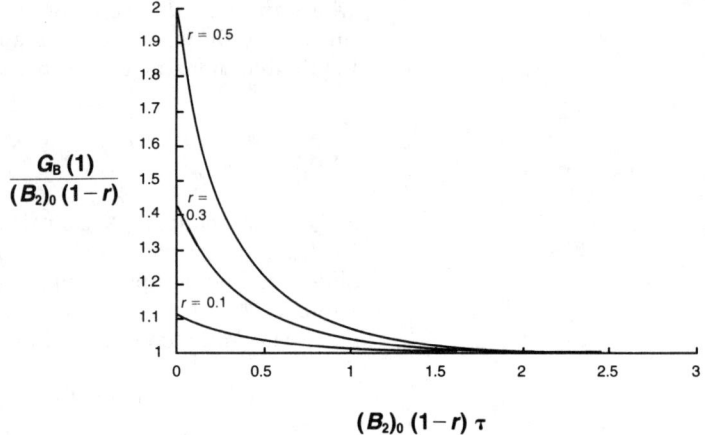

$$\frac{G_B(1)}{(B_2)_0(1-r)}$$

$(B_2)_0(1-r)\tau$

Analytical solution of equations (2.4.6) to find the distribution at any conversion is difficult; a statistical approach will prove simpler for obtaining the complete distribution at all conversions. The distribution at complete conversion is relatively simple, though[26]:

$$G_B(s,\infty) = (B_2)_0 \frac{(1-r)^2}{1-rs} \tag{2.4.10}$$

which is obviously a geometric distribution, not in chain length, but in the number of B_2 groups:

$$P_{i,i+1}(\infty) = (B_2)_0 (1-r)^2 r^i \qquad i \geq 0 \tag{2.4.11}$$

If we are looking for average degrees of polymerization (recalling that the degree of polymerization of a $P_{i,i+1}$ molecules is $2i + 1$), we find:

$$DP_n = \frac{1+r}{1-r} \tag{2.4.12}$$

Figure 2.6.
Total concentration of mixed polymers, or even n-mers for stoichiometrically imbalanced $A_2 + B_2$ polymerization.

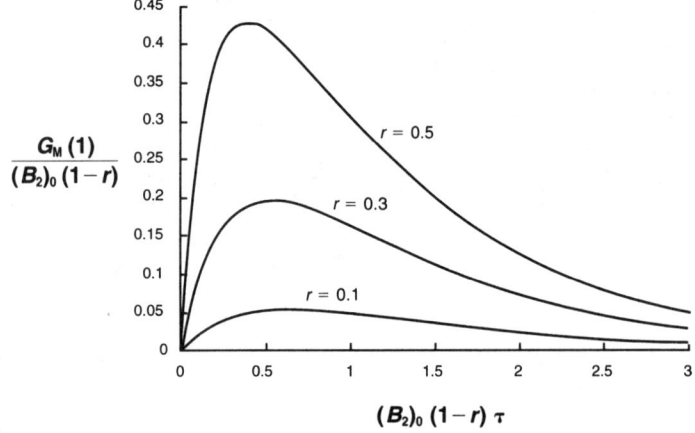

$$\frac{G_M(1)}{(B_2)_0(1-r)}$$

$(B_2)_0(1-r)\tau$

$$DP_w = \frac{1+r}{1-r} + \frac{4r}{1-r^2} \qquad (2.4.13)$$

The polydispersity is given as follows:

$$Q = 1 + \frac{4r}{(1+r)^2} \qquad (2.4.14)$$

which is a monotonically increasing function of r, from $r = 0$ ($Q = 1$, since no reaction is possible) to $r = 1$ ($Q = 2$).

2.4.2 Statistical Description: Combinatorial

Case[27] extended the combinatorial method presented in Section 2.2.2.1 in a straightforward, though laborious, manner to more complicated reaction schemes, such as step growth polymerization of three or more monomers. The first step is classifying polymers based on their end groups, as we did in Section 2.4.1 for the kinetic treatment. The distribution of each kind of polymer will have geometric character, and the entire chain length distribution (or molecular weight distribution) can be obtained from the sum of these subdistributions. Case gives a large catalog of molecular weight distributions calculated in this way.[27] Here, we are interested in the simple case $A_2 + B_2$. One can readily see that it is a straightforward way to calculate distributions, but extensions to more complex systems (e.g., $A_2 + B_2 + C_2$) are very tedious.

As in the kinetic derivation, we distinguish the three different kinds of molecule present in the system, but in the combinatorial method the third category, even n-mers, is split into two, depending on whether we start at the A end or the B end. Thus, we can identify four types of polymer molecule: type A (chains starting and ending in A groups), type B (chains starting and ending in B groups), type M-a (chains starting with an A group but ending in a B group), and type M-b (chains starting with B but ending in A). Despite the language used, no real sense of directionality (of a "start" or "end") exists on these chains.

The conversions of A and B functional groups are related through the stoichiometric ratio r by:

$$\frac{(A_2)_0}{(B_2)_0} = r = \frac{p_B}{p_A} = \frac{p_B}{p} \qquad (2.4.15)$$

where p_A, the conversion of the limiting reagent, is the proper measure of conversion in the system, p. The initial concentration of molecules was $\{(A_2)_0 + (B_2)_0\}$, or $\{(1 + r)(B_2)_0\}$, and so the total number of molecules remaining at a given conversion p is given by:

$$\text{total number of molecules} = r(B_2)_0(1 - p) + (B_2)_0(1 - rp) \qquad (2.4.16)$$

$$= (B_2)_0(1 + r - 2rp)$$

The numbers of molecules of the different types containing exactly iA_2 units are given as follows:

number of type A	$=$	$r(B_2)_0(1 - p)[p^{i-1}(rp)^{i-1}] \, (1 - p)$	$i \geq 1$	(2.4.17a)
number of type B	$=$	$(B_2)_0(1 - rp)[p^i(rp)^i] \, (1 - rp)$	$i \geq 0$	(2.4.17b)
number of type M-a	$=$	$r(B_2)_0(1 - p)[p^i(rp)^{i-1}] \, (1 - rp)$	$i \geq 1$	(2.4.17c)
number of type M-b	$=$	$(B_2)_0(1 - rp)[(rp)^i p^{i-1}] \, (1 - p)$	$i \geq 1$	(2.4.17d)

The formulation of these is fairly straightforward; we will take the first as an example. We think of the molecule as arranged from left to right; the left end is the end at which we start. The first term in the equation is the total number of A groups, the second the probability that that A functionality is unreacted (i.e., it is an end). The bracketed term represents the probability of each required step as we walk to the right. In this case, as we walk to the right we find that we require that $(i - 1)$ A groups $(i - 1)$ B groups be reacted; the two monomers alternate, of course. The final parenthetical term is the probability that the end A group is unreacted. The other three probabilities are fashioned similarly.

More important, though, we can now derive the chain length distribution at all times, by simple manipulation of the terms just defined. The results are:

$$P_{i,i-1} = (B_2)_0 \, (1 - p)^2 \, r^i p^{2i-2} \qquad\qquad i \geq 1 \qquad\qquad (2.4.18a)$$

$$P_{i,i+1} = (B_2)_0 \, (1 - rp)^2 \, r^i \, p^{2i} \qquad\qquad i \geq 0 \qquad\qquad (2.4.18b)$$

$$P_{i,i} \; = 2(B_2)_0 \, (1 - p) \, (1 - rp) \, r^i \, p^{2i-1} \qquad i \geq 1 \qquad\qquad (2.4.18c)$$

By summing over the permissible values of i, one may calculate the total concentrations of polymer of the three types, equivalent to $G_A(1)$, $G_B(1)$, and $G_M(1)$:

$$G_A(1) = r(B_2)_0 \frac{(1 - p)^2}{1 - rp^2} \qquad\qquad (2.4.19a)$$

$$G_B(1) = (B_2)_0 \frac{(1 - rp)^2}{1 - rp^2} \qquad\qquad (2.4.19b)$$

$$G_M(1) = 2r(B_2)_0 \frac{p(1 - p)(1 - rp)}{1 - rp^2} \qquad\qquad (2.4.19c)$$

The sum of these three is exactly equal to the number of molecules given in equation (2.4.16). Normalizing each subdistribution [equations (2.4.18)] by its total concentration [equations (2.4.19)] yields a geometric distribution with parameter (rp^2):

$$\frac{P_{i,i-1}}{G_A(1)} = \frac{P_{i-1,i}}{G_B(1)} = \frac{P_{i,i}}{G_M(1)} = (1 - rp^2)(rp^2)^{i-1} \qquad (2.4.20)$$

Each distribution is the same, but as p approaches 1 the concentrations of chains of type A and type M vanish, leaving only chains of type B, which are distributed as in equation (2.4.11).

Equations (2.4.19) are consistent with the earlier versions given in terms of time, equations (2.4.9), but to see this we must find the relationship between time and conversion. This is done with the equation for the disappearance of the limiting functional group

$$\frac{dA}{dt} = -kAB \tag{2.4.21}$$

with the initial condition that $A_0 = 2r(B_2)_0$. Given that

$$B = A + 2(B_2)_0(1 - r) \tag{2.4.22}$$

and that the conversion of the limiting reagent is defined as follows:

$$p = \frac{A_0 - A}{A_0} \tag{2.4.23}$$

integration of equation (2.4.21) yields:

$$p = \frac{1 - \exp[-(B_2)_0(1 - r)\tau]}{1 - r\exp[-(B_2)_0(1 - r)\tau]} \tag{2.4.24}$$

Given this, it is easy to show equivalence between the two versions of the equations for concentrations of the three molecular types. We could also show that equations (2.4.4) [or (2.4.6)] are satisfied by the distributions and thus the latter form the solution to the former.

From the foregoing equations we can also find the "regular" chain length distribution, the probability of finding a chain of length n, irrespective of its composition. As Flory showed,[25] this is given by:

$$\frac{P_n}{P} = \frac{2(1 - rp)(1 - p)}{1 + r - 2rp} p^{n-1} r^{n/2} \qquad n \text{ even}$$

$$= \frac{(1 - rp)^2 + r(1 - p)^2}{1 + r - 2rp} p^{n-1} r^{(n-1)/2} \qquad n \text{ odd} \tag{2.4.25}$$

Note that this is a normalized distribution. Figure 2.7 shows this distribution for various values of the stoichiometry parameter r. The cyclic or periodic behavior that appears when $r \neq 1$, which becomes more pronounced both at higher r and at higher p, arises from the lack of chains of even length. From this distribution one can find the average degrees of polymerization to be:

$$DP_n = \frac{1 + r}{1 + r - 2rp} \tag{2.4.26}$$

$$DP_w = \frac{(1 + r)(1 + rp^2) + 4rp}{(1 + r)(1 - rp^2)} \tag{2.4.27}$$

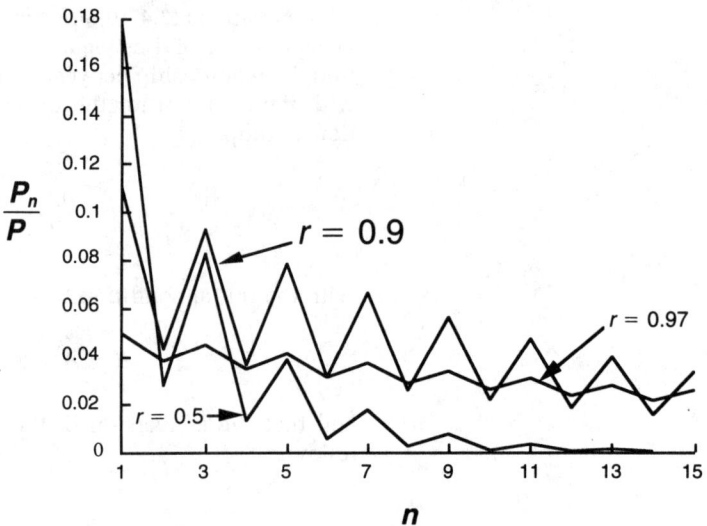

Figure 2.7.
Chain length distribution
for $A_2 + B_2$ polymerization
at stoichiometric
imbalance; $p = 0.97$.

which, for $p = 1$, reproduces equations (2.4.12) and (2.4.13). If $r = 1$, all these expressions are identical to those obtained in Section 2.2 for AB polymerization.

Example 2.2

Determine the number-average degree of polymerization for an $A_2 + B_2$ polymerization, starting with 49 mol % of A groups, at 95% fractional conversion of A groups. How much degree of polymerization has been lost by the slight stoichiometric imbalance?

Solution:

The ratio of initial concentrations of functional groups can be calculated from

$$r = \frac{A_0}{B_0} = \frac{0.49}{1 - 0.49} = 0.9608$$

For $p = 0.95$, equation (2.4.26) yields $DP_n = 14.5$. For $r = 1$, $p = 0.95$, $DP_n = 20$. Thus, there is a reduction of 27.5%.

2.4.3 Statistical Description: Formal Markov Chain Theory Approach

Obviously the combinatorial procedure will become unwieldy as the number of comonomers grows. A more formal approach would be better than such direct mechanistic reasoning. Thus we consider the polymer chain to be a Markov chain, a process that discretely evolves from state to state, with the transition probabilities dependent only on the identity of the present state.[28,29] It is possible to combine the various probabilities of reaction in a Markov chain transition matrix and then obtain the chain length distribution and its moments by matrix manipulation.[30]

Markov chains are often thought of as processes evolving in time through discrete time steps. When applying Markov chain theory to polymerizations, though, the most convenient approach is to envision as the Markovian process a walk along already formed chains (much as we have done for the combinatorial or recursive methods). The transient state is then identified with the kind of comonomer (e.g., A_2 or B_2) at some position along the chain, and the transition probabilities between those with the reaction probabilities (e.g., of an A group with a B group). A transition probability is zero if reaction (transition) between the particular pair of monomers is forbidden, as it would be for an A group with another A group in our present example. There are as many transient states as there are conomers in the reaction, N. The end of a chain (reached with a probability like $1 - p$), corresponds in Markov chain terminology to "absorption" into a state from which the process cannot exit.

The transition probability matrix P can be partitioned into submatrices in the following way:

$$P = \begin{pmatrix} I & 0 \\ R & Q \end{pmatrix}$$

(2.4.28)

The rank of the square matrix P is one greater than the number of transient states, N, in the Markov chain process that represents the polymerization. Q is an $N \times N$ matrix of probabilities of transition among the transient states. R is an $N \times 1$ matrix of probabilities of absorption (termination) from each of the transient states. 0 is a $1 \times N$ matrix containing only zeros, which indicates the impossibility for a transient state to be reached from an absorbing state (a chain end), and $I = 1$, which assures that once the system has entered the absorbed state (a chain end is reached), it will stay there. The sequential process must begin at a chain end if it is to count all the units and give us the chain length distribution and so, besides knowledge of the elements of P, we need to know q^0, the vector of probabilities of each transient state (type of comonomer unit) initiating the Markov chain (occurring at the end of a polymer chain). With these two arrays of information, Lowry shows that the chain length distribution is given by[30]:

$$\frac{P_n}{P} = q^0 \, Q^{n-1} \, (I - Q) \, 1$$

(2.4.29)

where I is the identity matrix of rank N and 1 is a column vector of length N consisting of ones; R need not appear, since it contains no information that Q does not have. Any moment of the distribution P_n, μ_i, may be calculated from:

$$\mu_i = q^0 t_i$$

(2.4.30)

where $\mathbf{t}_i$ are row vectors given by the following recursion formulas:

$$\mathbf{t}_1 = (I - Q)^{-1}\,\mathbf{1} \tag{2.4.31}$$

$$\mathbf{t}_i = \mathbf{t}_1 + [(I - Q)^{-1} - I]\sum_{j=1}^{i-1}\binom{i}{j}\mathbf{t}_j \tag{2.4.32}$$

This gives the degrees of polymerization:

$$DP_n = \mu_1 = \mathbf{q}^0\,(I - Q)^{-1}\,\mathbf{1} \tag{2.4.33}$$

$$DP_w = \frac{\mu_2}{\mu_1} = \frac{\mathbf{q}^0[2(I - Q)^{-1} - I]\,(I - Q)^{-1}\,\mathbf{1}}{\mathbf{q}^0\,(I - Q)^{-1}\,\mathbf{1}} \tag{2.4.34}$$

Notice that equation (2.4.29) is simply a normalized matrix generalization of equation (2.2.55). Also, DP_n is defined as the first moment, rather than the ratio of the first moment to the zeroth, because the latter is necessarily unity for a normalized distribution. For the case of AB polymerization, $N = 1$ and we have $Q = p$ and $\mathbf{q}^0 = 1$. Thus $(I - Q)^{-1} = 1/(1 - p)$, and by application of the foregoing equations the results of equations (2.2.43) and (2.2.44) are recovered for DP_n and DP_w. For more complicated cases, we diagonalize Q by a similarity transformation[31] and evaluate Q^{n-1} from:

$$Q^{n-1} = S^{-1}E^{n-1}S \tag{2.4.35}$$

where E^{n-1} is the matrix in which the eigenvalues of Q are raised to the $(n - 1)$st power (e_i^{n-1}) along the principal diagonal, and S is the matrix of eigenvectors a_i corresponding to the eigenvalue e_i. The resulting matrix formulas are compact, but it is very tedious to reduce them to the corresponding algebraic equations, especially for P_n. The matrix computations can be performed numerically on the computer, and this is the prime advantage of the Markov chain formulation. This approach has the following main disadvantage: it is difficult, if not impossible, to put the molecular weights of the individual comonomers into this scheme and thus calculate molecular weight distribution instead of chain length distribution. This poses a problem for copolymerization, since usually molecular weight is experimentally accessible, not degree of polymerization.

Let us now apply this formalism to the case of interest, $A_2 + B_2$ step growth polymerization. First we must find the transition probability matrix, Q, a 2×2 matrix, since $N = 2$. The elements of this matrix are denoted as p_{ij}, the probability that given being in state i, the next transition is to state j. In this case, the matrix of probabilities of transitions is

$$Q = \begin{pmatrix} p_{AA} & p_{AB} \\ p_{BA} & p_{BB} \end{pmatrix} = \begin{pmatrix} 0 & p_A \\ p_B & 0 \end{pmatrix} = \begin{pmatrix} 0 & p \\ rp & 0 \end{pmatrix} \tag{2.4.36}$$

since A_2 and B_2 do not react with themselves. The matrix Q is not 'stochastic," since the row elements do not sum to unity. That is because of the possibility of being at a chain end, expressed in the matrix vector $\mathbf{R}$. The initial transient state probability vector $\mathbf{q}^0$ comprises the probabilities that an end is an A group or a B group. A simple inventory of unreacted group suffices. If there are N_0 initial monomer units $\{2rN_0/(1 + r)\}$ A functionalities and $\{2N_0/(1 + r)\}$ B functionalities), then

concentration of unreacted A groups = $2r(B_2)_0(1 - p)$ (2.4.37)

concentration of unreacted B groups = $2(B_2)_0(1 - rp)$ (2.4.38)

so that

$$\text{fraction unreacted A groups} = q_A^0 = \frac{r(1 - p)}{1 + r - 2rp} \qquad (2.4.39)$$

$$\text{fraction unreacted B groups} = q_B^0 = \frac{1 - rp}{1 + r - 2rp} \qquad (2.4.40)$$

Note that in this case these probabilities are not equal to the mole fractions of the two groups, as was the case in AB polymerization, except in the case where $r = 1$, in which case $q_A^0 = q_B^0 = x_A = x_B = 1/2$. This subtlety arises in many situations—for example, when calculating DP_n by the recursive method, as we will do below.

Using equation (2.4.33):

$$DP_n = \mu_1 = \frac{1}{1 + r - 2rp} \left(r(1 - p) \quad 1 - rp \right) \begin{pmatrix} 1 & -p \\ -rp & 1 \end{pmatrix}^{-1} \begin{pmatrix} 1 \\ 1 \end{pmatrix} = \frac{1 + r}{1 + r - 2rp} \qquad (2.4.41)$$

as we found before in equation (2.4.26). Equation (2.4.27) for DP_w may similarly be reproduced by use of equation (2.4.34). The chain length distribution itself can be found by equation (2.4.29). In most cases calculation of the chain length distribution is best performed by a computer, but in this case it is straightforward:

$$\frac{P_n}{P} = \frac{1}{1 + r - 2rp} \left(r(1 - p) \quad 1 + rp \right) \begin{pmatrix} 0 & p \\ rp & 0 \end{pmatrix}^{n-1} \begin{pmatrix} 1 & -p \\ -rp & 1 \end{pmatrix} \begin{pmatrix} 1 \\ 1 \end{pmatrix} \qquad (2.4.42)$$

Here, an analytical solution is possible. As could be expected from the combinatorial analysis, it turns out that Q^{n-1} is different depending on whether n is even or odd, and the final result for P_n is the same as in equation (2.4.25).

The related approach of Peller[32] has the same usefulness and elegance as the Markov chain approach, but also has the same drawbacks (e.g., no clear way to incorporate molecular weights of the monomers to facilitate the calculation of molecular weight distribution; increasing intricacy of the formalism as the reaction scheme becomes more elaborate). On the other hand, the method does give treatment of sequence distribution (see Section 4.3), but this is not relevant for the present case of A_2 + B_2 polymerization, which must be strictly alternating.

2.4.4 Statistical Description: Recursive

Other statistical methods could be brought to bear on this problem. Perhaps the simplest route to the average degrees of polymerization is a recursive derivation.[20] To derive the weight-average molecular weight, we grab a unit of mass (or repeat unit) at random and ask for the average mass of the polymer chain in which the chosen unit resides. Thus:

$$M_w = w_{A2}[M_{A2} + 2E(W_A^{out})] + (1 - w_{A2})[M_{B2} + 2E(W_B^{out})] \quad (2.4.43)$$

where w_{A2} is the weight fraction of A_2 monomers in the reaction mixture, and M_{A2} and M_{B2} are the masses of the two monomers. We will again neglect the mass of the condensate products. We can write the following recursive relations, avoiding the intermediate step of dealing with "in" expected quantities:

$$E(W_A^{out}) = p\{M_{B2} + E(W_B^{out})\} \quad (2.4.44)$$

$$E(W_B^{out}) = rp\{M_{A2} + E(W_A^{out})\} \quad (2.4.45)$$

These two equations can be solved to yield:

$$E(W_A^{out}) = \frac{pM_{B2} + rp^2 M_{A2}}{1 - rp^2} \quad (2.4.46)$$

$$E(W_B^{out}) = \frac{rpM_{A2} + rp^2 M_{B2}}{1 - rp^2} \quad (2.4.47)$$

The weight fraction of equation (2.4.43) can be written:

$$w_{A2} = \frac{rM_{A2}}{rM_{A2} + M_{B2}} \quad (2.4.48)$$

The weight-average molecular weight is thus given as follows:

$$M_w = \frac{rM_{A2}^2 + M_{B2}^2}{rM_{A2} + M_{B2}} + \frac{2rp^2\{rM_{A2}^2 + 2M_{A2}M_{B2} + M_{B2}^2\}}{(rM_{A2} + M_{B2})(1 - rp^2)} \quad (2.4.49)$$

The first term is the average mass of the chosen monomer, the second that of the attached groups; DP_w can be found by letting $M_{A2} = M_{B2} = 1$, which the reader can show reproduces equation (2.4.27).

Although care must be taken in properly calculating the probability of A and B end groups, M_n can also be found by recursive arguments. This difficulty was mentioned in connection with the calculation of the vector $\mathbf{q}^o$ for the Markov chain treatment. Since the elements of that vector give the probabilities of end groups of each type [see equations (2.4.39) and (2.4.40)], M_n is given as follows:

$$M_n = \frac{r(1 - p)}{1 + r - 2rp}\{M_{A2} + E(W_A^{out})\} + \frac{1 - rp}{1 + r - 2rp}\{M_{B2} + E(W_B^{out})\} \quad (2.4.50)$$

Substituting in the results from equations (2.4.46) and (2.4.47) gives M_n:

$$M_n = \frac{rM_{A2} + M_{B2}}{1 + r - 2rp} \tag{2.4.51}$$

which is exactly what one would expect from a stoichiometric argument (i.e., one based on the mass divided by the number of molecules) and which corroborates equation (2.4.26) for DP_n.

2.4.5 Determination of Rate Constant

We proceed to briefly discuss the topic of how to determine the rate constant k, which has been assumed to be known in the discussion up to now.[6,23] Polyamidation to form a nylon appears to be an uncatalyzed reaction, and thus the kinetics follow the simple form of equation (2.4.21), which in the case of balanced stoichiometry leads to:

$$\frac{dA}{dt} = \frac{dB}{dt} = -kA^2 \tag{2.4.52}$$

Thus, if one performs the experiment under stoichiometric balance, one obtains the result that the conversion of functional groups, p, is related to time as follows:

$$p = \frac{A_0 \int_0^t k\, dt'}{1 + A_0 \int_0^t k\, dt'} \tag{2.4.53}$$

The integral is equal to $\tau/2$. Note that we have not assumed the rate constant to be constant. Thus, DP_n, which is necessarily $1/(1 - p)$, is related to time as follows:

$$DP_n = 1 + A_0 \int_0^t k\, dt' \tag{2.4.54}$$

If the rate constant were actually a constant, the plot of DP_n versus time would be linear with an intercept of 1 and a slope proportional to the rate constant k through A_0. If k is not constant over the entire reaction but only over some period of the reaction, the value of k during that period can be measured from the slope during that period.

Polyesterification provides a more interesting case. In the rare cases that use direct esterification rather than transesterification [as for poly(ethylene terephthalate)], the reaction is often acid-catalyzed (different catalysts are used for transesterification). If the (constant) concentration of the acid is combined into the rate constant, the same equations as given earlier can be used. If, on the other hand, no added acid catalyst is present, the reaction is self-catalyzed through the carboxylic acid, and the rate is then given as:

$$\frac{dA}{dt} = \frac{dB}{dt} = -kA^3 \qquad (2.4.55)$$

from which one finds that here the square of DP_n is similar in its dependence on time to DP_n in the catalyzed esterification or the case of a nylon polymerization:

$$DP_n^2 = 1 + 2A_0^2 \int_0^t k \, dt' \qquad (2.4.56)$$

(This integral, although defined as was τ, differs because a different kinetic scheme is being used. This in turn implies different forms for all the rate equations in Section 2.4.1. It leaves the resulting distributions and the applicability of the statistical analyses intact, though; proof of this is left to the reader as a problem.)

We have intentionally written the preceding equations not assuming the rate constant to be truly constant. This is because even in isothermal polymerizations, deviations occur from the appropriate straight-line behavior at both low and high conversions.[2,6,23,33] The deviation at low conversions (< 80%) may be due to the effects of unequal reactivity (see Section 2.7) but may be for other more global reasons. In the early part of the reaction the polarity and pH of the solution are changing drastically, quite possibly affecting the apparent rate constant. This extremely large range of conversion at which the rate constant is changing should not worry the reader, because it corresponds to a relatively small range of time and degrees of polymerization not of interest (DP_n < 5). During the important range of conversion in which high polymer is formed, the rate constant does appear to be fairly constant, until such high conversions are reached that diffusion limitations probably play a role.

Both this kinetic analysis and the various derivations of molecular weights and distributions assume that no side reactions occur. This is of course a gross simplification. For example, polyurethanes and polyureas are subject to allophanate and biuret formation, which lead to branching.[6,23] Polymerization to produce poly(ethylene terephthalate) is accompanied by side reactions that produce diethylene glycol (which when incorporated into the polymer decreases the melting point) and acetaldehyde (which affects the flavor of beverages). Numerous works attempt fairly complete kinetic schemes of poly(ethylene terephthalate) production,[34] and comprehensive general models for polycondensation reactions (from a kinetic point of view) have been presented.[35] The nature of side reactions (e.g., whether they lead to branching, whether they lead to an inactive end) will determine their effect on kinetic and statistical treatments.

2.5 Effect of Monofunctional Agents

For step growth polymerizations of monomers of type AB or A_2 + B_2, where the functional groups react only with each other and are in stoichiometric balance, we have seen that the molecular weight distribution is a function uniquely defined by conversion. For the latter under stoichiometric imbalance, however, there is a second parameter, the degree of imbalance, which alters but does not destroy the geometric character of the distribution. More importantly, it serves to limit the molecular weight to a finite value. This may be problematic when high molecular weights are necessary, but it is useful when oligomers or polymers of low molecular weight are to be produced (examples of that were given in Section 2.3). Stoichiometric imbalance may then intentionally be established to limit the molecular weight. There is another method, however: namely, adding monofunctional "impurities." In an AB polymerization for which stoichiometry is ensured by purity, this is the most obvious way. The chain stoppers can be represented schematically as a XB molecule, where the X group is nonreactive. Note from the beginning that this imposes a stoichiometric imbalance between A and B functionalities, r, which we define as before. How this sort of stoichiometric imbalance affects the molecular weights is the subject of this section.[3,17] We examine only this one system, and by only one (kinetic) method; this suffices to show the sort of effect obtained.

For the reaction of an AB monomer with an XB monomer, we have two distinguishable kinds of polymer: those with two reactive ends:

$$P_i: \text{A}\text{---}(\text{---BA---})_{i-1}\text{B} \qquad (2.5.1)$$

and those with one end blocked by the monofunctional agent:

$$P_{ix}: \text{X}\text{---}(\text{---BA---})_{i-1}\text{B} \qquad (2.5.2)$$

There are, therefore, two different polymerization reactions:

$$P_i + P_j \xrightarrow{k} P_{i+j} \qquad (2.5.3)$$

$$P_i + P_{jx} \xrightarrow{k} P_{(i+j)x} \qquad (2.5.4)$$

for which it is reasonable to assume that the rate constants are the same. Since A and B groups do not react with themselves, x-terminated molecules cannot react with one another (although that would be possible in A_2 + B_2 + BX polymerization).

Assuming equal reactivity of functional groups, we can write mass balances on the two kinds of polymeric species:

$$\frac{dP_i}{dt} = k\sum_{j=1}^{i-1} P_j P_{i-j} - kP_i \left(2\sum_{j=1}^{\infty} P_j + \sum_{j=1}^{\infty} P_{jx} \right) \qquad (2.5.5a)$$

$$\frac{dP_{ix}}{dt} = k \sum_{j=1}^{i-1} P_{jx} P_{i-j} - k P_{ix} \sum_{j=1}^{\infty} P_j \qquad (2.5.5b)$$

where the production terms obviously do not exist for the monomeric equations $i = 1$. We can solve this set of equations using the generating function technique. Since we have molecules of two different kinds, though, we will need two generating functions, just as we needed three in Section 2.4 to deal with the different molecules occurring in an $A_2 + B_2$ polymerization.

$$G(s) = \sum_{i=1}^{\infty} s^i P_i \qquad (2.5.6a)$$

$$G_x(s) = \sum_{i=1}^{\infty} s^i P_{ix} \qquad (2.5.6b)$$

Further defining $d\tau = k\, dt$, we can write:

$$\frac{\partial G(s)}{\partial \tau} = G^2(s) - G(s)\,(2G(1) + G_x(1)) \qquad (2.5.7a)$$

$$\frac{\partial G_x(s)}{\partial \tau} = G_x(s)\,(G(s) - G(1)) \qquad (2.5.7b)$$

with the initial conditions that $G(s,0) = sP_0$, and $G_x(s,0) = sP_{x0}$. Setting $s = 1$ in equations (2.5.7) gives:

$$\frac{\partial G(1)}{\partial \tau} = -(G^2(1) + G(1)G_x(1)) \qquad G(1,0) = P_0 \quad (2.5.8a)$$

$$\frac{\partial G_x(1)}{\partial \tau} = 0 \qquad\qquad\qquad\qquad G_x(1,0) = P_{x0} \quad (2.5.8b)$$

These equations illustrate that as molecular weight builds, the total number of difunctional molecules, $G(1)$, decreases but the total number of monofunctional species, $G_x(1)$, remains constant at a concentration of P_{x0}, although these chains do grow in length. The solution for the total amount of difunctional polymer (equation 2.5.8a) is given by:

$$G(1) = \frac{P_0 \exp\,(-P_{x0}\tau)}{1 + (P_0/P_{x0})\,[1 - \exp\,(-P_{x0}\tau]} \qquad (2.5.9)$$

and for $\tau \to \infty$, $G(1) \to 0$, meaning that eventually all polymer molecules will become unreactive, "capped" (at one end) by the monofunctional agent.

The distributions $G(s)$ and $G_x(s)$ are best obtained by normalized generating functions[17]:

$$y = \frac{G(s)}{G(1)} \qquad (2.5.10a)$$

$$y_x = \frac{G_x(s)}{P_{x0}} \qquad (2.5.10b)$$

Equations (2.5.7) thus become:

$$\frac{\partial y}{\partial \tau} = G(1)y(y - 1) \tag{2.5.11a}$$

$$\frac{\partial y_x}{\partial \tau} = G(1)y_x(y - 1) \tag{2.5.11b}$$

Since the logarithms of y and y_x change at the same rate, the two generating functions (and therefore the two distributions) satisfy the proportionality:

$$G(s) = \frac{G(1)}{P_{x0}} G_x(s) \tag{2.5.12}$$

This makes it plain that $y = y_x$. The normalized generating function for difunctional molecules is given by the solution to equation (2.5.11a), subject to $y = s$ at $\tau = 0$:

$$y_x = y = \frac{\dfrac{G(1) + P_{x0}}{P_0 + P_{x0}} s}{1 - \left(\dfrac{P_0 - G(1)}{P_0 + P_{x0}}\right) s} \tag{2.5.13}$$

which is a normalized geometric distribution with the parenthetical term in the denominator as the parameter. The distribution is thus given as follows:

$$\frac{P_i}{G(1)} = \frac{P_{ix}}{P_{x0}} = \left(\frac{G(1) + P_{x0}}{P_0 + P_{x0}}\right)\left(\frac{P_0 - G(1)}{P_0 + P_{x0}}\right)^{i-1} \tag{2.5.14}$$

For $P_{x0} = 0$, equation (2.5.14) is exactly the same as equation (2.2.36), as should be. The entire distribution of the polymeric mixture is given by adding these last two equations, and it is also a geometric distribution. Ultimately, as $\tau \to \infty$, the distribution of the mixture will be given by equation (2.5.14) with $G(1) = 0$.

To compare with earlier (statistical) results for $A_2 + B_2$ polymerization under stoichiometric imbalance, we may rewrite the distributions in terms of conversion p of the limiting reagent A, which in this case is given as follows:

$$p = \frac{A_0 - A}{A_0} = \frac{P_0 - G(1)}{P_0} \tag{2.5.15}$$

the latter equality being assured because the concentration of bifunctional chains is also the concentration of A groups. The stoichiometric ratio r, as before, equals the ratio A/B, which here equals $\{P_0/(P_0 + P_{x0})\}$. Thus, equation (2.5.14) becomes:

$$\frac{P_i}{G(1)} = \frac{P_{ix}}{P_{x0}} = (1 - rp)(rp)^{i-1} \tag{2.5.16}$$

The average degrees of polymerization are obviously given by:

$$DP_n = \frac{1}{1 - rp} \qquad (2.5.17)$$

$$DP_w = \frac{1 + rp}{1 - rp} \qquad (2.5.18)$$

which in the limit of complete conversion ($p = 1$) gives:

$$DP_n(p = 1) = \frac{1}{1 - r} \qquad (2.5.19)$$

$$DP_w(p = 1) = \frac{1 + r}{1 - r} \qquad (2.5.20)$$

Here again we see that the monofunctional impurity limits the molecular weights attainable, as did stoichiometric imbalance in the case of $A_2 + B_2$ polymerization.

While the stoichiometric imbalance is defined the same way, it does not have precisely the same effect on the degree of polymerization. To obtain the same equation for DP_n as in the $A_2 + B_2$ case, one must define a parameter r' given as follows:

$$r' = \frac{P_0}{P_0 + 2P_{x0}} = \frac{r}{2 - r} \qquad (2.5.21)$$

This parameter plays the same basic role as does r in $A_2 + B_2$ polymerization, although there are some slight differences in the equation for DP_w.

2.6 Reversible Polymerization

In all the systems analyzed so far, the resulting set of moment equations (whether we wrote them or not) is closed and consequently readily solved using either analytical or numerical means. In certain situations, however, the set of moment equations is not closed; that is, the equation for the kth moment depends on moments of higher order. The physical situation leading to this generally occurs whenever a polymer participates in a reaction that can occur not only at the ends but at every monomer unit along the chain (or at least some fraction thereof). Reactions leading to branching or crosslinking, to be discussed in Chapter 5, fit into this scheme but do not entail moment closure problems. Reversibility and degradation, however, do lead to moment closure problems, because here smaller chains are being made from larger ones (making it impossible to solve the equations for P_i sequentially). Esterification and amidation, the most common stepwise reactions, are both reversible (the former more than the latter), so this is clearly important.

For the case of a reversible AB polymerization, a simple kinetic mechanism may be written:

$$(AB)_i + (AB)_j \underset{k'}{\overset{k}{\rightleftharpoons}} (AB)_{i+j} + W \tag{2.6.1}$$

In this notation, the free W species is the condensation product—for example, water, as in the polymerization of the 11-aminoundecanoic acid shown in equation (2.2.1). The material balances on P_i is:

$$\frac{dP_i}{dt} = k\sum_{j=1}^{i-1} P_{i-j}P_j - 2kP_i\sum_{j=1}^{\infty} P_j - k'W(i-1)P_i + 2k'W\sum_{j=i+1}^{\infty} P_j \tag{2.6.2}$$

As usual the first term, indicating production by polymerization, does not exist for the monomer $i = 1$, but contrary to before the equation for P_1 does have a positive term, the depolymerization term shown last. If $W = 0$, the depolymerization terms vanish and equation (2.2.7) is regained. Equation (2.6.2) does not imply that for stepwise polymerizations lacking a condensation product that reversibility is necessarily unimportant. The presence of W in this equation is due to its assumed participation in the depolymerization step; the analogous equation for a depolymerization of a urethane, for example, might be a first-order decomposition.[36]

The depolymerization reaction depends not on the polymer concentration, but on the number of A—B linkages. Species P_{i+j} has $i + j$ repeat units and $i + j - 1$ linkages, so the polymer can break in $i + j - 1$ places. Since every molecule has two sites at which it can break to form two molecules (one of size i, the other of size j), the fraction of events leading to such scission is $2/(i + j - 1)$. (For $i = j$, a chain of even length splitting in the middle, there is of course only one site, but the factor of 2 persists because two chains of length i are formed.) The reverse reaction for step growth is proportional to the number of linkages, so that the rate expression includes the term:

$$k'W\sum_{j=1}^{\infty} \frac{2}{i+j-1}(i+j-1)P_{i+j} = 2k'W\sum_{j=1}^{\infty} P_{i+j} = 2k'W\sum_{j=i+1}^{\infty} P_j \tag{2.6.3}$$

We use the generating function formalism to derive the moment equations. Using Table 2.1, the generating function representation of equation (2.6.2) is:

$$\frac{\partial G(s)}{\partial t} = k\,G^2(s) - 2kG(1)G(s) - k'W\left\{s\frac{\partial G(s)}{\partial s} - G(s)\right\} - 2k'W\left\{\frac{G(s) - sG(1)}{1 - s}\right\} \tag{2.6.4}$$

This equation is a nonlinear partial differential equation. Although one might struggle to obtain an exact analytical solution, it is far easier to be content with the information afforded by the moment representation of equation (2.6.4). Taking care in the differentiation (e.g., using l'Hôpital's rule) gives the following moment equations:

$$\frac{d\mu_0}{dt} = -k\mu_0^2 + k'W(\mu_1 - \mu_0) \qquad (2.6.5)$$

$$\frac{d\mu_1}{dt} = 0 \qquad (2.6.6)$$

$$\frac{d\mu_2}{dt} = 2k\mu_2^2 + \frac{k'W}{3}(\mu_1 - \mu_3) \qquad (2.6.7)$$

Note that the equation for μ_2 depends on μ_3. Likewise, the equation for μ_3 depends on μ_4, and so on with the higher moments. Such a system of moment equations is not closed. This lack of closure obviously breaks down with the first moment, so that it is possible to calculate DP_n. An analytic solution is available of course for the uninteresting case of W equal to a constant. In the case for which no condensate is removed, $W = (\mu_1 - \mu_0)$, and DP_n of polymerization can be found analytically. As in the cases of stoichiometric imbalance and monofunctional agents, though, the degree of polymerization or conversion is no longer found to be simply a function of time, but also a function of a second variable, in this case the equilibrium constant K, equal to (k/k'):

$$DP_n = \frac{K - 1 - (\sqrt{K} - 1)^2 \exp(-2kP_0t/\sqrt{K})}{(\sqrt{K} - 1)\,[1 + \exp(-2kP_0t/\sqrt{K})]} \qquad (2.6.8)$$

The conversion, p, is given as follows[3]:

$$p = \frac{K}{K - 1} - \frac{\sqrt{K}}{K - 1}\frac{K - 1 + (\sqrt{K} - 1)^2\exp(-2kP_0t/\sqrt{K})}{K - 1 - (\sqrt{K} - 1)^2\exp(-2kP_0t/\sqrt{K})} \qquad (2.6.9)$$

since necessarily $DP_n = 1/(1 - p)$. At equilibrium $(t \to \infty)$, the conversion is found to be:

$$p = \frac{\sqrt{K}}{\sqrt{K} + 1} \qquad (2.6.10)$$

corresponding to a degree of polymerization given by:

$$DP_n = 1 + \sqrt{K} \qquad (2.6.11)$$

Equation (2.6.10) can also be derived from the definition of the equilibrium constant. Since

$$K = \frac{W\,[A - B]}{A\,B} \qquad (2.6.12)$$

where $[A - B]$ is the concentration of the linkages that constitute the noncondensate product. In terms of conversion p this can be written:

$$K = \frac{p^2}{(1-p)^2} \qquad (2.6.13)$$

from which equation (2.6.10) is easily obtained.

Equilibrium thus limits the degree of polymerization attainable, just as did stoichiometric imbalance and the presence of monofunctional impurities. The limit is quite severe, as seen in equation (2.6.11); an equilibrium constant of 10^4 is needed for a DP_n of 100, and this requirement is generally much higher than the equilibrium constants manifested by polyester or nylon systems. Polyesterifications generally have equilibrium constants on the order of 1 to 10, while transesterifications are often an order of magnitude less. Even though polyamidation reactions have much higher K values (10^2–10^3), this still falls short of our requirement.[6,23] Thus, high polymer is difficult to obtain in closed systems from which the product is not removed.

In practice, therefore, polymerizations such as polyesterification and polyamidation are run at low pressures (down to several mm Hg) and elevated temperatures ($\geq 250°C$) to force the reaction to completion and obtain higher molecular weight polymer for film, fiber, or packaging applications; the low molecular weight condensation product (water or glycol) is extracted from the reacting mass because the vapor pressure of these compounds is higher than that of the polymer chains. In this case the dynamics of condensate removal is also a significant factor, and integration of equation (2.6.5) must be performed numerically.

To solve for DP_w, we need to devise a closure procedure. One such procedure[37] uses associated Laguerre polynomials, $L_m^\lambda(x)$, discussed in Appendix 2B, and relates the third moment to the first three in the following way:

$$\mu_3 = \frac{\mu_2}{\mu_0 \mu_1} (2\mu_0\mu_2 - \mu_1^2) \qquad (2.6.14)$$

This allows for a closed equation for the second moment, written as follows:

$$\frac{d\mu_2}{dt} = k\mu_1^2 + \frac{k'W}{3} \left(\mu_1 - 2\frac{\mu_2^2}{\mu_1} + \frac{\mu_1\mu_2}{\mu_0} \right) \qquad (2.6.15)$$

a Bernoulli equation that can be solved numerically. Despite successful closure of the equations, however, there is no guarantee that the resulting solution will be a good one. The applicability of the solution depends on the polydispersity and the shape of the real distribution; the further from geometric generally, the worse it will become.[38] An alternative procedure is to simply assume that the distribution is geometric, so that the problem reduces to calculating p, which is straightforward.

The assumption of a geometric distribution is supported by the tendency of the reversibility of polymerization to maintain the polymer product in a geometric distribution. At equilibrium, statistical thermodynamics shows that the geometric distribution (or "most probable," here a quite applicable term) holds.[39,40] This is not the case when equal reactivity (see Section 2.7) does not hold.[41] The ubiquity of the geometric distribution is an important result. Random scission processes, like thermal or radiation degradation, give products that will tend toward a geometric distribution, even if starting from a monodisperse product. Nonuniform scission processes, however, such as shear degradation (which may tend to break a chain near its middle), do not lead to a geometric distribution.[42]

The elevated temperatures employed for the more efficient removal of the condensate in the (semibatch) operation also promote interchange reactions between the formed polymer molecules, such as amide and ester interchange.[4,24] The interchange reaction is a simultaneous unlinking and relinking by two polymers

$$P_i + P_j \underset{k_i}{\overset{k_i}{\rightleftarrows}} P_{i+j-k} + P_k \tag{2.6.16}$$

Reversible polymerization including interchange reactions has been studied.[43-46] In some cases the geometric (or near-geometric) distribution has been assumed,[45] but it has also been shown[46] to hold at all times, even far from equilibrium, as long as the initial feed is geometrically distributed. If the distribution is other than geometric, it will of course tend toward that as equilibrium is approached. The interested reader should also refer to the more recent work in this area.[47,48]

2.7 Violation of the Equal Reactivity Assumption

All the complexities dealt with up to this point have been details that do not disturb the basic geometric character of the resulting distributions. They thus leave intact the applicability of the statistical approaches. Now we turn to an effect that limits our repertoire of mathematical techniques: a violation of the equal reactivity assumption on which all the foregoing treatments have been based. This is very different from the change in apparent rate constant due to changing polarity in polyesterifications, discussed in Section 2.4.5. The effect, rather than a global change as would be brought about by a polarity or temperature change, is a local change or difference affecting specific functional groups.

Here we must use terms carefully, making a distinction more common in the literature of network formation than in that of linear stepwise polymerization, namely, the difference between *unequal reactivity* and *substitution effect*. The former refers to intrinsically different reactivity of the same kind of functional group—for example, two different amine functionalities with different neighboring groups. These may be on different molecules (as, say, in an $A_2 + B_2 + C_2$ polymerization, where the B and C groups are of the same type but are not equally reactive) or on the same molecule ($A_2 + BC$ polymerization). A monomer that shows such an effect is 2,4-toluene diisocyanate (TDI),[23] commonly used in the synthesis of polyurethanes:

$$
\begin{array}{c}
CH_3 \\
\end{array}
$$

$$(2.7.1)$$

The isocyanate group ortho to the methyl is approximately 2.7 times less reactive than its para counterpart, since the nearby methyl group increases its electronegativity. Such an effect is a complicating factor, but it does not wreck the ability to analyze the system statistically. The evolution of the different probabilities of reaction (p_A, p_B, p_C) is determined by the different kinetic rate constants, but the problem is easily solvable. Statistical treatments of unequal reactivity[21,22,49,50] have generally been for nonlinear, network-forming systems, but the methods of solution carry over to linear polymerizations.

Substitution effect, although perhaps springing from the same physical sources (such as the steric hindrance or the electron-donating or, -withdrawing capacity of nearby groups), is nonetheless qualitatively different in terms of how it must be viewed. For this difference in reactivity is not intrinsic, rather, it is induced by the reaction of other groups on the monomer. Thus, whereas in the former case the reactivity depends only on an identity (B or C), established at the beginning of the reaction and unchanging, in this latter case the identity of a group changes upon the reaction of a neighboring group (B_2 changes to BC upon reaction of the first B). In our example of TDI, the reaction of the more reactive paraisocyanate further decreases the reactivity of the remaining ortho group by a factor of approximately 4.[23] Thus TDI suffers from both unequal reactivity and substitution effect. The latter apparently does not destroy the ability to perform correct statistical analysis as long as it is confined to one of the comonomers.[51] However, in homopolymerizations (A_2 or AB) or in copolymerizations of the type $A_2 + B_2$, where both monomers suffer from substitution effects, a number of studies have shown the inapplicability of simple statistical ideas.[52-56] (Again, most of these investigations are in the context of network-forming polymers, but the conclusions still apply. We also hasten to point out that this restrictive conclusion applies only to simple statistical ideas; more complex statistical derivations correctly account for the substitution effect.[57] Such intricate treatments lie beyond the scope of this text.)

The equal reactivity assumption, as the term is used in linear polymerizations, refers to a prohibition of a substitution effect rather than a difference in intrinsic reactivity. The question here is whether the reactivity of a group depends on the size of the molecule to which it is attached—in other words, whether the reaction should be written as follows:

$$P_i + P_j \underset{k_{i+j}}{\overset{k_{i,j}}{\rightleftarrows}} P_{i+j} + W \qquad (2.7.2)$$

with a dependence of the rate constants on the length of the participating molecules. In general, then, this appears as an infinite-shell substitution effect, depending not just on the state of the partner functional group on the same monomer (a first-shell substitution effect) but on the state of monomers far away along the contour length of the polymer, or more precisely, on the length of the chain. Although it is hard to imagine this effect arising from electronic or steric sources for chains more than a few units long, other physical sources (e.g., diffusion) may produce it.

The prevalence of the assumption that reactivities are equal is in sharp contrast to the prevailing opinion in the initial years of polymer science, when it was held that reactivity strongly decreased with molecular size.[4] That belief implied a severe difficulty in obtaining high molecular weights. The introduction of the principle of equal reactivity implied that high molecular weights could be attained and that enormous simplifications in both the kinetic and statistical analyses were present. This is not merely wishful thinking, for there exists evidence for the validity of equal reactivity. The most commonly cited evidence is that on reactions of homologous series, such as that on the rate constant data for the esterification of a series of homologous carboxylic acids[58]:

$$H{-}(CH_2)_n{-}\overset{\displaystyle O}{\overset{\|}{C}}{-}OH + HOC_2H_5 \xrightarrow{HCl} H_2O + H{-}(CH_2)_n{-}\overset{\displaystyle O}{\overset{\|}{C}}{-}OC_2H_5 \qquad (2.7.3)$$

The data are shown in Table 2.2. The reactivity does decrease with an increase in molecular size, but when n is 3 or more,

TABLE 2.2 / A Test of Equal Reactivity Assumption

Molecular Size (n)	$k \times 10^4 \ [mol/L \cdot s)]$
1	22.1
2	15.3
3	7.5
4	7.5
5	7.4
8	7.5
9	7.4
11	7.6
13	7.5
15	7.7
17	7.7

the rate constant reaches a limiting value that is approximately constant and independent of molecular weight in the range examined. Similar conclusions were reached for esterification of dicarboxylic acids with ethanol,[58] for saponification of esters,[59] and for etherification.[60] The same principle has been demonstrated for polymerizing systems, specifically the reaction of sebacoyl chloride with α,ω-alkane diols[61]:

$$ i\text{HO}-(\text{CH}_2)_n-\text{OH} + i\text{Cl}-\overset{\overset{\text{O}}{\|}}{\text{C}}-(\text{CH}_2)_8-\overset{\overset{\text{O}}{\|}}{\text{C}}-\text{Cl} \rightarrow $$

$$ i\text{HCl} + \text{H}\left[\text{O}-(\text{CH}_2)_n-\text{O}-\overset{\overset{\text{O}}{\|}}{\text{C}}-(\text{CH}_2)_8-\overset{\overset{\text{O}}{\|}}{\text{C}}\right]_i\text{Cl} \qquad (2.7.4) $$

The data for these systems are shown in Table 2.3. Moreover, a theoretical basis for equal reactivity was supplied early on by Rabinowitch.[6,23,62]

It may be argued that the experiments on homologous series, especially for reactions of monofunctional reactants in which no polymerization occurs, are not entirely applicable to the case at hand. Similarly, evidence against the equal reactivity assumption in similar experiments[63] may consist of manifestations of a global change in rate constant because of changes in the medium. A direct test of a lack of dependence of reactivity on the polymer chain length would be more satisfying. Such a study was done on the reaction of epichlorohydrin with sodium salts of branched monocarboxylic acids.[64] Here, the oligomer distribution obtained compared favorably with that predicted by a kinetic scheme assuming equal reactivity. Other studies, however, contradict this result. In a study of the reversible reaction in the production of poly(ethylene terephthalate),[65] the equilibrium constant was shown to be a function of the extent of polycondensation (perhaps a global effect), but the monomer content at various degrees of reaction exceeded that calculated from the geometric distribution. In this case, the results could be explained by assuming that the monomer had a lower reactivity. Other studies also support the view that the reactivity of the monomer, at least, may differ from that of longer chains.[66–70]

Because of the variety of different forms violation of the equal reactivity assumption may take, we attempt no general

TABLE 2.3 / Test of Equal Reactivity Assumption with Polymerizing System

Molecular Size (n)	$k \times 10^3$ [(L/mol · s)]
5	0.60
6	0.63
7	0.65
8	0.62
9	0.65
10	0.62

treatment here, but rather give reference to the work that has been done. For intrinsically unequally reactive functional groups, in addition to the above-cited references, Durand and Bruneau[71] have applied Markov chain theory, while the more basic combinatorial or statistical approaches were used by Case,[27] Peebles,[72] and Gandhi and Babu.[73] For induced unequal reactivity or substitution effect, the better treatments are those based on kinetic methods. A number of these studies suggest a reactivity for the monomeric species differing from that of larger species. Thus, the most relevant work is that of the Indian group based on just such a model,[3,74,75] which has even been extended to reversible polymerizations in nonbatch polymerizations.[76] Nanda and Jain[77] earlier assumed a linear chain length dependence of the rate constants to obtain analytical results; however, such a dependence is not supported by the evidence, and mathematical problems arise when the rate constant decreases with chain length.

2.8 Cyclization

In the treatment up to this point we have excluded one possible "side reaction" that is chemically identical to the kind of polymerization reaction considered thus far, but different topologically: cyclization. There is nothing, after all, excluding the possibility of the reacting groups on the opposite ends of the chain from meeting in space and reacting, given appropriate flexibility of the chain. In fact, this was the reason given for not producing most monadic nylons by polymerization of amino acids. If cyclization does occur, a cyclic species is formed in addition to the linear polymer generally of interest, which is unreactive (barring reversibility).

To see the analytical difficulties cyclization leads to, let us briefly reconsider an irreversible AB polymerization. We are forced to use a kinetic description because the long-range nature of the cyclization reaction prohibits statistical approaches of the kind given in this chapter. We denote the linear polymer product by P_i and the cyclic by-product by C_i. If the monomer AB is large enough to permit the end-to-end distance to be described by a Gaussian distribution,[4] the concentration of one end about the other will scale as $i^{-3/2}$ for a chain of length i. Cyclization can then be treated as a first-order reaction the rate constant of which decreases as $i^{-3/2}$. Thus:

$$\frac{dP_1}{dt} = -2kP_1P - k_{\text{cyc}}\, P_1 \tag{2.8.1}$$

$$\frac{dP_i}{dt} = k \sum_{j=1}^{i-1} P_j P_{i-j} - 2kP_iP - k_{\text{cyc}} i^{-3/2} P_i \tag{2.8.2}$$

$$\frac{dC_i}{dt} = k_{\text{cyc}}\, i^{-3/2} P_i \tag{2.8.3}$$

In terms of the generating functions of the desired linear product, $G(s)$, and the cycles, $H(s)$:

$$\frac{dG(s)}{d\tau} = G^2(s) - 2G(s)G(1) - \frac{k_{cyc}}{k} \sum_{i=1}^{\infty} s^i i^{-3/2} P_i \qquad (2.8.4)$$

$$\frac{dH(s)}{d\tau} = \frac{k_{cyc}}{k} \sum_{i=1}^{\infty} s^i i^{-3/2} P_i \qquad (2.8.5)$$

The remaining summation might be expressed in terms of a fractional integral of the generating function $G(s)$ (if convergent), resulting in a partial integral–differential equation. The moment equations exhibit a moment closure problem in which, contrary to the reversibility problem, moments are defined in terms of successively lower nonintegral moments, such as $\mu_{-1/2}$. Thus the easiest method is to follow Gordon and Temple,[78] who solved the problem by sequential numerical solution; we do not do this here.

It should be noted that an analytical result is available if it is assumed that cycle formation is too rare to perturb the geometric distribution of the linear species.[79] In this case:

$$C_i = \frac{k_{cyc}}{k} i^{-5/2} \left(\frac{kP_0 t}{1 + kP_0 t}\right)^i \qquad (2.8.6)$$

The distribution, in other words, is a product of a power law (with an exponent of $-5/2$) and an exponential cutoff. At long times, thus, the distribution is power law, decreasing rapidly with size. Note that the concentration of these cyclic species is not proportional to P_0; this is a peculiar but physically reasonable result of the assumption of an unperturbed linear chain distribution.

While equation (2.8.6) may not form the solution we would desire, it does give us the qualitative features we should expect. The distribution of cyclic species should rapidly decrease with length as a power law, and also with an exponential cutoff. It should be noted that for ring–chain equilibrium one also gets a power law distribution with an exponent of $-5/2$, $(-3/2)$ contributed from the Gaussian distribution and (-1) contributed from the decomposition of a ring of size i into a linear chain of i units at a rate proportional to i.[80]

APPENDIX 2A

AN EXPLANATION OF THE RATE EQUATIONS FOR STEP POLYMERIZATIONS

2A.1 AB Polymerization

2A1.1 Balance Equation for an *i*-mer

If we define the rate constant by the equation:

$$\frac{dA}{dt} = \frac{dB}{dt} = -kAB \tag{2A.1}$$

the balance for *i*-mers is given as follows:

$$\frac{dP_i}{dt} = k \sum_{j=1}^{i-1} P_j P_{i-j} - 2k P_i \sum_{j=1}^{\infty} P_j \tag{2A.2}$$

Let us look at each term more closely.

2A.1.2 Disappearance (Negative) Terms

Let us call the rate at which an *i*-mer *disappears* $(dP_i/dt)_{dis}$ and the rate at which it disappears *by reaction with a j-mer* $(dP_i/dt)_{dis,j}$. So, from equation (2A.2), we have

$$\left(\frac{dP_i}{dt}\right)_{dis} = 2kP_i \sum_{j=1}^{\infty} P_j = 2kP_i(P_1 + P_2 + P_3 + \cdots P_i + \cdots) \tag{2A.3}$$

$$\left(\frac{dP_i}{dt}\right)_{dis,j} = 2kP_i P_j \tag{2A.4}$$

The factors of 2 in these equations calls for some explanation, which requires us to look at how an *i*-mer reacts with a *j*-mer. For $i \neq j$, we have:

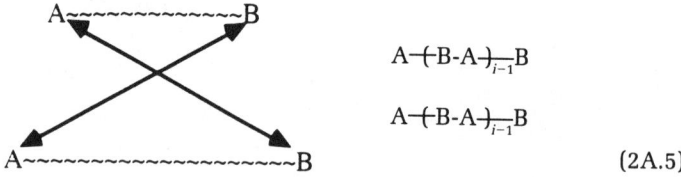

$$\tag{2A.5}$$

There are *two* possible reactions (indicated by the arrows), and these are *distinguishable* because $i \neq j$; in each reaction a B and an A are lost—but in one the B is lost on the *i*-mer, and in the other on the j-mer. Thus, since there are two distinguishable ways in which an i-mer and a j-mer can react, a factor of 2 is introduced.

However, for $i = j$, these two reactions are indistinguishable, and so the factor of 2, which remains in equation (2A.4), would seem to overcount the number of reactions. The factor of 2 is appropriate, however, because two i-mers disappear. For the disappearance of an i-mer by reaction with a j-mer ($i \neq j$) only one i-mer disappeared, but in the disappearance of an i-mer with an i-mer, *two* i-mers disappear.

2A.1.3 Appearance (Positive) Terms

Now let us look at the *appearance* terms. Let us call the rate at which an i-mer *appears* $(dP_i/dt)_{app}$ and the rate at which it appears *by reaction of a j-mer* [with an $(i - j)$-mer necessarily] $(dP_i/dt)_{app,j}$ ($j \leq i/2$). Thus:

$$\left(\frac{dP_i}{dt}\right)_{app} = k \sum_{j=1}^{i-1} P_j P_{i-j} \tag{2A.6}$$

$$\left(\frac{dP_i}{dt}\right)_{app,j} = k\, g(j, i - j)\, P_j P_{i-j} \tag{2A.7}$$

where

$$g(j, i - j) = 1 \quad \text{if } j = i - j \quad (\text{i.e., if } j = i/2)$$
$$= 2 \quad \text{if } j \neq i - j$$

Let us look at a few special cases to understand how these terms relate to the disappearance terms.

First, for $i = 2$:

$$\left(\frac{dP_2}{dt}\right)_{app} = \left(\frac{dP_2}{dt}\right)_{app,1} = kP_1^2 \tag{2A.8}$$

Now, the rate at which 2-mer appears because of reaction between 1-mer and 1-mer should be half the rate at which 1-mers disappear by reaction with 1-mers, since one 2-mer appears for each pair of 1-mers that disappear. And indeed it is, since

$$\left(\frac{dP_1}{dt}\right)_{dis,1} = 2kP_1^2 \tag{2A.9}$$

For the case $i = 3$, we have

$$\left(\frac{dP_3}{dt}\right)_{app} = \left(\frac{dP_3}{dt}\right)_{app,1} = 2kP_1P_2 \tag{2A.10}$$

and again:

$$\left(\frac{dP_3}{dt}\right)_{app} = (1/2)\left[\left(\frac{dP_1}{dt}\right)_{dis,2} + \left(\frac{dP_2}{dt}\right)_{dis,1}\right] = (1/2)\,4kP_1P_2 \tag{2A.11}$$

A 4-mer can form in two different ways:

$$\left(\frac{dP_4}{dt}\right)_{app} = \left[\left(\frac{dP_4}{dt}\right)_{app,1} + \left(\frac{dP_4}{dt}\right)_{app,2}\right] = 2kP_1P_3 + kP_2^2 \tag{2A.12}$$

and here:

$$\left(\frac{dP_4}{dt}\right)_{app,1} = (1/2)\left[\left(\frac{dP_1}{dt}\right)_{dis,3} + \left(\frac{dP_3}{dt}\right)_{dis,1}\right] = (1/2)\,4kP_1P_3 \tag{2A.13}$$

$$\left(\frac{dP_4}{dt}\right)_{app,2} = (1/2)\left(\frac{dP_2}{dt}\right)_{dis,2} = (1/2)\,2kP_2^2 \tag{2A.14}$$

Thus, by writing the equation the way we did, we achieve the proper balance of the appearance and disappearance terms.

2A.1.4 Balance Equations for A, B

As a final show of consistency, we will see that the balance equation for an i-mer is consistent with the rate constant definition of equation (2A.1). Note that

$$A = B = \mu_0 \tag{2A.15}$$

since each molecule bears one A group and one B group. Summing equation (2A.2) from 1 to ∞ gives the equation for μ_0:

$$\frac{d\mu_0}{dt} = k\mu_0^2 - 2k\mu_0^2 = -k\mu_0^2 \tag{2A.16}$$

which by equation (2A.15) grants us consistency with equation (2A.1).

2A.2 A$_2$ Homopolymerization

2A.2.1 Balance Equation for an i-mer

The rate constant is defined by the following equation:

$$\frac{dA}{dt} = -kA^2 \tag{2A.17}$$

that is, again by functional group disappearance. The indistinguishability of A groups does imply that a factor of (1/2) is needed to prevent overcounting, but this is canceled by a factor of 2, which appears because two A's disappear in each reaction. Thus equation (2A.17) is the correct definition of the rate constant. The balance equation for an i-mer is then given as follows:

$$\frac{dP_i}{dt} = 2k \sum_{j=1}^{i-1} P_j P_{i-j} - 4k P_i \sum_{j=1}^{\infty} P_j \qquad (2A.18)$$

The factors of 2 and 4 require explanation; we will proceed as in the AB case.

2A.2.2 Disappearance (Negative) Terms

One might be tempted to think that the factor of 4 in the disappearance term comes from the existence of four possible ways to react:

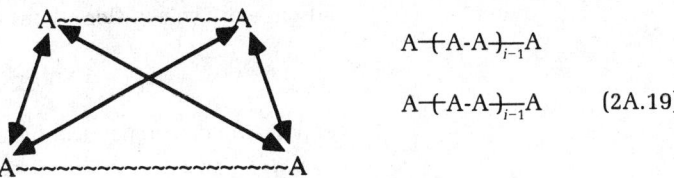

$$A\text{---}(A\text{-}A)_{\overline{i-1}}A$$

$$A\text{---}(A\text{-}A)_{\overline{i-1}}A \qquad (2A.19)$$

until we realize that these are indistinguishable. The factor of 4 arises for a less exotic reason: if k is the rate constant for functional group reaction, then we must realize that

$$\left(\frac{dP_i}{dt} \right)_{\text{dis},j} = k 2 P_i 2 P_j \qquad (2A.20)$$

since an i-mer and a j-mer each bear two A groups. Then, for the same reasons as in Section 2A.1, the appearance term has $2k$.

2A.2.3 Balance Equation for A

Here, we realize that

$$A = 2\mu_0 \qquad (2A.21)$$

The rate equation for μ_0 is found by summing equation (2A.18):

$$\frac{d\mu_0}{dt} = 2k\mu_0^2 - 4k\mu_0^2 = -2k\mu_0^2 \qquad (2A.22)$$

which, because of equation (2A.21), easily leads to equation (2A.17). So again we have written our balance equation for an i-mer consistent with the definition of the rate constant.

2A.3 A_f Homopolymerization

A case we will encounter in Chapter 5 as a paradigm of branched polymer formation is A_f homopolymerization. The chemistry is the same as in A_2 polymerization: like groups react, as in a polyetherification. The only significant alteration here is the number of hydroxy groups initially present on a monomer. This leads to vastly different structures, of course; here we are only concerned with the form of the kinetic equation, which is altered somewhat.

In the balance equation for an i-mer, we maintain the definition of the rate constant k of equation (2A.17). Remembering that in the case of A_2, each concentration P_i was weighted by the number of functional groups, we can immediately write:

$$\frac{dP_i}{dt} = \frac{k}{2} \sum_{j=1}^{i-1} [(f-2)j + 2] P_j [(f-2)(j-i) + 2] P_{i-j} - k[(f-2)i + 2] P_i \sum_{j=1}^{\infty} [(f-2)j + 2] P_j \qquad (2A.23)$$

since an i-mer bears $[(f-2)i + 2]$ unreacted groups (in the absence of cycles). Here then,

$$A = (f-2)\mu_1 + 2\mu_0 \qquad (2A.24)$$

and again consistency can be shown.

2A.4 Smoluchowski's Coagulation Equation

We can write all the foregoing balance equations for i-mers in a general form, which is called *Smoluchowski's coagulation equation*:

$$\frac{dP_i}{dt} = (1/2) \sum_{j=1}^{i-1} K_{j,i-j} P_j P_{i-j} - P_i \sum_{j=1}^{\infty} K_{i,j} P_j \qquad (2A.25)$$

where $K_{i,j}$ is the *coagulation kernel*, which describes the reaction between an i-mer and a j-mer. Thus we summarize the cases as follows:

A-B polymerization: $K_{i,j} = 2k$
A-A homopolymerization: $K_{i,j} = 4k$
A_f homopolymerization: $K_{i,j} = k[(f-2)i + 2][(f-2)j + 2]$

2A.5 Comments with Regard to the Termination Step of Free-Radical Polymerization

In most of the reactions in free-radical polymerization (to be discussed in Chapter 3), the reacting groups are distinguishable (one is a free radical, the other a monomer, transfer agent, etc.) and can happen in only one way, so that these problems do not arise. However, the termination step does involve two similar groups. When an R_i terminates with an R_j, distinguishability becomes a question only when $i = j$. But here again, the factor of (1/2) is canceled by the factor of 2, which arises because two R_i's disappear. Thus the termination rate is correctly written:

$$\left(\frac{dR_i}{dt}\right)_{term} = k_t R_i R \qquad (2A.26)$$

where k_t is the rate constant for radical disappearance.

APPENDIX 2B

LAGUERRE POLYNOMIALS

In general, reconstruction of a distribution or its transform $G(s)$ from moments requires an infinite number of moments. However, the geometric distribution we have encountered is a one-parameter distribution. Hence, only one (nonunity) moment or a molecular weight is necessary to reconstruct that distribution. Other distributions we will encounter in this text are characterized by two or more parameters. Simple mathematical cases have a small number of parameters; this recurring feature suggests that in practical cases the use of moments is a straightforward and valuable means of characterizing a chain length distribution.[81] Once the use of moments in the analysis of polymerization kinetics had been suggested,[82] it was shown that a distribution may be reconstructed using a finite number of moments.[37,83] There are, however, limitations of such procedures.[38]

The number of moments necessary to provide a sufficient description of the distribution is generally unknown, although one can say that the more complex the polymerization mechanism, the greater the number needed. If N_k kinetic parameters are required for the characterization, one can evaluate or measure a number of moments N_m, then use the rate equation to determine the set of kinetic parameters. Clearly we require that N_m equal or exceed N_k. The accuracy of the moments dictates whether more moments are essential for complete characterization of the distribution.

Generalized Laguerre polynomials provide a suitable method of obtaining the chain length distribution from the moments.[82] The distribution is expanded in orthogonal series of Laguerre polynomials, the coefficients of which are given in terms of the moments of the original distribution. The mth-order associated Laguerre polynomial is given as follows:

$$L_m^\lambda(x) = \sum_{n=0}^{m} (-1)^n \frac{m!}{n!(m-n)!} \frac{\Gamma(m+\lambda)}{\Gamma(m+\lambda-n)} x^{m-n} \qquad (2B.1)$$

For example,

$$L_0^\lambda(x) = 1 \qquad (2B.2)$$

$$L_1^\lambda(x) = x - \lambda \qquad (2B.3)$$

$$L_2^\lambda(x) = x^2 - 2(\lambda + 1)x + (\lambda + 1)\lambda \qquad (2B.4)$$

$$L_3^\lambda(x) = x^3 - 3(\lambda + 2)x^2 + 3(\lambda + 2)(\lambda + 1)x - (\lambda + 2)(\lambda + 1)\lambda \qquad (2B.5)$$

One can express any function that ranges from zero to infinity (and satisfying certain conditions of continuity) as an expression in these polynomials. In particular, the distribution P_i can be expressed as follows:

$$P_i = \frac{\lambda}{a} w^\lambda \frac{\lambda i}{a} \sum_{m=0}^{\infty} k_m L_m^\lambda \frac{\lambda i}{a} \qquad (2B.6)$$

where $w^\lambda(x)$ is a weighting function for the orthogonal polynomial. In this case it is a Γ-distribution weighting function given by

$$w^\lambda(x) = \frac{x^{\lambda-1}e^{-x}}{\Gamma(\lambda)} \tag{2B.7}$$

The parameters λ and a in equation (2B.6) are, in essence, adjustable parameters to be determined from a knowledge of the distribution P_i or the moments of the distribution, and they can be chosen to provide either a simple representation of the expansion or a simple relation between the moments themselves, in cases as moment closure techniques discussed in Section 2.6.

The coefficients of the expansion, k_m, can be found by letting $x = \lambda i/a$ and making use of the orthogonality property of the Laguerre polynomials. Thus

$$k_m = \sum_{n=0}^{m} (-1)^n \frac{1}{n!(m-n)!} \frac{\Gamma(\lambda)}{\Gamma(m+\lambda-n)} \left(\frac{\lambda}{a}\right)^{m-n} \mu_{m-n} \tag{2B.8}$$

These coefficients are given in terms of the moments μ_k, of the distribution P_i, and it is this relationship that permits representation of the whole chain length distribution from a few moments, provided the series represented by equation (2B.6) converges. The accuracy of this approximation depends on the number of terms used in equation (2B.6) or, equivalently, the number of moments used. For a chain length distribution arising from a free-radical polymerization (see Chapter 3), five terms in the expansion were found to give 1% error in some cases,[82] but since accuracy depends on the relative magnitude of the competing rates, in other cases as many as ten terms may be required to obtain 5% error.

The first three such coefficients of the expansion are given by:

$$k_0 = \mu_0 \tag{2B.9}$$

$$k_1 = \frac{1}{a}\mu_1 - \mu_0 \tag{2B.10}$$

$$k_2 = \frac{1}{2a^2}\frac{\lambda\mu_2}{\lambda+1} - \frac{\mu_1}{a} + \frac{\mu_0}{2} \tag{2B.11}$$

An arbitrary choice of a and λ can be taken to make $k_1 = k_2 = 0$, so that the expansion of equation (2B.6) can be simplified. So we make the following choices:

$$a = \frac{\mu_1}{\mu_0} \quad \text{and} \quad \lambda = \frac{a^2}{\mu_2/\mu_0 - a^2} \tag{2B.12}$$

Then equation (2B.6) becomes

$$P_i = \frac{\lambda}{a} w^\lambda \frac{\lambda i}{a} \left(\mu_0 + \sum_{m=3}^{\infty} k_m L_m^\lambda \frac{\lambda i}{a}\right) \tag{2B.13}$$

One can effectively approximate the distribution by truncating the expansion say, at $m = q$, so that

$$P_i = \frac{\lambda}{a} w^\lambda \frac{\lambda i}{a} \left(\mu_0 + \sum_{m=3}^{q} k_m L_m^\lambda \frac{\lambda i}{a} \right) \tag{2B.14}$$

Actually the approximation uses all the leading moments $\mu_0, \mu_1, \ldots, \mu_q$. We can now express any moment of P_i in terms of the coefficients $k_0, k_1, \ldots, k_q$ using equation (2B.14):

$$\mu_k = \frac{\Gamma(k + \lambda)}{\Gamma(\lambda)} \frac{\mu_0}{(\lambda/a)^k} + \sum_{m=3}^{q} \frac{k_m}{(\lambda/a)^k} \sum_{j=0}^{m} (-1)^j \frac{m!}{j!(m-j)!} \frac{\Gamma(m + \lambda)\Gamma(m + \lambda + k - j)}{\Gamma(\lambda)\Gamma(m + \lambda - j)} \tag{2B.15}$$

If we choose to truncate expression (2B.14) at $m = 2$, an expression for μ_3 could be obtained from equation (2B.15) by substituting the chosen relations for a and λ. The result is given in the body of the chapter.

Problems

2.1. Prove that:

$$\mu_k = \lim_{s \to 1} \left\{ \frac{\partial^k G(s)}{\partial(\ln(s))^k} \right\}$$

2.2. Prove equations (2.2.40) through (2.2.42).

2.3. Evaluate the following sums:

(a) $\sum_{i=1}^{\infty} P_0(1 - p)^2 p^{i-1}$

(b) $\sum_{i=1}^{\infty} P_0(1 - p)^2 i p^{i-1}$

(c) $\sum_{i=1}^{\infty} P_0(1 - p)^2 i^2 p^{i-1}$

(d) $\sum_{i=1}^{\infty} i p^{i-7}$

(e) $\sum_{i=83}^{\infty} p^{i-1}$

(f) $\sum_{i=2}^{\infty} \sum_{j=1}^{i-1} P_j P_{i-j}$

When would a double summation like that of (f) appear in the analysis of a polymerization reaction? When are sums (a)–(c) important?

2.4. Toward the end of this book, we will encounter the following generating function:

$$G(s) = \frac{\sqrt{1 + 4P_0 k\theta} - \sqrt{1 + (1 - s)4P_0 k\theta}}{2k\theta}$$

where P_0 is the initial concentration of monomer, k is a rate constant, and θ is a time variable. Derive from this generating function DP_n and DP_w, as well as the polydispersity Q. Is there any obvious qualitative difference between this distribution and the distributions encountered so far? Derive also the chain length distribution P_i.

2.5. Prove equations (2.2.33) and (2.2.34).

2.6. Derive both DP_n and DP_w from recursive arguments for the case of AB polymerization with a monofunctional agent XB.

2.7. Derive both DP_n and DP_w from recursive arguments for the case of $A_2 + B_2 + AX$, where the last is a monofunctional agent. Assume the molar concentration of A_2 and AX monomers equals that of B_2 monomers. Define the answer in terms of the stoichiometric imbalance $r = A/B$.

2.8. Consider the irreversible step polymerization of AB monomers, beginning not from monomer, but from polymer of the following initial distribution:

$$P_i(0) = P_0 e^{-\tau} \frac{\tau^{i-1}}{(i-1)!}$$

Find DP_n and DP_w. How does the breadth of the polymer formed compare to that of the most probable distribution? Go as far as you can toward finding the chain length distribution.

2.9. Polyesterification is generally acid catalyzed with an added acid, and so the rate equation for the disappearance of both carboxylic acid and alcohol groups is:

$$\frac{d[COOH]}{dt} = \frac{d[OH]}{dt} = -k[COOH][OH]$$

If no acid catalyst is added, then the carboxylic acid acts as a catalyst (i.e., the reaction is self-catalyzed) and the rate equation becomes:

$$\frac{d[COOH]}{dt} = \frac{d[OH]}{dt} = -k[COOH]^2[OH]$$

For the case of an uncatalyzed AB polymerization (or equivalently an $A_2 + B_2$ polymerization at balanced stoichiometry)
(a) Find [COOH] and [OH] as a function of time.
(b) Substantiate the claim that the chain length distribution is unaffected by the difference in the kinetics. Is this surprising? Give reasons for why this occurs.

2.10. In an acid-catalyzed polymerization of diethylene glycol and adipic acid at 109°C, the degree of polymerization DP_n was followed by end group titration and was found to increase from 20 at $t = 200$ minutes to 77 at t = 650 minutes, all conditions being held constant. Both alcohol and acid groups are present in equimolar amounts, the initial concentration of each being $9.8M$ (i.e., the concentration of each monomer initially was $4.9M$).
(a) Determine the rate constant k' as defined by:

$$\frac{d[COOH]}{dt} = -k'[COOH][OH]$$

where the concentrations are in moles per liter and time t in seconds.

(b) Why might we prefer to use these data (at 200 and 650 min) over data at $t = 0$ minutes (where we know that $DP_n = 1$)?

(c) The equilibrium constant K,

$$K = \frac{[-C(O)O-][H_2O]}{[-COOH][-OH]}$$

is approximately 1 at 109°C (this is typical for polyesters).

(1) What is the maximum DP_n accessible if no water is removed?

(2) What molar ratio of dissolved water of condensation to carboxyl end group concentration would lead to an equilibrium value of $DP_n = 77$ at 109°C?

(3) How must the DP_n value given in (2) have been achieved? That is, what percentage of the condensate must have been removed?

(4) How would the reversibility of the reaction qualitatively affect the plot of DP_n versus time?

(5) What can you conclude from the fact that the data of DP_n versus time between 200 and 650 minutes is linear?

(d) Adipic acid has a boiling point of 265°C at 100 mm Hg, whereas diethylene glycol boils at 245°C at 1 atm. Thus, it is reasonable that the diol has a much higher vapor pressure than the diacid, so that the removal of water results in the loss of diethylene glycol, which in turn induces a stoichiometric imbalance. Suppose we lose 1% of the diethylene glycol. What maximum degree of polymerization can be achieved?

(e) For polyamides (nylons), reversibility also occurs except that the equilibrium constant is much higher: in the range of 250–900 for the polymerization of adipic acid and hexamethylene diamine. It is found that the melt or bulk viscosity of a nylon 6/6 sample was a function of the season (the melt viscosity was lower in August than in January). Explain this result. Would this effect be as extreme for polyesters?

2.11. Consider the system $A_2 + B-B'$, where B' is the same type functional group as B but has a different reactivity, as indicated:

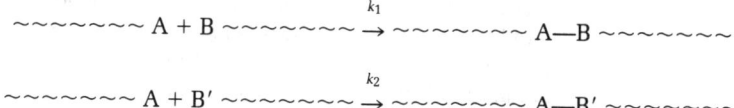

Consider the case of equal stoichiometry: $A = B + B'$.

(a) Set up the kinetic equations needed for obtaining the conversion of the three functional groups, p_A, p_B, and $p_{B'}$. Go as far as you can toward solving these.

(b) Derive DP_w in terms of p_A, p_B, and $p_{B'}$.

References

1. W. H. Carothers, *J. Am. Chem. Soc.*, **51**, 2548 (1929).

2. D. H. Solomon, Ed., *Step-Growth Polymerizations*. Dekker, New York (1972).

3. S. K. Gupta and A. Kumar, *Reaction Engineering of Step Growth Polymerization*. Plenum, New York (1987).

4. P. J. Flory, *Principles of Polymer Chemistry*. Cornell University Press, Ithaca, New York (1953).

5. P. J. Flory, *J. Am. Chem. Soc.*, **61**, 3334 (1939).

6. G. Odian, *Principles of Polymerization*, 2nd ed., Wiley, New York (1981).

7. M. B. Huglin, *Eur. Polym. J.*, **27**, 875 (1991).

8. J. P. Howe, *J. Chem. Phys.*, **23**, 899 (1955).

9. A. T. Bharucha-Reid, *Elements of the Theory of Markov Processes and Their Applications*. McGraw-Hill, New York (1960).

10. W. Feller, *An Introduction to Probability Theory and Its Applications*, 3rd ed., Wiley, New York (1968).

11. J. Scanlan, *Trans. Faraday Soc.*, **52**, 1286 (1956).

12. E. I. Jury, *Theory and Application of the Z-Transform Method*. Wiley, New York (1964).

13. W. H. Abraham, *Ind. Eng. Chem. Fundam.*, **2**, 221 (1963).

14. J. R. Ragazzini and G. F. Franklin, *Sampled-Data Control Systems*. McGraw-Hill, New York (1958).

15. B. C. Kuo, *Analysis and Synthesis of Sampled Data Control Systems*. Prentice-Hall, Englewood Cliffs, NJ (1963).

16. R. V. Churchill, *Complex Variables and Applications*. McGraw-Hill, New York (1960).

17. H. Kilkson, *Ind. Eng. Chem. Fundam.*, **3**, 281 (1964).

18. A. E. Taylor, *Advanced Calculus*. Ginn, Boston (1955).

19. M. Gordon, *Proc. R. Soc. London, Ser. A*, **268**, 240 (1962).

20. C. W. Macosko and D. R. Miller, *Macromolecules*, **9**, 199 (1976).

21. D. R. Miller and C. W. Macosko, *Macromolecules*, **11**, 656 (1978).

22. F. Lopez-Serrano, J. M. Castro, C. W. Macosko, and M. Tirrell, *Polymer*, **21**, 263 (1980).

23. G. Odian, *Principles of Polymerization*, 3rd ed., Wiley, New York (1991).

24. R. W. Lenz, *Organic Chemistry of Synthetic High Polymers*. Wiley-Interscience, New York (1967).

25. P. J. Flory, *J. Am. Chem. Soc.*, **58**, 1877 (1936).

26. J. J. Hermans, *Makromol. Chem.*, **87**, 21 (1965).

27. L. C. Case, *J. Polym. Sci.*, **29**, 455 (1958).

28. S. M. Ross, *Introduction to Probability Models*, 3rd ed. Academic Press, Orlando, FL (1985).

29. G. G. Lowry, Ed., *Markov Chains and Monte Carlo Calculations in Polymer Science*. Dekker, New York (1970).

30. G. G. Lowry, Molecular weight distributions, in *Markov Chains and Monte Carlo Calculations in Polymer Science*, G. G. Lowry, Ed. Dekker, New York (1970).

31. N. R. Amundson, *Mathematical Methods in Chemical Engineering. Matrices and Their Application.* Prentice-Hall, Englewood Cliffs, NJ (1966).

32. L. Peller, *J. Chem. Phys.*, **36**, 2976 (1962).

33. D. H. Solomon, *J. Macromol. Sci.: Rev. Macromol. Chem.*, **C1**, 179 (1967).

34. For example, see C. Laubriet, B. LeCorre, and K. Y. Choi, *Ind. Eng. Chem. Res.*, **30**, 2 (1991), and references therein.

35. (a) L. L. Jacobsen and W. H. Ray, *J. Macromol. Sci., Rev. Macromol. Chem. Phys.*, **C32**, 407 (1992). (b) L. L. Jacobsen and W. H. Ray, *AIChE J.*, **38**, 911 (1992).

36. W. P. Yang, C. W. Macosko, and S. T. Wellinghoff, *Polymer*, **27**, 1235 (1986).

37. H. M. Hulburt and S. Katz, *Chem. Eng. Sci.*, **19**, 555 (1964).

38. H. Tobita and K. Ito, *Polym. React. Eng.*, **1**, 407 (1992–1993).

39. P. J. Flory, *J. Chem. Phys.*, **12**, 425 (1944).

40. A. V. Tobolsky, *J. Chem. Phys.*, **12**, 402 (1944).

41. (a) S. K. Gupta, N. L. Agarwalla, P. Rajora, and A. Kumar, *J. Polym. Sci., Polym. Phys.*, **20**, 933 (1982). (b) A. Kumar, P. Rajora, N. L. Agarwalla, and S. K. Gupta, *Polymer*, **23**, 222 (1982).

42. R. M. Ziff and E. D. McGrady, *Macromolecules*, **19**, 2513 (1986).

43. W. H. Abraham, *Chem. Eng. Sci.*, **21**, 327 (1966).

44. T. T. Szabo and J. F. Leathrum, *J. Appl. Polym. Sci.*, **13**, 477, 487, 561 (1969).

45. D. A. Mellichamp, *Chem. Eng. Sci.*, **24**, 125 (1969).

46. W. H. Abraham, *Chem. Eng. Sci.*, **25**, 331 (1970).

47. A. Kumar, *Polym. Eng. Sci.*, **28**, 1240 (1988).

48. A. Kumar and A. Khanna, *J. Appl. Polym. Sci.*, **41**, 2077 (1990).

49. M. Gordon and G. R. Scantlebury, *Trans. Faraday Soc.*, **60**, 604 (1964).

50. D. Durand and C.-M. Bruneau, *Polymer*, **24**, 587 (1983).

51. D. R. Miller and C. W. Macosko, *Macromolecules*, **13**, 1063 (1980).

52. S. I. Kuchanov, *Kinetic Calculation Methods in Polymer Chemistry.* Khimiya, Moscow (1978).

53. S. I. Kuchanov, *Dokl. Akad. Nauk SSSR*, **249**, 1014 (1979).

54. S. I. Kuchanov and Ye. S. Povolotskaya, *Polym. Sci. USSR*, **A24**, 2499, 2512 (1982) (*Vysokomol. Soyed. A*, **A24**, 2179, 2190).

55. J. Mikes and K. Dusek, *Macromolecules*, **15**, 93 (1982).

56. (a) H. Galina, *Europhys. Lett.*, **3**, 1155 (1987). (b) H. Galina and A. Szustalewicz, *Macromolecules*, **22**, 3124 (1989). (c) H. Galina and A. Szustalewicz, *Macromolecules*, **23**, 3833 (1990). (d) H. Galina, K. Kaczmarski, B. Para, and B. Sanecka, *Makromol. Chem., Theory Simulations*, **1**, 37 (1992).

57. C. Sarmoria and D. R. Miller, *Macromolecules*, **24**, 1833 (1991).

58. B. V. Bhide and J. J. Sudborough, *J. Indian Inst. Sci.*, **8A**, 89 (1925).

59. D. P. Evans, J. J. Gordon, and H. B. Watson, *J. Chem. Soc.*, 1439 (1938).

60. P. C. Haywood, *J. Chem. Soc.*, **121**, 1904 (1922).

61. K. Ueberreiter and M. Engel, *Makromol. Chem.*, **178**, 2257 (1977).

62. E. Rabinowitch, *Trans. Faraday Soc.*, **33**, 1225 (1937).

63. S. I. Kuchanov, M. L. Keshtov, P. G. Halatur, V. A. Vasnev, S. V. Vinogradova, and V. V. Korshak, *Makromol. Chem.*, **184**, 105 (1983).

64. W. J. M. Rootsaert and J. G. van de Vusse, *Chem. Eng. Sci.*, **21**, 1067 (1966).

65. G. Challa, *Makromol. Chem.*, **38**, 105, 123, 138 (1960).

66. J. H. Hodgkin, *J. Polym. Sci., Polym. Chem.*, **14**, 409 (1976).

67. I. Vancso-Szmercsany and E. Makay-Bödi, *J. Polym. Sci.*, **C16**, 3709 (1968).

68. K. Weisskopf and G. Meyerhoff, *Makromol. Chem.*, **187**, 411 (1986).

69. V. N. Ignatov, V. A. Vasnev, S. V. Vinogradova, V. V. Korshak, and H. M. Tseitlin, *Makromol. Chem.*, **189**, 975 (1988).

70. V. N. Ignatov, V. A. Vasnev, and S. V. Vinogradova, *Polym. Sci. USSR*, **29**, 993 (1987) (*Vysokomol. Soyed. A.*, **A29**, 899).

71. D. Durand and C.-M. Bruneau, *Eur. Polym. J.*, **17**, 707, 715 (1981).

72. L. H. Peebles, *Molecular Weight Distributions in Polymers.* Wiley, New York (1971).

73. K. S. Gandhi and S. V. Babu, *AIChE J.*, **25**, 266 (1979).

74. R. Goel, S. K. Gupta, and A. Kumar, *Polymer*, **18**, 851 (1977).

75. (a) S. K. Gupta, A. Kumar, and A. Bhargava, *Eur. Polym. J.*, **15**, 557 (1979). (b) S. K. Gupta, A. Kumar, and A. Bhargava, *Polymer*, **20**, 305 (1979).

76. (a) A. Kumar and A. Khanna, *Polymer*, **30**, 1733, 1742 (1989). (b) A. Kumar and A. Khanna, *Macromolecules*, **22**, 866 (1989).

77. V. S. Nanda and S. C. Jain, *J. Chem. Phys.*, **49**, 1318 (1968).

78. M. Gordon and W. B. Temple, *Makromol. Chem.*, **152**, 277 (1972); *ibid.*, **160**, 263 (1972).

79. X.-F. Yuan, A. J. Masters, C. V. Nicholas, and C. Booth, *Makromol. Chem.*, **189**, 823 (1988).

80. (a) H. Jacobson and W. H. Stockmayer, *J. Chem. Phys.*, **18**, 1600 (1950). (b) H. Jacobson, C. O. Beckman, and W. H. Stockmayer, *J. Chem. Phys.*, **18**, 1607 (1950).

81. W. H. Ray, *J. Macromol. Sci.: Rev. Macromol. Chem.*, **C8**, 1 (1972).

82. C. H. Bamford and H. Tompa, *Trans. Faraday Soc.*, **50**, 1097 (1954).

83. C. H. Bamford, W. G. Barb, A. D. Jenkins, and P. F. Onyon, *The Kinetics of Vinyl Polymerization by Radical Mechanisms.* Academic Press, New York (1958).

3

CHAIN GROWTH POLYMERIZATION

3.1 Introduction

Chainwise polymerization occurs by the successive addition of monomers to the ends of growing chains, in contrast to stepwise polymerization, in which species of all sizes react with one another. This growth by the addition of monomer units alone, which is the primary distinction of chain growth polymerization from step growth polymerization, earns it the alternative name "addition" polymerization.[1] The root of the difference between these two types of polymerization lies in the chemistry; the types of reaction occurring in the one differ from those occurring in the other. The monomers used in chainwise polymerization are, for the most part, not the same in step growth and chain growth polymerizations; by-products of low molecular weight are extremely rare for chain growth polymerizations; and the reactions are much more exothermic.

By far the most common chain growth mechanism involves the successive opening of carbon–carbon double bonds. Table 3.1 lists common monomers and the uses of the polymers derived therefrom; note how well represented these are in Tables 1.1 through 1.4. In all these cases (except formaldehyde), the resulting polymer has a carbon backbone, in contrast to the heteroatomic backbones of most stepwise polymers. In most cases the carbon backbone is saturated, but there are exceptions. Conjugated 1,3-dienes are often reacted to form 1,4-polymers as in formula (1.3.9), so that an unsaturation remains in the backbone rather than pendant to the backbone.[2] Acetylenic monomers can be polymerized to form polymers with conjugated backbones that can be made conductive by doping. There are exceptions to the pure carbon backbone as well. The carbon–oxygen double bond of formaldehyde can be polymerized to form polyoxymethylene (as shown in Table 3.1), but more importantly, ring-opening polymerizations are commonly used to produce polyethers, nylons, and polysiloxanes; the most common of these are listed in Table 3.2.

The simplest step growth polymerizations can be described by one reaction, such as that between an amine and a carboxylic acid. At least two reactions are needed for even the simplest chain growth polymerization, though: an initiation reaction

TABLE 3.1 / Typical Unsaturated Monomers Used in Chainwise Polymerization

Category	Name	Monomer	Uses (Examples)
Olefins	Ethylene	$H_2C{=}CH_2$	Plastics, films
	Propylene	$H_2C{=}CH{-}CH_3$	Plastics, fibers
Dienes	Butadiene	$H_2C{=}CH{-}CH{=}CH_2$	Rubber (in copolymers often)
	Isoprene	$H_2C{=}C(CH_3){-}CH{=}CH_2$	"Natural rubber"
	Chloroprene	$H_2C{=}CCl{-}CH{=}CH_2$	Neoprene rubber, resistant to high temperatures and oil
Vinyls	Vinyl chloride	$H_2C{=}CHCl$	Packaging, piping, tubing
	Styrene	$H_2C{=}CHC_6H_5$	Packaging, foams
	Vinyl acetate	$$H_2C{=}\underset{\underset{H}{\mid}}{C}{-}O{-}\overset{\overset{O}{\parallel}}{C}{-}CH_3$$	Intermediates, paint latices
	Acrylonitrile	$H_2C{=}CHCN$	Acrylic fiber
(Meth)acrylates	Methyl methacrylate	$$H_2C{=}\underset{\underset{CH_3}{\mid}}{C}{-}\overset{\overset{O}{\parallel}}{C}{-}O{-}CH_3$$	Plexiglas, hard contact lenses
	Ethyl acrylate	$$H_2C{=}\underset{\underset{H}{\mid}}{C}{-}\overset{\overset{O}{\parallel}}{C}{-}O{-}\underset{\underset{CH_3}{\mid}}{CH_2}$$	Paint latices
	Hydroxyethyl methacrylate	$$H_2C{=}\underset{\underset{CH_3}{\mid}}{C}{-}\overset{\overset{O}{\parallel}}{C}{-}O{-}\underset{\underset{\underset{OH}{\mid}}{CH_2}}{\overset{}{CH_2}}$$	Soft contact lenses (lightly crosslinked)
Fluoro compound	Tetrafluoroethylene	$F_2C{=}CF_2$	Teflon
Aldehyde	Formaldehyde	$H_2C{=}O$	Plastics

(which produces an active species) and a propagation reaction (by which the active species adds monomer successively to form polymer). The reactive species may be free radicals, anions, or cations, and they are produced from compounds called *initiators*. (The initiator may even be the monomer itself, as in thermal polymerization wherein the monomer undergoes a reaction to produce the free radicals.[2]) A number of other reactions may occur, and may even be desirable, but for now we focus on the two required steps.

The *initiation* reaction generally comprises two parts: the production of an active species R_0, which subsequently attacks monomer, M, to produce a "polymer" of length *1*, R_1:

$$R_0 + M \rightarrow R_1 \qquad (3.1.1)$$

TABLE 3.2 / Typical Monomers Used in Ring-Opening Polymerization

Category	Name	Monomer → Polymer	Uses (Examples)
Epoxides	Ethylene oxide	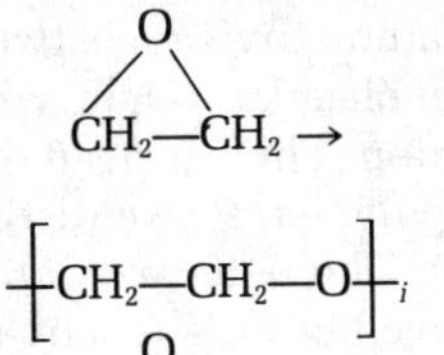	Adhesives, polymer additive
	Propylene oxide	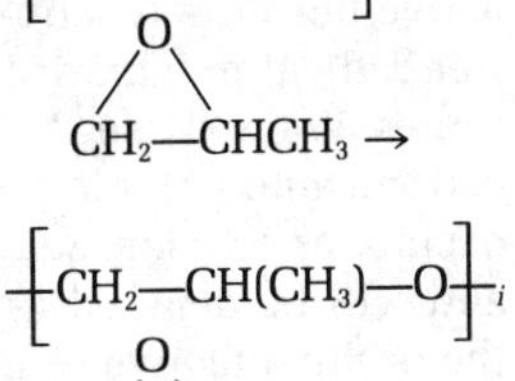	Prepolymer
Other cyclic ethers	Trioxane (→ polyoxy- methylene or polyacetal)	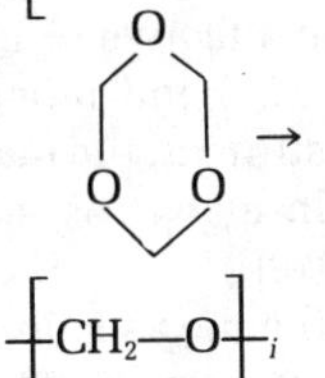	Plastic
Cyclic amines	ε-Caprolactam (→ nylon 6)	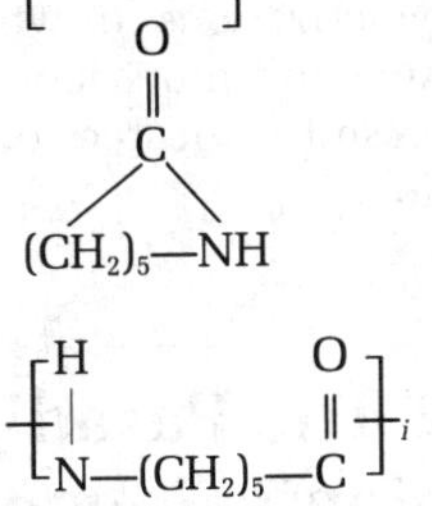	Fibers
Cyclic siloxanes	Tetramethylsiloxane	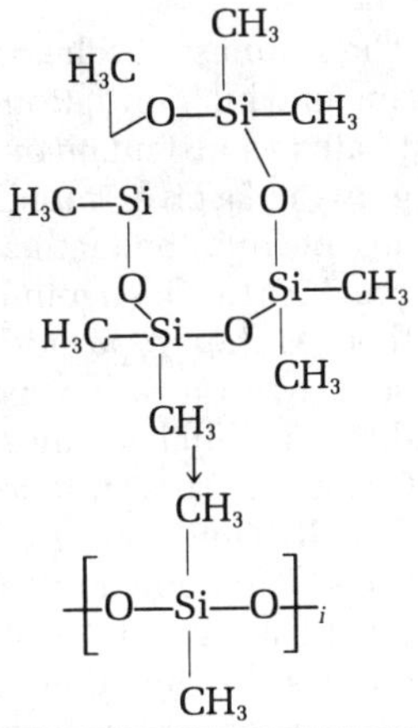	Silicone rubbers and gums

Ethylene oxide:

$$\underset{\displaystyle CH_2-CH_2 \;\rightarrow}{\overset{\displaystyle O}{\triangle}}$$

$$\left[CH_2-CH_2-O \right]_i$$

Propylene oxide:

$$\underset{\displaystyle CH_2-CHCH_3 \;\rightarrow}{\overset{\displaystyle O}{\triangle}}$$

$$\left[CH_2-CH(CH_3)-O \right]_i$$

Trioxane:

$$\left[CH_2-O \right]_i$$

ε-Caprolactam:

$$\begin{array}{c} O \\ \parallel \\ C \\ (CH_2)_5-NH \end{array}$$

$$\left[\begin{array}{cc} H & O \\ | & \parallel \\ N-(CH_2)_5-C \end{array} \right]_i$$

Tetramethylsiloxane:

$$\left[O-\underset{\displaystyle CH_3}{\overset{\displaystyle CH_3}{Si}}-O \right]_i$$

The rate-limiting step may either be the production of R_0, or the first addition of monomer shown in reaction (3.1.1). The subsequent *propagation* reaction actually produces the polymer and can be written:

$$R_i + M \rightarrow R_{i+1} \tag{3.1.2}$$

The sequential addition of monomer, rather than the random reaction of all species, means that what we call "polymer" when we model chain growth polymerization differs from that for stepwise polymerizations. As indicated in Table 1.5, there is a natural division between monomer and polymer, based largely on molecular weight, which was not present in stepwise polymerization. Thus, instead of considering all species, even monomer, as polymer, P_i, we distinguish between monomer (M) and polymer. The latter we divide into the active species, R_i, and once-active but now inactive species P_i. The "R" obviously suggests free-radical polymerization, the most common of the chainwise routes. We will use "R," however, to denote the active species in cationic and anionic polymerizations as well. If no deactivation occurs, as Section 3.2, the active species is our final product, and will be denoted as P_i. This particular notation thus allows the product (active or not) to be denoted as P, have a generating function $G(s)$ and moments μ_i, just as in Chapter 2, while the new constituent, the (active) intermediate, is denoted by R, $H(s)$, and λ_i. (In many cases this involves a change of notation from the source.)

In the next section we introduce the simplest of chainwise polymerizations, in which only the two requisite reactions, initiation and propagation, occur, to see what effects this has on the resulting degrees of polymerization and chain length distribution.

3.2 The Paradigm: Ideal Anionic Polymerization

The simplest of schemes involving initiation and propagation is one in which initiation happens very rapidly. If initiation can be treated as instantaneous, the situation reduces to the concurrent growth of a given number of chains. The kinetics of initiation are absent from the mathematical analysis, but the fact of initiation provides the fixed number of chains. The only reaction of interest, then, is the propagation reaction shown in (3.1.2), assumed to occur with a rate constant independent of the chain length, a direct parallel to the so-called equal reactivity assumption of Chapter 2. This situation may sound somewhat contrived, but it actually can be approached, most easily in anionic polymerizations.[3] This polymerization process is important in practice because it allows a level of control over the chain length distribution not possible by other mechanisms.

Balances for the concentrations of various polymer species in a batch reactor result in an infinite set of equations for the monomer M and the polymeric species P_i in the reactor:

$$\frac{dM}{dt} = -kM \sum_{i=1}^{\infty} P_i \qquad (3.2.1)$$

$$\frac{dP_1}{dt} = -kP_1 M \qquad (3.2.2)$$

$$\frac{dP_i}{dt} = k(P_{i-1} - P_i)M \qquad (3.2.3)$$

Initial conditions at $t = 0$ are: $M(0) = M_0$, $P_1(0) = I_0$, and $P_i(0) = 0 (i > 1)$. Since there are no termination reactions and all the growing chains are created at $t = 0$, the concentration of polymer molecules remains constant. Summing equations (3.2.2) and (3.2.3) for all i verifies this:

$$\frac{d}{dt}\left(\sum_{i=1}^{\infty} P_i\right) = 0 \qquad (3.2.4)$$

The number of polymer molecules, μ_0, is constant and equal to the number of initiators:

$$\sum_{i=1}^{\infty} P_i = \mu_0 = I_0 \qquad (3.2.5)$$

Because a constant concentration of chains is consuming the monomer, the monomer concentration, M, decreases in simple exponential fashion, as seen from solving equation (3.2.1):

$$M = M_0 \exp\left\{-I_0 \int_0^t k\, dt'\right\} \qquad (3.2.6)$$

The rate constant k has been left to the time- (though not chain length-) dependent, to allow, for example, for the possibility of varying temperature during the reaction. This flexibility in modeling can be achieved because the simplicity of the solution is not destroyed by it. The chain length distribution, in fact, is expressed in terms of the "eigenzeit" transform variable, τ[4]:

$$\tau = \int_0^t kM\, dt' \qquad (3.2.7)$$

This change of time variables allows for easier solution, since τ contains the complications due to the need for monomer concentration to decrease in a batch reactor, as well as to the possibility that k will change. Because $d\tau = kM\, dt$, equation (3.2.3) can thus be written in the following alternate form:

$$\frac{dP_i}{d\tau} = -(P_i - P_{i-1}) \qquad (3.2.8)$$

where of course P_0 is defined to be zero. Equation (3.2.8) can be solved by any of the kinetic methods discussed in Chapter 2, plus two others shortly.

3.2.1 Direct Sequential Solution

As we saw for step growth polymerizations, one can solve these equations sequentially and, if the solution is analytic, determine the chain length distribution by induction. Here again, for this simple case, the general form of the solution for P_i becomes recognizable in very short order. In more complex cases for which the solution is not analytic, it can still be done numerically.[5]

The evolution equation for chains of length *1* is given as follows:

$$\frac{dP_1}{d\tau} = -P_1 \tag{3.2.9}$$

Thus P_1 exponentially decreases with τ:

$$P_1 = I_0 e^{-\tau} \tag{3.2.10}$$

The equations for the larger species P_i, $i > 1$, can be expressed in the following form:

$$P_i + \frac{dP_i}{d\tau} = P_{i-1} \tag{3.2.11}$$

which, by way of an exponential integrating factor, can be expressed as follows:

$$\frac{d}{dt}(e^\tau P_i) = e^\tau P_{i-1} \tag{3.2.12}$$

This has the general solution:

$$P_i = e^{-\tau} \int_0^\tau e^{\tau'} P_{i-1} d\tau' \tag{3.2.13}$$

Successive solution yields:

$$P_2 = I_0 e^{-\tau} \tau \tag{3.2.14}$$

$$P_3 = I_0 e^{-\tau} \frac{\tau^2}{2} \tag{3.2.15}$$

$$P_4 = I_0 e^{-\tau} \frac{\tau^3}{6} \tag{3.2.16}$$

This progression leads to a suggestion:

$$P_i = I_0 e^{-\tau} \frac{\tau^{i-1}}{(i-1)!} \tag{3.2.17}$$

Inserting this into equation (3.2.13) gives:

$$P_i = I_0 e^{-\tau} \int_0^\tau \frac{(\tau')^{i-2}}{(i-2)!} d\tau' = I_0 e^{-\tau} \frac{\tau^{i-1}}{(i-1)!} \tag{3.2.18}$$

Thus the distribution of equation (3.2.17) is proven by induction. It is the *Poisson distribution*, the properties of which are discussed in more detail later.

3.2.2 Laplace Transform

The system introduced in Section 3.2.1 is amenable to solution by the Laplace transform,[6] a route that was not available to the stepwise paradigm examined in Section 2.2 because of the convolution term representing creation. We first define the Laplace transform variable $\overline{P}_i$.

$$\overline{P}_i(\lambda) = \int_0^\infty P_i(\tau)e^{-\lambda\tau}d\tau \tag{3.2.19}$$

Applying the Laplace transform to equation (3.2.8) and to the initial condition gives:

$$\lambda\overline{P}_i + \overline{P}_i = \overline{P}_{i-1} \tag{3.2.20}$$

or

$$\overline{P}_i = \frac{1}{1+\lambda}\overline{P}_{i-1} \tag{3.2.21}$$

The general solution to this difference equation is:

$$\overline{P}_i = K\left(\frac{1}{1+\lambda}\right)^i \tag{3.2.22}$$

where K is a constant determined from the normalization condition of equation (3.2.5). This condition is expressed in Laplace transform space as follows:

$$\sum_{i=1}^{\infty} \overline{P}_i = \frac{I_0}{\lambda} \tag{3.2.23}$$

leading to:

$$K\sum_{i=1}^{\infty}\left(\frac{1}{1+\lambda}\right)^i = \frac{K}{\lambda} = \frac{I_0}{\lambda} \tag{3.2.24}$$

so that $K = I_0$. Inserting this into equation (3.2.22) and inverting the Laplace transform, using standard tables, gives equation (3.2.17), our previous result.

3.2.3 Continuous Variable Approximation

As the complexity of a kinetic scheme increases, solution of the infinite set of differential equations sequentially becomes increasingly difficult, and the generating function approach can be used, or numerical approaches resorted to. Before proceeding in that direction, we note a convenient and effective scheme for *approximate* solution that, for reasons that will become obvious, is particularly useful for chain growth polymerization. The method consists of replacing the discrete variable, chain length, by a continuous variable, so that the difference-differential equations become partial differential equations. The method was first applied to free-radical polymerization[6–8] and can be applied to both batch and continuous polymerizations.[9] More recently it has been applied for modeling[10] and control[11] of high conversion free-radical polymerizations. Reviews of the numerical techniques and approximations are available.[12,13] The method has also been applied to ideal anionic polymerization, the case at hand.[14]

If the chain length i is considered a continuous variable, the Taylor series expansion for P_{i-1} about P_i gives

$$P_{i-1} = P_i - \frac{\partial P_i}{\partial i} + \frac{1}{2} \frac{\partial^2 P_i}{\partial i^2} + \cdots \qquad (3.2.25)$$

Inserting this approximation, truncated at the second-order term, into equation (3.2.8) gives:

$$\frac{\partial P_i}{\partial \tau} + \frac{\partial P_i}{di} = \frac{1}{2} \frac{\partial^2 P_i}{\partial i^2} \qquad (3.2.26)$$

The initial condition remains the same: $P_1(0) = I_0$, $P_i(0) = 0$, $i > 1$. Boundary conditions, however, need to be added: that P_i remains finite as $i \to \pm \infty$. The solution is:

$$P_i(\tau) = \frac{I_0}{(2\pi\tau)^{1/2}} \exp\left\{ \frac{[i - (\tau + 1)]^2}{2\tau} \right\} \qquad (3.2.27)$$

This distribution is Gaussian with a mean of $\tau + 1$ and a variance of τ, and is only an approximation* to the true solution, a Poisson distribution. The two become equivalent as i becomes very large, indicating that a continuous variable approximation is valuable at large chain lengths.[15] This limitation must be stressed; only for long chains can we validly assume small $1/i$, upon which the usefulness of the Taylor series (3.2.25) rests. This virtually eliminates the usefulness of the continuous variable approximation for step growth polymerization, for which chain length becomes large only very near complete conversion. The accuracy of the approximation depends not only on the polymer chain length, but also on the number of terms retained in the expansion.[14] In the preceding example, we have used a second-order approximation because it is known that a first-order approximation is insufficient.[16] It should be kept in mind that a second-order approximation is required at the very least.

3.2.4 Application of the Generating Function

As we saw in Chapter 2, use of the moment generating function is often the best way of solving the kinetic equations. Applying the operator G to equation (3.2.8), we obtain:

$$\frac{\partial G(s)}{\partial \tau} = -(1 - s)\, G(s) \qquad (3.2.28)$$

---

*That it is an approximation is obvious, since by the distribution of equation (3.2.27), the existence of chains of negative length is predicted.

while the initial condition becomes $G(s,0) = I_0 s$. Integration of the equation (3.2.28) yields:

$$G(s,\tau) = I_0 s e^{-(1-s)\tau} \tag{3.2.29}$$

The transform can be readily inverted by expanding the solution (3.2.29) in a MacLaurin series in s or equivalently by use of property 13 in Table 2.1, again yielding equation (3.2.17).

The generating function $G(s)$ allows easy examination of some of the properties of the Poisson distribution. The zeroth moment of the distribution is the total polymer concentration I_0, as shown in equation (3.2.5). The next two moments are obtained by successive differentiation:

$$\mu_1 = \lim_{s \to 1} \left(\frac{\partial G(s,\tau)}{d(\ln s)} \right) = I_0(1 + \tau) \tag{3.2.30}$$

$$\mu_2 = \lim_{s \to 1} \left(\frac{\partial^2 G(s,\tau)}{d(\ln s)^2} \right) = I_0[(1 + \tau)^2 + \tau] \tag{3.2.31}$$

For this case μ_0 is constant while μ_1 increases, in contrast to stepwise polymerization for which μ_0 decreased while μ_1 was constant. This is because of the different conventions chosen for what is considered to be polymer; here, monomer is excluded.

The degrees of polymerization DP_n and DP_w are given by:

$$DP_n = \frac{\mu_1}{\mu_0} = 1 + \tau \tag{3.2.32}$$

$$DP_w = \frac{\mu_2}{\mu_1} = 1 + \tau + \frac{\tau}{1 + \tau} \tag{3.2.33}$$

which yield a polydispersity Q, given as follows:

$$Q = \frac{\mu_2 \mu_0}{\mu_1^2} = 1 + \frac{\tau}{(1 + \tau)^2} \tag{3.2.34}$$

Equation (3.2.34) shows us the most important property of the Poisson distribution: that as τ becomes large, the polydispersity approaches unity. In contrast to the geometric distribution, which becomes broader ($Q \to 2$) as conversion increases, this distribution becomes narrower after an initial maximum. This narrowness can be seen in the chain length and molecular weight distributions shown in Figures 3.1 and 3.2. At first glance, these figures may seem to contradict the idea that narrowing of the distribution occurs as conversion of τ increases. Clearly, the standard deviation of these distributions increases as they travel to the right. This is expected from the continuous variable analysis, which showed that the standard deviation was $\sim \tau^{1/2}$. The polydispersity, however, is not a measure of the absolute breadth of the distribution, but the breadth relative to the mean or average. Thus, while the mean increases as τ, the breadth increases, but only as $\tau^{1/2}$. The polydispersity thus decreases as τ increases.

Large τ is obtained at large times t or high conversion of monomer. We have been using τ as some measure of conversion,

Figure 3.1.
Poisson chain length
distribution arising from
ideal anionic
polymerization.

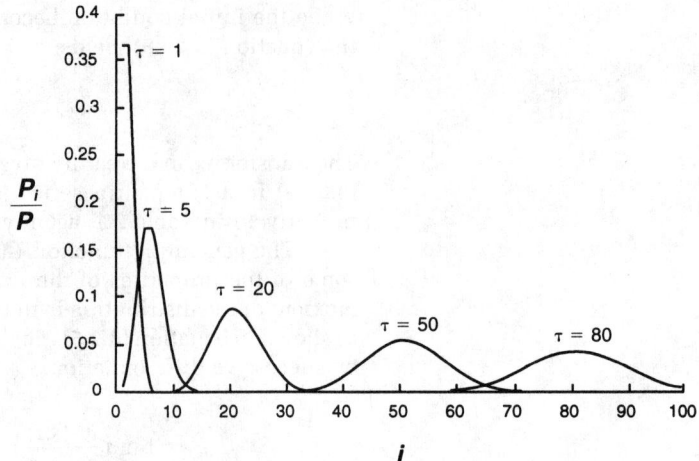

but have yet to make the relation explicit. Given equations (3.2.6) and (3.2.7), it is easy to show that:

$$\tau = \frac{M_0 - M}{I_0} \tag{3.2.35}$$

The variable τ is thus a scaled conversion; the conversion p of the monomer remaining after initiation is $\tau(I_0/M_0)$. As monomer is depleted, τ approaches M_0/I_0, and high molecular weights are attainable only if M_0/I_0 is sufficiently large. The stoichiometry thus determines the chain length and the narrowness of the distribution.

The Poisson distribution, like the geometric distribution, is a single-parameter distribution. Knowledge of only one moment (other than μ_0) is sufficient to completely specify the chain length distribution.[17] For the geometric distribution, the parameter was the conversion p, whereas the parameter here is τ, a scaled conversion. The two show marked differences, however, supporting the notion that anyone who uses a finite number of moments to

Figure 3.2.
Molecular weight
distribution corresponding
to a Poisson distribution
arising from an ideal
anionic polymerization.

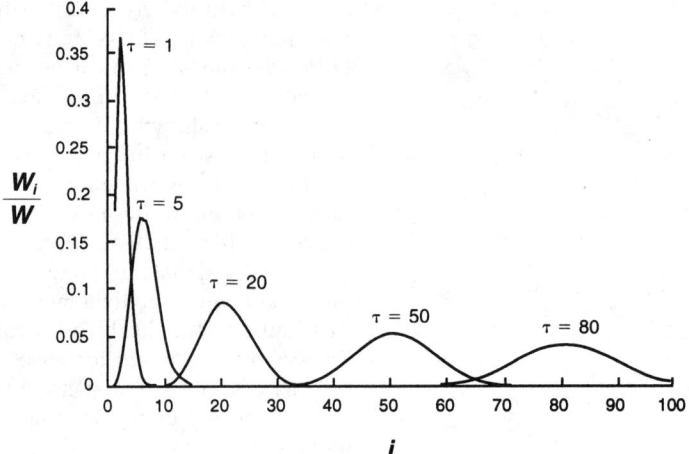

characterize a distribution should have some knowledge of what kind of distribution to expect.

Example 3.1

Derive the molecular weight distribution for ideal anionic polymerization, normalized like P_i/P so that the distribution sums to unity.

Solution:

The relation between the chain length distribution and the molecular weight distribution is shown in equation (1.4.8). Neglect of the mass of the initiator fragment gives m_i exactly proportional to i. The proportionality constant, equal to the mass of a monomer unit, is unimportant for homopolymers because it will cancel out in the normalization. Thus, we let

$$W_i = i\, I_0 e^{-\tau} \frac{\tau^{i-1}}{(i-1)!} \qquad (3.2.36)$$

The normalization constant W is merely the first moment, μ_1, of the chain length distribution, or $I_0(1 + \tau)$. The normalized molecular weight distribution is thus:

$$\frac{W_i}{W} = \frac{i}{1 + \tau} \frac{\tau^{i-1}}{(i-1)!} \qquad (3.2.37)$$

3.2.5 Why Not Use Statistical Methods?

Although the kinetic approach has been as useful here as for stepwise polymerizations, no use has been made of any statistical approach. This is because statistical approaches such as the recursive method are inapplicable*; a walk along these polymer chains does not lead to a statistically equivalent point. Anionic polymerization provides a clearer illustration of this point than the substitution effects described in Section 2.7. In ideal stepwise polymerization, functional groups react randomly, so that whenever one walks from one monomer to the next, one can still apply the same probability of reaction of functional groups. In the combinatorial method, it is what allows each $i - 1$ reacted functional groups to share the identical probability p of having reacted; in the recursive approach, it is what allows the same expectation to be repeatedly encountered when walking down the chain.

Ideal anionic polymerization affords no such simplicity. There is a real sense of direction on these chains not found with stepwise polymers, and it derives from the initiation step—that

---

*One can derive the Poisson distribution as a limit of the binomial distribution obtained by the random tossing of $(M_0 - M)$ objects into I_0 boxes, as $(M_0 - M)$ and $I_0 \to \infty$ with $(M_0 - M)/I_0 = $ constant. This does constitute a statistical argument, although of a different kind.[18]

is, the chain begins at an initiator and ends elsewhere. Upon choosing a monomer, one can either look "up" toward the initiator or "down" to the active end. While this sense of direction is not sufficient to destroy the recursive nature of the chains (as we shall see), it does mean that if one walks down the chain, for example, one necessarily walks toward monomers that reacted later in time, and the probability of not having reacted grows as a result of monomer depletion. Likewise, if one walks "up" the chain, one walks toward monomers that reacted earlier in time, and one is progressively more likely to reach the initiator. The probability of reaction is not constant along the chain. There is no parallel to this ordering of events in step growth, at least in the absence of substitution effects and cyclization. For anionic polymerization, this ordering means that simple statistical methods fail.

3.3 Nonideal Ionic Chain Growth Polymerization

The kind of polymerization described in Section 3.2 is often referred to as a "living" polymerization[2,3] because of the absence of reactions that result in inactive or "dead" chains. The term was coined for anionic polymerization of styrene initiated by sodium naphthalene,[19] but there were earlier reports of absence of termination reactions in the polymerization of dienes and styrene with organometallic initiators free from impurities like water.[20] Likewise, that absence of termination reactions implied the possibility of achieving narrow molecular weight distributions had been recognized by Flory[21] for the polymerization of ethylene oxide free of termination.[22]

The remarkably narrow molecular weight distributions obtained for living polymerization with rapid initiation provides the basis for a number of the uses of polymers made by this route (e.g., in the calibration of size exclusion chromatography columns). But the term "living" suggests other possibilities. The livingness of these chains allows the quantitative addition of new monomer in semibatch fashion. In this way, well-designed block copolymers can be made by the successive addition of different monomers, as in the thermoplastic elastomers made of styrene–butadiene block copolymers. Similarly, one can quantitatively "kill" these chains with small molecules to provide well-characterized end groups (e.g., reactive groups for the so-called macromonomers). Other tailor-made polymers, such as the star polymers discussed in Section 5.1.3, are also best synthesized through anionic means.[23]

"Livingness" has been put in the context of anionic polymerization because that is where it is most easily achieved. Cationic polymerization has usually been contrasted with anionic as not being living, but rather being plagued with reactions leading to dead chains. This is no longer strictly true; the progress of recent years has led to living cationic polymerizations,[24] but history nonetheless testifies to the relative difficulty of achieving this.

Thus for most cationic polymerizations, and perhaps many anionic, the scheme of Section 3.2 is oversimplified; indeed, the rather stringent requirements make it difficult to achieve the narrow distributions. Most complications broaden the chain length distribution, although cases can be contrived that narrow the distribution yet further, as when the rate constant decreases with chain length.* The more important additions to the scheme come from other reactions, specifically from termination and from transfer reactions. A termination reaction both produces a dead chain and results in the loss of an active site; a transfer reaction produces a dead chain, the active site surviving by transfer to some agent to form another chain. While it is not within the scope of this book to elaborate on the fine details of the chemistry of ionic polymerization (for which excellent sources exist[2,3,26-30]), it is worthwhile to briefly discuss the kinds of reactions that occur for anionic, cationic, and ring-opening polymerizations, and for simple cases to observe how these reactions affect the chain length distribution.

It should be borne in mind that there are many uncertainties concerning the mechanisms, side reactions, and rate constants of ionic polymerizations. Anionic polymerizations are not as well understood as free-radical polymerizations, to which we will turn after this, because of very high reaction rates, extreme sensitivity to minor impurities, participation of heterogeneous inorganic catalysts, and the existence of the active species in different states. These features obscure the mechanism and the kinetic phenomena.

3.3.1 Anionic Polymerization

Anionic polymerization is carried out most commonly in solvent, at low temperature, with nucleophilic initiators (e.g., alkyllithiums), and on monomers with carbon–carbon double bonds with electron-withdrawing groups or on the carbon–oxygen bond of, for example, formaldehyde.[31] Ion pairs in anionic polymerization may be intimate or distant, separated by solvent, or in many intermediate forms, all existing in an equlibrium determined in part by the polarity of the medium.[32] The apparent propagation rate constant is thus some composite of the rate constants for each of these states, the more loosely bound being much more reactive. In nonpolar solvents, the situation can be further complicated by various states of aggregation between chain ends which decrease the rate of polymerization yet more and alter (lower) the order of dependence on initiator concentration.[2,33] A more nonpolar solvent also slows the initiation step, making it noninstantaneous, a case we examine shortly.

Dead chains are most often caused by transfer to some impurity present, such as oxygen, carbon dioxide, alcohols, or (as a good example) water:

$$R_i^- + \text{HOH} \overset{k_{\text{tr}}}{\to} P_i + \text{OH}^- \qquad (3.3.1)$$

---

*Violations of the equal reactivity assumption for anionic polymerization have been discussed in the literature.[9,25]

(Counterions are not shown). The hydroxide ion is unable to reinitiate polymerization, and thus even though chemically a transfer reaction, (3.3.1) acts as a termination reaction. The effect of this, which we will see, testifies to the importance of rigorous purification to obtain high molecular weight polymers with a narrow Poisson distribution. In polar monomers such as (meth)-acrylates, there can be competing reactions (e.g., with the carbonyl group).* Even anionic living systems do not live forever; they generally slowly expire over days or weeks, due to mechanisms not completely understood.[2,36]

Reversibility, which is so significant for the common stepwise systems such as polyesters and nylons, is often ignored in chainwise polymerizations; and for the more common monomers this is justified, at least at polymerization temperatures less than 100°C. *Depropagation* reactions, leading to an equilibrium between polymer chains and monomer, are possible, however, and are significant for certain monomers polymerized anionically, such as α-methylstyrene and formaldehyde. Reversibility in exothermic chainwise polymerizations is most often described by a *ceiling temperature* T_c, above which no polymerization can occur. It is thus a function of the monomer concentration but is most often given for pure monomer.[2,37] The low temperature of anionic polymerizations clearly facilitates polymerization in these cases, but subsequent use at higher temperatures is made possible by end capping with groups prohibiting depropagation (in the anionic polymerization of formaldehyde, acetic anhydride[2,31])—an excellent example of the importance of end groups on the properties of the polymer, particularly stability.

3.3.2 Cationic Polymerization

Cationic polymerizations are best characterized by contrast to anionic polymerizations. They are initiated most commonly by protonic or Lewis acids, such as BF_3, and the latter rely on an impurity such as water for initiation. They prefer monomers with electron-donating groups [providing routes to polymerize α-olefins and vinyl ethers, but not, e.g., (meth)acrylates], although this still allows for a bit of overlap with anionic systems (e.g., styrene and formaldehyde).

Cationic polymerizations are, as a rule, much less ideal than anionic. Even if living, they generally last for a much shorter time. Side reactions are greater in number and frequency; the most important of these is chain transfer to monomer, which

---

*Another related chainwise mechanism can be used to polymerize methacrylates (especially) and acrylates: the so-called group transfer polymerization. Narrow molecular weights can also be obtained by this route. Initiation requires an initiator (a silyl ketene acetal) and a catalyst (a nucleophilic anion or Lewis acid). Purity requirements are similar to those for anionic polymerization, proton-donating compounds being particularly detrimental. The mechanism of group transfer polymerization is under debate but some anionic character may be present.[34,35]

occurs by the transfer of a proton from the carbocation to the monomer:

$$R_i^+ + M \overset{k_{tr,s}}{\rightarrow} P_i + M^+ \qquad (3.3.2)$$

This transfer reaction, which leaves an unsaturation at the end of the preceding chain, does not terminate the propagation of the active center. Various first-order termination reactions also occur; these may regenerate the initiator, or not (as in counterion combination). Olefin polymerizations are characterized by an array of rearrangement and isomerization reactions, leading to complicated structures.[2,27,38,39] Reversibility can also affect cationic polymerizations, as in the important case of isobutylene, which has a T_c of 50°C for bulk monomer.[2]

3.3.3 Ring-Opening Polymerization

Ring-opening polymerization provides a route to some of the same kinds of polymer discussed in Chapter 2, such as nylons, polyesters, and siloxanes; typical monomers polymerized by ring-opening polymerization were shown earlier in Table 3.2. These polymerizations can be either cationic or anionic, and the same sorts of generalization made for ionic polymerizations of unsaturated monomers apply to these cyclic monomers. Even though anionic ring-opening polymerization is generally more specific than cationic polymerization, it is quite important.[40] For example, cyclic ethers can be polymerized only cationically, except for important epoxides such as ethylene oxide and propylene oxide, which can be polymerized anionically. Polymerization of epoxides can be living; but exchange reactions with water or alcohols can lead to chains that are temporarily dormant, transfer to monomer may occur, and initiation may not be instantaneous. Copolymers of polyoxymethylene are made by cationic polymerization of 1,3,5-trioxane with other cyclic ethers such as ethylene oxide, a polymerization marked by equilibration effects.[2,31]

Equally important is the polymerization of cyclic amides (lactams), such as that of ε-caprolactam to form nylon 6. These polymerizations are performed most often by hydrolytic (cationic) polymerization with water, which involves various equilibria and thus tends to give polymers of geometric distribution. Polymerization of cyclic siloxanes by cationic or anionic routes generally have the characteristics of living polymerixations; this is industrially important in the production of high-molecular-weight polydimethylsiloxane, those of lower molecular weight being formed by stepwise mechanisms (see Section 2.3.6).

3.3.4 Departures from Ideal Living Polymerization

Many side reactions and nonidealities have been suggested by the preceding discussions; it is instructive to examine what happens to the chain length distribution if some of the conditions of ideal living polymerization are not achieved. Numerous variants, many of which are well documented, lead to broadening of a molecular weight distribution.[41] Here we restrict ourselves to two simple cases: noninstantaneous initiation and first-order termination. More detailed treatments of more intricate schemes of slow initiation[33] and of termination and transfer[42] are available.

3.3.4.1 Noninstantaneous Initiation

Slow initiation can occur in anionic, cationic, and ring-opening polymerizations. It can be idealized by the following two second-order reactions:

$$I + M \overset{k_{in}}{\rightarrow} P_1 \tag{3.3.3}$$

$$P_i + M \overset{k}{\rightarrow} P_{i+1} \tag{3.3.4}$$

As can easily be imagined, because all chains do not start at the same time, a broadening of the chain length distribution, relative to the Poisson distribution, occurs.

The species balances are:

$$\frac{dI}{dt} = -k_{in}IM \tag{3.3.5}$$

$$\frac{dM}{dt} = -k_{in}\,IM - kM\sum_{j=1}^{\infty} P_j \tag{3.3.6}$$

$$\frac{dP_i}{dt} = k_{in}IM\,\delta(i-1) - kM\,(P_i - P_{i-1}) \tag{3.3.7}$$

where the Dirac delta function $\delta(i-1)$ designates a term that occurs only when the argument is zero, in this case when $i = 1$. Using the generating function and the eigenzeit transformation, along with initial conditions $I(0) = I_0$, $M(0) = M_0$ and $P_i(0) = 0$, one gets, in the s-domain,

$$G(s, \tau) = \frac{\alpha I_0 s}{1 - \alpha - s}\{e^{-\alpha\tau} - e^{-(1-s)\tau}\} \tag{3.3.8}$$

where $\alpha = k_{in}/k$.

The inversion of $G(s)$ is straightforward, if lengthy, and is left as an exercise to the reader. The final expression, however, is[43,44]:

$$P_i = \alpha I_0 e^{-\tau}(1 - \alpha)^{-i} \sum_{j=i}^{\infty} \frac{[(1 - \alpha)\tau]^j}{j!} = \alpha I_0 e^{-\tau}(1 - \alpha)^{-i}\left(e^{(1-\alpha)\tau} - \sum_{j=0}^{i-1} \frac{[(1 - \alpha)\tau]^j}{j!}\right) \tag{3.3.9}$$

the second form being computationally more convenient. The distribution of equation (3.3.9) is often termed the Gold distribution,[44] and for values assigned to the parameter α, the distribution can vary from broad at small α, to narrow at large α. For $\alpha = \infty$, equation (3.3.9) yields the Poisson distribution, as it must. Also, as $\alpha \rightarrow 0$, the distribution resembles a step function. This is

Figure 3.3.
Effect of noninstantaneous
initiation on chain length
distribution for anionic
polymerization: $\tau = 25$.

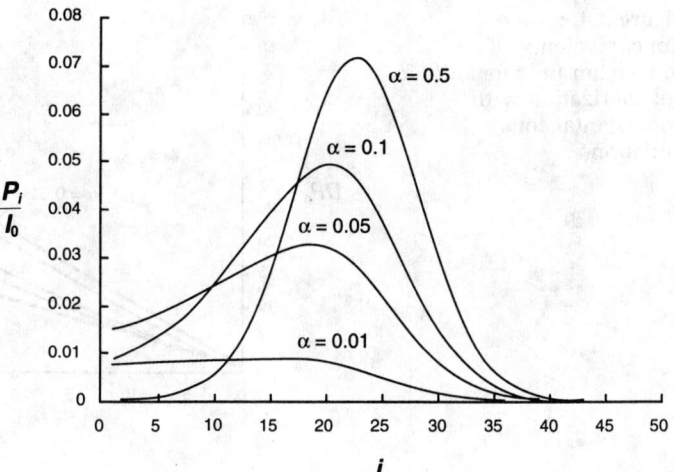

clearly seen in Figure 3.3 where the chain length distribution is
plotted as a function of α.

The zeroth moment is the concentration of polymer chains,
which for this mechanism is given by the solution to equation
(3.3.5):

$$\sum_{i=1}^{\infty} P_i = \mu_0 = I_0 - I = I_0(1 - e^{-\alpha\tau}) \qquad (3.3.10)$$

The first moment gives the number of monomer units incorporated in the polymer chains, and so is given by the soluton
to (3.3.6):

$$\sum_{i=1}^{\infty} i\,P_i = \mu_1 = M_0 - M = I_0\left\{\tau + \left(1 - \frac{1}{\alpha}\right)(1 - e^{-\alpha\tau})\right\} \qquad (3.3.11)$$

Hence, the number-average degree of polymerization can be written as follows:

$$DP_n = 1 - \frac{1}{\alpha} + \frac{\tau}{1 - e^{-\alpha\tau}} \qquad (3.3.12)$$

Here we may note two distinctions from the description
of ideal anionic polymerization. The first difference is that the
initial monomer concentration in Section 3.2 was defined as
the monomer concentration left over after initiation; here it is the
true initial monomer concentration. This difference is minor,
since for high molecular weights we must have M_0 much greater
than I_0. The more important point is that the eigenzeit variable
τ no longer is a scaled conversion, as is clear from equation
(3.3.11). In Figure 3.4, we see that the DP_n decreases as α approaches zero. In the Poisson distribution limit, DP_n is $1 + \tau$, but
for $\alpha = 0$, DP_n is $1 + (\tau/2)$; all intermediate values of α give
degrees of polymerization between these two limits.

Figure 3.4.
Nonequivalence of conversion and τ for anionic polymerization with noninstantaneous initiation.

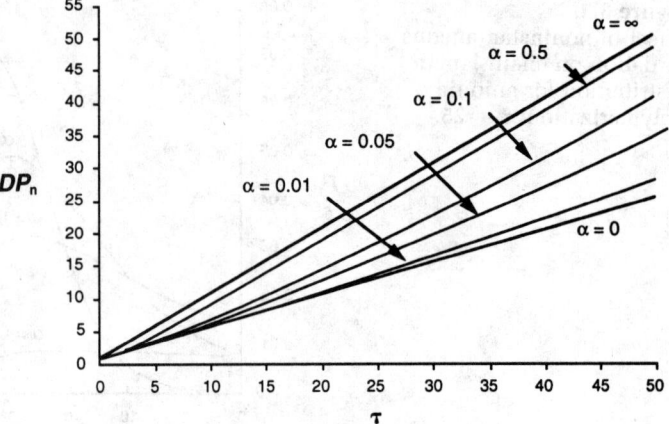

It can be shown that the second moment is given by

$$\mu_2 = I_0 \left\{ \tau \left(3 + \tau - \frac{2}{\alpha} \right) + \left(\frac{2}{\alpha^2} - \frac{3}{\alpha} + 1 \right)(1 - e^{-\alpha\tau}) \right\} \qquad (3.3.13)$$

so that

$$DP_w = 1 - \frac{2}{\alpha} + \frac{\tau(\tau + 2)}{\tau + (1 - 1/\alpha)(1 - e^{-\alpha\tau})} \qquad (3.3.14)$$

Figure 3.5 shows how noninstantaneous initiation tends to broaden the chain length distribution, but within bounds. The greatest polydispersity occurs for $\alpha = 0$, as expected. For this case, the maximum polydispersity occurs at $\tau = 6$, which gives $Q = 1.375$. Thereafter, the distribution narrows as it does for a Poisson distribution. The limiting polydispersity for the $\alpha = 0$ case as $\tau \to \infty$ is 4/3, and the limiting form for DP_w is given as follows:

$$DP_w(\alpha = 0) = 1 - \frac{2\tau}{3} \frac{\tau + 3}{\tau + 2} \qquad (3.3.15)$$

Figure 3.5.
Effect of noninstantaneous initiation on polydispersity for anionic polymerization.

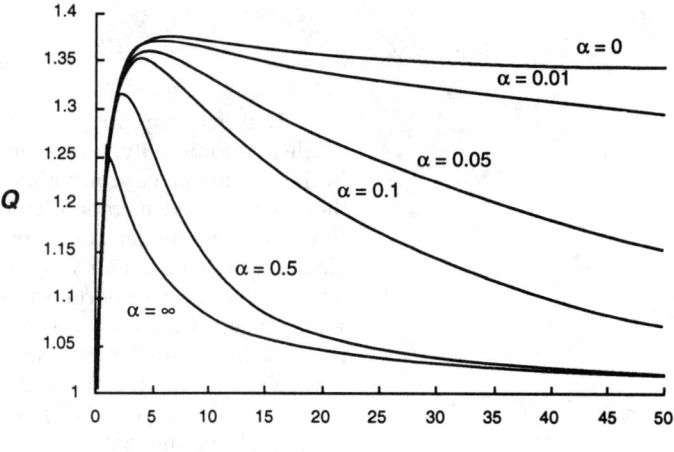

For less extreme cases of slow initiation, the polydispersities of course fall between the Poisson limit and the limit of equation (3.3.15).

3.3.4.2 Ionic Polymerization with First-Order Termination

Ionic polymerization with terminating mechanisms is treated in several sources.[42,45-48] In the present formulation of the problem, transfer does not occur but first-order termination as discussed earlier does. This includes the case of ionic polymerization wherein chains are terminated by the presence of impurities X, the concentration of which is constant (k_t in that case being equated with $k_t X$). Because we now have living chains and dead, product chains, we alter our notation as discussed in Section 3.1.

The following balances can be written for this system if initiation is intantaneous:

$$\frac{dM}{dt} = -k_p M \sum_{i=1}^{\infty} R_i \tag{3.3.16}$$

$$\frac{dR_i}{dt} = -k_p M(R_i - R_{i-1}) - k_t R_i \tag{3.3.17}$$

$$\frac{dP_i}{dt} = k_t R_i \tag{3.3.18}$$

with initial conditions: $M(0) = M_0$, $P_i(0) = 0$, $R_i(0) = I_0 \delta(i - 1)$. These equations can be reformulated as generating function equations as follows:

$$\frac{dM}{dt} = -k_p M \, H(1) \tag{3.3.19}$$

$$\frac{\partial H(s)}{\partial t} = -k_p M(1 - s)H(s) - k_t H(s) \tag{3.3.20}$$

$$\frac{\partial G(s)}{\partial t} = k_t H(s) \tag{3.3.21}$$

where $H(s)$ is the live chain generating function and $G(s)$ is the (product) dead chain generating function. These two, respectively, have moments λ_i and μ_i. The initial conditions are now transformed into $H(s,0) = I_0 s$, $G(s,0) = 0$.

From equation (3.3.18) it is clear that the concentration of living chains, $H(1)$ or λ_0, decays exponentially:

$$H(1) = \lambda_0 = I_0 e^{-k_t t} \tag{3.3.22}$$

The monomer concentration thus decays in a more complicated fashion:

$$M = M_0 \exp\left[-\frac{k_p I_0}{k_t} (1 - e^{-k_t t}) \right] \tag{3.3.23}$$

The eigenzeit transform cannot be used because the monomer concentration M does not appear in every term of equation (3.3.17). Nonetheless, equation (3.3.20) does form a separable equation because $H(s)$ appears in every term, so $H(s)$ can be solved for analytically to yield:

$$H(s) = I_0 s \, e^{-k_t t} \exp\left\{ -(1-s)\frac{k_p M_0}{k_t} e^{-k_p I_0 / k_t} \left[k_t t + \sum_{n=1}^{\infty} \frac{\left(\frac{k_p I_0}{k_t}\right)^n (1 - e^{-k_t t n})}{(n)(n!)} \right] \right\} \tag{3.3.24}$$

The complicated form of this equation masks the simplicity of the resulting distribution. All the complexity is in the time dependence, not in the dependence on the transform variable s. Indeed, the distribution is still Poisson (although with a complicated "τ," which is again not a scaled conversion), and the overall concentration of these chains decays exponentially with time, indicated by the first exponential term. That the Poisson character of the live chain distribution should survive the introduction of termination is, in hindsight, obvious inasmuch as the termination reaction, since it has no preference for chain of certain lengths, "kills" chains at random, which can in no way disturb the shape of the distribution.

The moments of the live chain distribution are analogous to those found in Section 3.2, with the appropriate replacement for τ. The concentration of the dead chains grow at the expense of the living chains:

$$G(1) = \mu_0 = I_0(1 - e^{-k_t t}) \tag{3.3.25}$$

The distribution of the chains will not be Poisson, but will instead be broader. This is because the dead chain distribution is a *cumulative* one:

$$G(s) = k_t \int_0^t H(s) \tag{3.3.26}$$

At any time, the increment to the dead chain distribution, the *instantaneous* dead chain distribution, is given by the time derivative of equation (3.3.25) [or equation (3.3.21)] and is Poisson. The cumulative distribution, however, is not Poisson. Rather, it is an accumulation of many Poisson distributions of different "τ" values. We do not attempt to solve equation (3.3.26).

Note that a linear dependence between DP_n and conversion is maintained in this system. This comes about because the number of polymer molecules is constant, and since DP_n is just the number of converted monomers divided by the number of chains, a linear relation is necessary. Thus the claim that a linear relationship between DP_n and conversion constitutes a criterion for living polymerization is false.[49] Rather it only assures that reactions that increase the number of polymer molecules with time, such as transfer or slow initiation, do not occur appreciably.

The reaction schemes, especially for cationic polymerization, thus can become increasingly intricate, and already, at this simple case, we have been kept from an analytical solution. However, some simplifications do occur for these complicated cases when initiation is sufficiently slow. Although we could analyze this limit in the context of cationic polymerization, we instead analyze it in the context in which it is more familiar: free-radical polymerization. In free-radical polymerizations, initiation, propagation, transfer reactions to solvent, agent, or monomer, and termination all occur simultaneously. We first turn to the chemistry of these systems, since it differs from that of ionic systems in several significant ways.

3.4 Chemistry of Free-Radical Polymerization

A brief review of the chemistry of free-radical polymerization is recommended by its commercial (and scientific) significance, if not by better understanding the resulting math. The greatest share of the commodity polymers are made by this mechanism, and all the monomers listed earlier in Table 3.1 (except propylene, or α-olefins in general, and formaldehyde) can be and often are polymerized in this way. We do not pretend that this treatment of the chemistry is at all complete; other sources[2,26,50] (on which the following discussion draws) should be consulted for a deeper understanding of the chemistry.

3.4.1 Initiation

The first, necessary step is the production of free radicals. This can be done in a number of ways. The most common of these is *thermal decomposition* of some compound, generally a peroxide (e.g., benzoyl peroxide), an azo [e.g., azobis(isobutyronitrile) (AIBN)], or a disulfide. With AIBN, for example, the decomposition proceeds as follows:

$$
\begin{array}{ccc}
\mathrm{CH_3} & \quad & \mathrm{CH_3} \\
| & & | \\
\mathrm{H_3C-C-N{=}N-C-CH_3} & \longrightarrow & 2\ \mathrm{H_3C-C\cdot} + \mathrm{N_2} \\
| & & | \\
\mathrm{CN} \qquad \mathrm{CN} & & \mathrm{CN}
\end{array}
\tag{3.4.1}
$$

Decomposition is actually much more complicated than this, but for the moment the important point is that two radicals are formed from the compound. One of these may react with a monomer, such as methyl methacrylate:

$$
\begin{array}{ccccc}
\mathrm{CH_3} & & \mathrm{CH_3} & & \mathrm{H_3C\ \ H\ \ CH_3} \\
| & & | & & |\ \ \ \ |\ \ \ | \\
\mathrm{H_3C-C\cdot} & + & \mathrm{H_2C{=}C} & \longrightarrow & \mathrm{H_3C-C\ -C-C\cdot} \\
| & & | & & |\ \ \ \ |\ \ \ | \\
\mathrm{CN} & & \mathrm{C{=}O} & & \mathrm{NC\ \ H\ \ C{=}O} \\
& & | & & | \\
& & \mathrm{O} & & \mathrm{O} \\
& & | & & | \\
& & \mathrm{CH_3} & & \mathrm{CH_3}
\end{array}
\tag{3.4.2}
$$

The two reactions in sequence read:

$$I \xrightarrow{k_d} 2R_0 \tag{3.4.3}$$

$$R_0 + M \xrightarrow{k_1} R_1 \tag{3.4.4}$$

The decomposition of the initiator generally follows a first-order rate law:

$$\frac{dI}{dt} = -k_d I \tag{3.4.5}$$

Thus, at a given temperature the initiator decays from its initial concentration of I_0 exponentially:

$$I = I_0 e^{-k_d t} \tag{3.4.6}$$

The use temperatures for thermal indicators generally yield k_d on the order of 10^{-6} to 10^{-4} s^{-1}, so that half-lives $\ln(2)/k_d$ are generally long. Initiation is thus a continuing process throughout the reaction, contrary to ideal anionic polymerization. The range of use temperatures is relatively small, since the activation energies are quite high (100–200 kJ/mol).

The reaction scheme implied in equations (3.4.3) and (3.4.4) does not fairly represent what fates may befall a radical, or even what may happen to the initiator. The variety of side reactions that plague benzoyl peroxide decomposition are discussed in Odian.[2] The somewhat simpler case of AIBN is discussed, for example, by Moad et al.[51] We will not go into these complexities here, except to say that the overall effect of these is that some fraction f of the radicals survive to initiate chains. The initiation process is thus inefficient and f is termed the efficiency. A good share of the reactions that waste radicals occur because of the "cage effect,"[52] arising from the confinement of the radical pair for some time in the close space of a "solvent cage." The "solvent" in this case comprises the environment in general (e.g., monomer, inert diluent, polymer), and the nature of this "solvent" will affect the lifetime of a cage and thus the efficiency.

The decomposition represented by equation (3.4.3), rather than initiation step proper (3.4.4), is generally rate-limiting, allowing us to collapse the kinetic scheme into:

$$I \xrightarrow{k_d} 2fR_1 \tag{3.4.7}$$

Combining this result with the fact of initiator inefficiency allows us to write that the rate of initiation is:

$$\text{rate of initiation} = 2fk_d I \tag{3.4.8}$$

A second method of producing radicals is *photoinitiation*,[53,54] in which ultraviolet or visible light absorbed by a molecule leads to an excited species that either dissociates on its own or interacts with another species in order to dissociate. The common photoinitiator α,α-dimethoxy-α-phenylacetophenone is an example of the first kind:

$$\underset{\text{(3.4.9)}}{}$$

(3.4.9)

Most of these are aromatic ketones, such as benzophenone,

(3.4.10)

which requires interaction with a proton donor. Some thermal initiators, such as AIBN, can also be used as photoinitiators. Photoinitiation offers the advantages of spatial resolution, control of the reaction through the intensity of light, and a temperature-independent process owing to the absence of an activation energy. The technique is generally limited to surface application such as coatings, because of Beer's law.

Redox reactions provide a third route to the production of radicals. Examples of common redox systems include (1) peroxides with a reducing agent (e.g., ferrous ions, *N,N*-dialkylaniline), (2) a combination of inorganic reductants and oxidants, such as persulfates with ferrous ions, and (3) organic–inorganic pairs. The monomer itself may also participate. The reaction may be induced by light or by heat. Because of their more moderate activation energy, such systems can be used over a broader temperature range. Other advantages include the possibility for use in aqueous systems, such as in the polymerization of acrylamide, or in emulsion polymerizations (see Chapter 7).

Additional methods for the generation of radical species include the use of *radiation* or *plasma,* and *electroinitiation.* Monomers can themselves react to form radicals in so-called *pure thermal initiation.* In many cases impurities (particularly O_2, which may form peroxides) have been the source of alleged pure thermal initiation, but it does seem that methyl methacrylate and styrene polymerize, albeit slowly, by pure thermal initiation. It is important to bear all these possibilities in mind; in this chapter we will concentrate on thermally decomposing initiators, keeping in mind that the rate expression for initiation will be different from $2fk_dI$ under other modes of initiation.

3.4.2 Propagation

Once the monomeric radical R_1 has been produced, the true polymerization reaction, propagation, occurs, as in the following example of methyl methacrylate:

$$(3.4.11)$$

The reaction has been drawn indicating a head-to-tail arrangement of the monomers, which is generally favored both sterically and for stabilization. Note that the tacticity of the end group (i.e., its stereoconfiguration) is determined only upon addition of the next monomer.

The propagation step can be generalized symbolically as follows:

$$R_i + M \overset{k_p}{\to} R_{i+1} \qquad (3.4.12)$$

leading to the rate of propagation (or polymerization):

$$\text{rate of propagation} = \text{rate of polymerization} = k_p RM \quad (3.4.13)$$

Values of k_p are generally more uncertain than those for k_d, but range from 10^3 to 10^5 L/(mol · s) at typical reaction temperatures, with activation energies from 10 to 40 kJ/mol.

3.4.3 Termination

Termination, the mechanism by which radical species die, consists of the mutual annihilation of two radicals (very different from, e.g., the termination mechanisms of cationic polymerization). This may occur by two routes: combination and disproportionation. *Combination* is the more common route, and the reaction may be written as follows:

$$R_i + R_j \overset{k_{tc}}{\to} P_{i+j} \qquad (3.4.14)$$

Two radical chains are destroyed, and one inactive chain remains. Monomers such as styrene and ethyl acrylate terminate mainly by this mechanism. *Disproportionation*, on the other hand, leaves two inactive polymer chains:

$$R_i + R_j \overset{k_{td}}{\to} P_i + P_j \qquad (3.4.15)$$

One of these two will bear a saturated end, the other an unsaturated end. We have used no notation to distinguish between these two, because we intend to ignore the reactivity of this group (but see Chapter 5). Methyl methacrylate when polymerized is thought to terminate predominantly by this route. Direct evidence in support of the dominance of one mechanism or the other generally comes from some measure of the number of initiators (radioactively labeled, e.g.) per polymer chain. Such a strategy does not work when transfer to monomer (see Section 3.4.4) is a significant factor, so that the mechanism is often unclear, as in the case of vinyl acetate.

A composite termination rate constant k_t ($= k_{tc} + k_{td}$) describes the rate of termination*:

$$\text{rate of termination} = k_t R^2 \qquad (3.4.16)$$

Values of k_t, which are even more uncertain than those for k_p, are in the range of 10^7–10^9 L/(mol · s) with activation energies similar to that for propagation. It should also be noted that the mechanism of termination may change depending on other reactions occurring in the system, especially when the system is initiated with a redox reaction.[2]

3.4.4 Transfer

Transfer reactions involve the transfer of a group (often a proton) from some species to the active chain, so that the radical is transferred to that species. The agent of transfer may be the monomer itself, solvent, some added agent, initiator, or the polymer product (see Chapter 5). Transfer breaks the production of a polymer chain but not the *kinetic chain,* which describes the longevity of a given radical; thus transfer agents are often said to affect chain length but not rate of polymerization.

Transfer to monomer M or solvent or agent S can be written as follows:

$$R_i + X \overset{k_{tr,x}}{\rightarrow} P_i + R_1 \qquad (3.4.17)$$

and so follows the rate law:

$$\text{rate of transfer to monomer} = k_{tr,x} RX \qquad (3.4.18)$$

where $X = M$ or S. Transfer reactions are generally described by ratios of k_{tr} to k_p, C_x, since these more naturally describe the effect on polymer chain length. Values of C_m ($= k_{tr,m}/k_p$) are important because they place a limit on the polymer chain length that can be achieved; these are generally low, on the order of 10^{-5}, but higher for very reactive monomers such as vinyl acetate (on the order of 10^{-4}) and vinyl chloride (on the order of 10^{-3}).

Identification of all small-molecule radicals resulting from transfer reactions as R_1 is not necessarily valid. If transfer agents do not reinitiate chains as rapidly as initiated monomers, they

---

*Another common convention introduces a factor of 2 into this equation,[2] but we will use (3.4.16) for reasons discussed in Appendix 2A. Values of k_t must be consistent with the convention used.

are known as retarders, and both chain length and the rate of polymerization will be affected.

3.4.5 Inhibition

Retardation of polymerization in the extreme results in inhibition, in which radical species react with some species, Z, to yield completely inactive product. Although the chemistry involved may be different, retardation and inhibition differ more in degree than in kind. Inhibition is important because compounds such as hydroquinone are usually added to commercial monomer to prevent polymerization. Good reviews of the chemistry of inhibition (and retardation)[55,56] and the resulting effects on kinetics[57] are available; we do not deal with inhibition in this chapter.

3.5 Molecular Weight Distributions in Free-Radical Polymerization

3.5.1 Active or Living Polymer Chains

Calculating the chain length distribution arising from the entire system of reactions presented thus far may at first seem too difficult, so for the sake of illustration we begin with a simple scheme involving only initiation, propagation, and termination.[58] We first describe the evolution of the distribution of the active polymer species R_i and the kinetics of the disappearance of monomer. (The rate of disappearance of initiator was discussed in Section 3.4.1.)

The material balance equations on the species of interest are:

$$\frac{dM}{dt} = -2fk_dI - k_pM\sum_{j=1}^{\infty}R_j \qquad (3.5.1)$$

$$\frac{dR_i}{dt} = 2fk_dI\delta(i-1) + k_pM(R_{i-1} - R_i) - k_tR_i\sum_{j=1}^{\infty}R_j \qquad (3.5.2)$$

Where R_0 is defined as zero. Here we have assumed that the reaction rate constants are independent of molecular size and that initiation is by thermal decomposition.

This infinite set of equations can be reduced to three by defining $H(s)$ as the generating function of the active polymer species. Equations (3.5.1) and (3.5.2) are then rewritten as follows:

$$\frac{dM}{dt} = -2fk_dI - k_pMH(1) \qquad (3.5.3)$$

$$\frac{\partial H(s)}{\partial t} = 2fk_dIs - k_pM(1-s)H(s) - k_tH(s)H(1) \qquad (3.5.4)$$

We can solve first for $H(1)$, then for M, and finally for $H(s)$, subject to initial conditions: $M(0) = M_0$, and $H(s,0) = 0$. The total radical concentration, $H(1)$ or R, is found from equation (3.5.4):

$$\frac{dH(1)}{dt} = 2fk_dI - k_tH^2(1) \tag{3.5.5}$$

If rate constants are truly constant, we may define a new time variable u[59-62]:

$$u = \left(\frac{8fI_0k_t}{k_d}\right)^{1/2} e^{-k_dt/2} \tag{3.5.6}$$

Equation (3.5.5) then becomes:

$$\frac{dH(1)}{du} - \frac{2k_t}{k_du} H^2(1) = -\frac{k_d}{2k_t} u \tag{3.5.7}$$

This can be solved with the further substitution,

$$H(1) = -\frac{uk_d}{2k_ty} \frac{dy}{du} \tag{3.5.8}$$

which yields a modified Bessel equation of zero order[63]:

$$\frac{d^2y}{du^2} + \frac{1}{u}\frac{dy}{du} - y = 0 \tag{3.5.9}$$

The solution for $H(1)$ as a function of time reads:

$$H(1) = \left(\frac{2fI_0k_d}{k_t}\right)^{1/2} e^{-k_dt/2} \left[\frac{K_1(u)I_1(u_0) - K_1(u_0)I_1(u)}{K_1(u_0)I_0(u) + K_0(u)I_1(u_0)}\right] \tag{3.5.10}$$

where K_i and I_i are the modified Bessel functions of order i. Were an analytical solution for M possible, it would be rather cumbersome and complicated, and the solution for $H(s)$ would be impossible in this rigorous approach. Limited solution for the case of constant M are available but of little use, especially given their complexity.[41,60]

3.5.1.1 The Quasi–Steady State Approximation (QSSA)

We have gone into some detail on the math to foster an appreciation of the benefits of a simplification known as the quasi–steady state approximation. The high reactivity of species such as radicals implies that the rate of change of radical concentration is negligible compared to the rate of production or destruction,[7] because adjustments to changes in environment are rapid. Thus:

$$\frac{dH(1)}{dt} = 0 \tag{3.5.11}$$

This quasi–steady state approximation means that initiation and termination are balanced so that

$$H(1) = \left(\frac{2fk_dI}{k_t}\right)^{1/2} \tag{3.5.12}$$

Equation (3.5.11) is not strictly true, for time dependence may enter through f, k_d, I, or k_t, but the derivative of equation (3.5.12) is small. An example of the rapidity with which the radical concentration adjusts is seen at the beginning of the reaction, for which $H(1)$ approaches its steady-state value in the following fashion:

$$\frac{H(1)}{H(1)_{ss}} = \tanh\{(2fk_dI_0k_t)^{1/2}t\} \tag{3.5.13}$$

where $H(1)$ reaches its steady-state value in a few chain lifetimes (often less than a second).[64] In terms of the preceding rigorous analysis, the QSSA implies that the bracketed term in equation (3.5.10) comprising the Bessel functions is unity, a reasonable assumption for values of u typical for radical chain growth polymerization.[60,61]

Of course there are situations in which the QSSA does not hold. When there are rapid changes in reagent concentration or, more importantly, rate constant (as when a diffusive mechanism changes), the QSSA will be violated. For these cases, non–steady-state analyses (usually numerical) are necessary.

We could now proceed to solve the monomer balance, but instead examine the chain length distribution $H(s)$, for the introduction of the QSSA allows this. Another way of phrasing the QSSA is that there is a vast separation in the time scale for changes in reagent and radical concentration and the lifetime of a chain, so that a chain during its life sees a constant environment. If this is so, we can solve for $H(s)$ at any given time, because the QSSA also applies to the distribution $H(s)$ itself:

$$\frac{dH(s)}{dt} = 0 \tag{3.5.14}$$

Through equation (3.5.4) we obtain:

$$H(s) = \frac{2fk_dIs}{k_pM(1-s) + (2fk_dk_tI)^{1/2}} \tag{3.5.15}$$

We should immediately recognize this as a most probable distribution:

$$H(s) = H(1)\frac{(1-q)s}{1-qs} \tag{3.5.16}$$

where q is the probability of propagation:

$$q = \frac{k_pM}{k_pM + (2fk_dk_tI)^{1/2}} = \frac{k_pM}{k_pM + k_tH(1)} \tag{3.5.17}$$

Thus the distribution is:

$$R_i = H(1)(1 - q)q^{i-1} \qquad (3.5.18)$$

and the average degrees of polymerization are:

$$DP_n^L = \frac{1}{1 - q} \qquad (3.5.19)$$

$$DP_w^L = \frac{1 + q}{1 - q} \qquad (3.5.20)$$

the superscript L indicating active or "living" chains. For long chains (q near unity), polydispersity of the active chains will be nearly 2; the narrow distributions possible by some ionic polymerizations cannot be achieved by free-radical polymerization.*

Contrary to the chainwise cases considered earlier in this chapter, here we can successfully apply statistical reasoning, as suggested by the use of the term "probability of propagation." The formation of a chain of length i requires $i - 1$ propagation steps, each of which occurs with a probability q, and one termination step, which occurs with a probability $1 - q$; this reasoning gives equation (3.5.18) directly. The statistical approach was applicable to stepwise case because the same probability of reaction p applied to all functional groups; we can use it here because at each step in a given chain, propagation occurs with the same probability. That the probability of propagation is constant during the lifetime of a chain is a direct result of the QSSA.[66,67] If the QSSA is violated, a geometric distribution will not result, and statistical derivation will not be possible. One can walk along the chain and always reach a statistically equivalent point (something not possible during non-QSSA situations such as those in an anionic polymerization) because during the lifetime of a chain, no changes occur in the environment. The situation here is thus comparable to a step polymerization, the probability of propagation q replacing the conversion p.

3.5.1.2 Inclusion of Other Reactions

The foregoing scheme is quite simple, not nearly as rich as could be expected from Section 3.4. The main neglect is that of transfer reactions; we will restrict ourselves to reactions with small and ideal transfer agents (monomer, solvent, or added agent). The concentration of the jth transfer agent S_j depletes as follows in a batch polymerization:

$$\frac{dS_j}{dt} = -k_{tr,sj}S_jH(1) \qquad (3.5.21)$$

---

*An exception arises if fast initiation is combined with an agent that effectively prevents termination, so as to achieve "living" free-radical polymerization.[65]

The counterpart to equation (3.5.4) is:

$$\frac{\partial H(s)}{\partial t} = 0 = \left[2fk_dI + k_{tr,m}MH(1) + \sum_{j=1}^{n} k_{tr,sj}S_jH(1) \right] s - k_pM(1 - s)H(s) - k_{tr,m}MH(s) \quad (3.5.22)$$

$$- \sum_{j=1}^{n} k_{tr,sj}S_jH(s) - k_tH(s)H(1)$$

This picture is much more complicated, but the geometric distribution survives. The propagation probability is merely redefined as follows:

$$q = \frac{k_pM}{k_pM + k_{tr,m}M + \sum_{j=1}^{n} k_{tr,sj}S_j + k_tH(1)} \quad (3.5.23)$$

The rate of initiation, note, now has the added terms due to the initiation of new chains by the various transfer mechanisms [see the bracketed term in equation (3.5.22)]. If transfer is nonideal, a reduction in $H(1)$ occurs because the radicals are in a dormant state or "on hold." The QSSA would have to be redefined, then, since two different kinds of radical would exist. As long as the QSSA holds, though, the distribution of living chains will be geometric, independent of the number of reactions occurring.

3.5.1.3 Long-Chain Hypothesis (LCH)

Under the more general kinetic scheme, which includes transfer to monomer, the monomer balance equation must be rewritten as follows:

$$\frac{dM}{dt} = -2fk_dI - (k_p + k_{tr,m})MH(1) \quad (3.5.24)$$

Generally, chain transfer to monomer results in a terminal double bond available for polymerization (see Section 5.1.1); here we ignore this possibility for branching and instead consider the monomer to have been consumed. This equation can be solved under the assumption of a constant radical concentration $H(1)$, which in turn implies a constant initiator concentration $I = I_0$, not unusual given the long half-lifes of these initiators, even at their use temperatures. The solution is:

$$\frac{M}{M_0} = e^{-t/\lambda} - \frac{2fk_dI_0}{k_p(1 + C_m)H(1)M_0}(1 - e^{-t/\lambda}) \quad (3.5.25)$$

where the time constant for monomer disappearance, λ, is given by

$$\lambda = \frac{1}{k_p(1 + C_m)H(1)} \quad (3.5.26)$$

Equation (3.5.25) yields negative values of the monomer concentration at sufficiently long times, because by assuming that decomposition was the rate-determining step in initiation, we imply that monomer is consumed through initiation after all monomer has been depleted. In fact, decomposition must not be rate-limiting when there is little monomer present; rather, the initiation step proper, equation (3.4.4), must become important.

The initiation term can be neglected entirely if the chain length is long, so that monomer is primarily consumed in propagation steps (large kinetic chain length). Equation (3.5.25) then simplifies to:

$$\frac{M}{M_0} = e^{-t/\lambda} \tag{3.5.27}$$

This *long-chain hypothesis* is based on the observation that free-radical polymerization produces polymer of high molecular weight from the start of the reaction. If initiator concentration decays exponentially with time, the monomer concentration decays as follows:

$$\ln\left(\frac{M}{M_0}\right) = \left(\frac{8fk_p^2 I_0}{k_t k_d}\right)^{1/2} (e^{-k_d t/2} - 1) \tag{3.5.28}$$

This exhibits the interesting feature of predicting a *dead-end polymerization* in which not all monomer is consumed before the initiator is depleted.[68,69] The ultimate monomer concentration M_∞ is given by:

$$M_\infty = M_0 \exp\left[-\frac{8fk_p^2 I_0}{k_t k_d}\right] \tag{3.5.29}$$

Appropriate choice of initiator type and amount can yield $M_\infty \approx 0$, but note that under such conditions the long-chain hypothesis on which this anaysis is based may not be valid.

3.5.2 The Effects of Concentration and Temperature

We can now begin to address the dependences of the rate of polymerization and the chain length on the concentrations of the reagents and temperature. We assume that rate constants are dependent solely on reaction temperature, through the usual Arrhenius relation; any dependence on the state of the reaction medium is deferred until Section 3.6.

Under the long-chain hypothesis, the rate of polymerization R_p can be identified with the rate of propagation, and so

$$R_p = -\frac{dM}{dt} = \left(\frac{2fk_d k_p^2}{k_t}\right)^{1/2} MI^{1/2} \propto MI^{1/2} (fk_d)^{1/2} k_p k_t^{-1/2} \tag{3.5.30}$$

The *kinetic chain length, ν,* which we mentioned in Section 3.4.4, quantifies the lifetime of a radical in terms of the number of propagation steps in which it participates:

$$\nu = \frac{\text{rate of polymerization}}{\text{rate of termination}} = \frac{k_p M}{\sqrt{2fk_d I k_t}} \propto MI^{-1/2} (fk_d)^{-1/2} k_p k_t^{-1/2} \tag{3.5.31}$$

It is thus very nearly DP_n, except when transfer reactions occur, for then DP_n is less than ν, since the *kinetic chain* is not broken by transfer as the polymer chain is.

The linear dependence of each of these on M and k_p is expected; the other dependences are slightly deeper and result from the QSSA. Both R_p and ν depend on the square root of the initiator concentration, although the rate increases with this (because of the greater number of active radicals as I increases) while the kinetic chain length decreases (because of the greater termination rate at the higher concentration of radicals). Hence, too, the dependences on k_d and f. When the termination rate constant k_t increases, however, both the rate of polymerization and the chain length diminish. The dependences on M and I moreover suggest how the chain length may change with conversion. As the monomer concentration M decreases or "drifts" with conversion, the chain length decreases. This may be countered by depletion of initiator I, which tends to increase chain length.

Of course, the scalings of equations (3.5.30) and (3.5.31) will not hold if the kinetic scheme deviates from the simple one given in this section.[2,50] Complications generally enter in because of differences in the initiation or termination steps; there are many examples. Photoinitiation thus introduces a dependence on the intensity of light, and redox polymerization introduces complications into both initiation and termination steps. The dependence of R_p on M may increase to M^2 if termination occurs predominantly by the action of initiator radicals (primary termination) or to $M^{3/2}$ if monomer assists in the homolysis of the initiator. Likewise, the dependence on M may appear different if the initiator efficiency f is a strong function of M (e.g., through viscosity).

The dependences on rate constants suggest the comparison between reactions at two different temperatures, or what will happen if the temperature increases with conversion, for example. The rate constants individually increase with increasing temperature, but the effect on the rate and chain length depends on the relative values of the activation energies (E) for the various processes. We may thus define an effective activation energy for the chain length:

$$E_\nu = E_p - \frac{1}{2}\{E_d + E_t\} \tag{3.5.32}$$

An effective activation energy, unlike the activation energy for a basic reaction step, can be negative, and indeed it is in this case, because of the high activation energy for thermal decomposition of initiators. Thus the kinetic chain length decreases with increasing temperature because the dominant factor is the much greater rate of production of radicals, increasing the termination rate. For the same reason, the rate of polymerization does increase with increasing temperature because of the dominance of $k_d^{1/2}$. The decreasing kinetic chain length is a result of the mode of initiation; in the case of photoinitiation in which no activation energy exists, the kinetic chain length increases with increasing temperature.

Example 3.2

Compare the solution polymerization of styrene at 50 and 60°C by plotting R_p and v versus time. Initiation is by AIBN, for which[70] $k_d = 1.58 \times 10^{15} \, e^{-15,500/T} \, s^{-1}$ (with T in K) and f may be assumed to be 0.65. The best data for k_p suggest[71,72] $k_p = 1.26 \times 10^7 \, e^{-3490/T} \, L/(mol \cdot s)$ and $k_t = 4.97 \times 10^7 \, L/(mol \cdot s)$ at 50°C and $5.53 \times 10^7 \, L/(mol \cdot s)$ at 60°C.[72] The initial monomer concentration is 1.0 mol/L. For the initial initiator concentration I_0, we have (a) 0.01 mol/L and (b) 0.04 mol/L. Assume that all rate constants are really constant and that the polymerizations are isothermal.

Solution:

At 50°C: $k_d = 2.28 \times 10^{-6} \, s^{-1}$ and $k_p = 256 \, L/(mol \cdot s)$; at 60°C: $k_d = 9.63 \times 10^{-6} \, s^{-1}$ and $k_p = 354 \, L/(mol \cdot s)$. (Note the greater than fourfold increase in k_d with a 10°C rise relative to the weaker increases in k_p and k_t). Equation (3.4.6) can be used to find I as a function of time, while (3.5.28) can be used to find M as a function of time. Equations (3.5.30) and (3.5.31) are then applied to find R_p and v.

Figure 3.6 shows that increasing either temperature or initiator concentration acts to increase the R_p. Increasing I_0 by a factor of 4 causes the expected twofold increase in R_p, but only initially; the course of monomer and, to a lesser extent, initiator depletion alters the situation thereafter. Increasing the temperature causes R_p to drop more quickly, not only because the propagation is faster but because initiator is being depleted more rapidly. This can also be noted by the difference in conversions after 36 hours for the four cases: at 50°C, 53% at 0.01 mol/L and 78% at 0.04 mol/L; at 60°C, 80% at 0.01 mol/L and 96% at 0.04 mol/L.

Figure 3.7 shows that I_0 and T have opposite effects on v and R_p, but monomer depletion affects both in the same fashion. Some of these chain lengths are too short to be useful polymers, but working at higher T with lower I_0, as well as in bulk, should produce chains of appropriate length.

Figure 3.6.
Rate of polymerization versus time for Example 3.2: solid curve, 50°C, $I_0 = 0.01$ mol/L; short dashes, 50°C, $I_0 = 0.04$ mol/L; long dashes, 60°C, $I_0 = 0.01$ mol/L; long–short dashes, 60°C, $I_0 = 0.04$ mol/L.

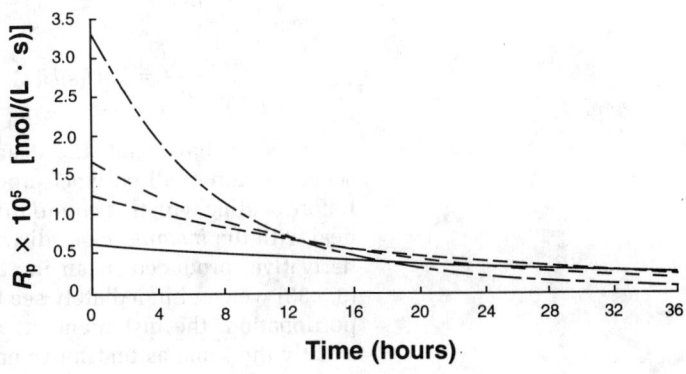

Figure 3.7.
Kinetic chain length versus time for Example 3.2; curves as in Figure 3.6.

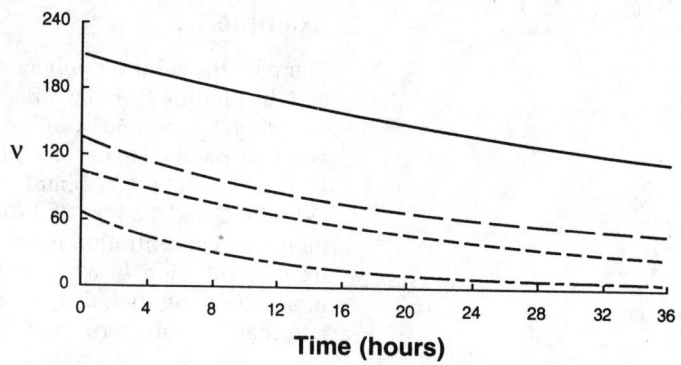

3.5.3 Instantaneous Dead Polymer

The active polymer chains we have dealt with so far are present in at most micromolar concentrations, as we saw in the preceding example. This clearly then is not our product! Our product, rather, consists of the inactive or "dead" chains that accumulate as a result of termination and transfer processes, given by P_i and its transform, $G(s)$.

For the active chains, we made no distinction with respect to the mode of termination. For disproportionation and combination alike, we used an overall termination constant k_t (= k_{tc} + k_{td}) because the rate of termination, not the mode, affects the living chain distribution. Since, however, the dead chain distribution does depend on the mode of termination, we must now distinguish between disproportionation and combination. The two will yield different distributions of polymers, so we first ignore the influence of transfer reactions and start with the rarer case of disproportionation, as might be applicable for methyl methacrylate.

3.5.3.1 Termination by Disproportionation
If termination is by disproportionation ($k_t = k_{td}$), the evolution equation for the individual species and the corresponding equation in s-space are:

$$\frac{dP_i}{dt} = k_{td}R_i \sum_{j=1}^{\infty} R_j = k_{td}H^2(1)\,(1 - q)q^{i-1} \qquad (3.5.33)$$

$$\frac{\partial G(s)}{\partial t} = k_{td}H(s)H(1) = k_{td}H^2(1)\frac{(1-q)s}{1-qs} \qquad (3.5.34)$$

As we have said, the dead polymer $G(s)$ represents the accumulation of all polymer produced up to that point in time. Before dealing with this *cumulative* distribution, though, we will deal with the *instantaneous* distribution, given by the foregoing derivative, produced in an increment of time. From equation (3.5.33) we can immediately see that for termination by disproportionation, the instantaneous distribution of dead chains is exactly the same as that for the active chains from which they are produced—that is, a geometric distribution with the same parameter q. So, we need not go through the formal machinery of expanding in power series, for we already know that the instantaneous degrees of polymerization are:

$$DP_n^{inst} = \frac{1}{1 - q} \tag{3.5.35}$$

$$DP_w^{inst} = \frac{1 + q}{1 - q} \tag{3.5.36}$$

Some treatments[14] do not include P_1 as polymer, hence arrive at slightly different equations for DP_n^{inst} and DP_w^{inst}.

3.5.3.2 Termination by Combination

For the more common case of combination (as for styrene or ethyl acrylate), we have termination predominantly by combination ($k_t = k_{tc}$). Here:

$$\frac{dP_i}{dt} = \frac{k_{tc}}{2} \sum_{j=1}^{i-1} R_j R_{i-j} \qquad i \geq 2 \tag{3.5.37}$$

$$\frac{\partial G(s)}{\partial t} = \frac{k_{tc}}{2} H^2(s) = \frac{k_{tc}}{2} H^2(1) \left(\frac{(1 - q)s}{1 - qs} \right)^2 \tag{3.5.38}$$

Unlike the case of pure disproportionation, the instantaneous dead chain distribution here is obviously not geometric, for the generating function characteristic of a geometric distribution has been squared. This is what one calls a self-convoluted geometric distribution, a particular example of a negative binomial distribution.[73] The distribution can be found from expanding equation (3.5.38) in a Taylor series in s, but it is even easier simply to deal with equation (3.5.37):

$$\frac{dP_i}{dt} = \frac{k_{tc}}{2} H^2(1) \sum_{j=1}^{i-1} (1 - q)q^{j-1} (1 - q)^{i-j-1} = \frac{k_{tc}}{2} H^2(1) (1 - q)^2 (i - 1) q^{i-2} \qquad j \geq 2 \tag{3.5.39}$$

This distribution can also be derived from simple combinatorial reasoning: the production of a dead chain by combination requires two active chains to experience a termination event [with probability $(1 - q)^2$], and a total of $i - 2$ propagation events (with probability q^{i-2}). This combination can be done in $i - 1$ different ways, with a factor of 2 accounting for indistinguishability. One can also derive this result by other statistical methods, such as by recursive arguments or similar methods.[66,67,74] One merely needs to account for the directionality on a chain arising because one end was initiated and the other terminated, hence possibly coupled to another chain.

The self-convoluted geometric has the following instantaneous degrees of polymerization:

$$DP_n^{inst} = \frac{2}{1 - q} \tag{3.5.40}$$

$$DP_w^{inst} = \frac{2+q}{1-q} \qquad (3.5.41)$$

This distribution is narrower than a geometric distribution, as is easily seen in the polydispersity:

$$Q = 1 + \frac{q}{2} \qquad (3.5.42)$$

Thus, for long chains, the polydispersity approaches 1.5 rather than 2. This in fact is a general result: the random combining of a fixed number of chains decreases polydispersity. This is not true for randomly combining a random number of chains, though, as we will see in Chapter 5.

3.5.3.3 Termination by Disproportionation and Combination and Transfer

For the general case discussed in Section 3.5.1, we have the following equations:

$$\frac{dP_i}{dt} = \{k_{td}H(1) + k_{tr,m}M + \sum_{j=1}^{n} k_{tr,sj}S_j\}H(1)\,(1-q)q^{i-1} + \frac{k_{tc}}{2}\,H^2(1)\,(1-q)^2\,(i-1)\,q^{i-2} \quad (3.5.43)$$

$$\frac{\partial G(s)}{\partial t} = k_{td}H(s)\,H(1) + \frac{k_{tc}}{2}\,H^2(s) + \{k_{tr,m}M + \sum_{j=1}^{n} k_{tr,sj}S_j\}\,H(s) \qquad (3.5.44)$$

$$= \{k_{td}H(1) + k_{tr,m}M + \sum_{j=1}^{n} k_{tr,sj}S_j\}\,H(1)\frac{(1-q)s}{1-qs} + \frac{k_{tc}}{2}\,H^2(1)\left(\frac{(1-q)s}{1-qs}\right)^2$$

The distribution is a combination of a geometric and self-convoluted geometric distribution. A second parameter is needed in addition to q. This may be either the fraction of chain-ending steps that are by combination, ξ, or the fraction of dead chains made by combination, ξ':

$$\xi = \frac{k_{tc}H(1)}{k_{td}H(1) + k_{tc}H(1) + k_{tr,m}M + \sum_{j=1}^{n} k_{tr,sj}S_j} \qquad (3.5.45)$$

$$\xi' = \frac{\dfrac{k_{tc}}{2}\,H(1)}{k_{td}H(1) + \dfrac{k_{tc}}{2}\,H(1) + k_{tr,m}M + \sum_{j=1}^{n} k_{tr,sj}S_j} = \frac{\xi}{2-\xi} \qquad (3.5.46)$$

The former arises more naturally in the statistical analyses,[66,67] the latter more naturally from the kinetic equations. The instantaneous degrees of polymerization are given as follows:

$$DP_n^{inst} = \frac{1}{(1-\xi/2)(1-q)} = \frac{1+\xi'}{1-q} \qquad (3.5.47)$$

$$DP_w^{inst} = \frac{1+q+\xi}{1-q} = \frac{1+q}{1-q} + \frac{2\xi'}{(1+\xi')(1-q)} \qquad (3.5.48)$$

The polydispersity is then given as follows:

$$Q = \frac{1+q}{1+\xi'} + \frac{2\xi'}{(1+\xi')^2} = \left(1 - \frac{\xi}{2}\right)(1 + q + \xi) \quad (3.5.49)$$

The polydispersity thus lies between 1.5 and 2 for long chains.

These two parameters, q and either ξ' or ξ, define the instantaneous distributions of dead chains. The accumulation of these chains to give our polymer product depends on how the two parameters change with conversion, and it is this question to which we turn next.

3.5.4 Cumulative Dead Polymer

The generating function of the cumulative polymer, $G(s)$, can be put into integral form for each of the cases we have studied so far. For the first case of dead chain production solely by disproportionation, we can write:

$$G(s) = \int_0^t k_{td} H^2(1) \frac{(1-q)s}{1-qs} dt' \quad (3.5.50)$$

All terms have been kept within the integral for the sake of generality. If the parameter q is a function of time (or conversion), the distribution will no longer be geometric, but some mixture of geometric distributions of different parameters q. The distribution will practically always be *broader*.

This broadening of the distribution has a source very different from that which has led to the polydispersity we have observed in all preceding cases. There, the breadth of the chain length distribution arose from *statistical dispersion*—that is, from the random aspects of the polymerization mechanism. The additional increase in polydispersity here, on the other hand, arises from a deterministic or nonrandom element, the combination of polymer of different distributions into a polymer product. An example would be the mixing of two samples of poly(methyl methacrylate), the first of $M_n = 100,000$, the second of $M_n = 500,00$, both of geometric distribution. Such a mixing is hardly a random event; it leads to a polydispersity greater than 2, the value depending on the amount of each sample added. The normal course of a free-radical polymerization, in which there are changes in q (and ξ') due to reagent depletion, temperature changes, and diffusion limitations (see Section 3.6), essentially performs a mixing analogous to our example, except that it does it "naturally" and on a continuous, rather than a discrete, basis. This kind of dispersion is referred to as *drift dispersion*. It is important to distinguish between these forms of dispersion because they originate from different sources and thus are affected by circumstances (e.g., reactor configuration) in different ways (see Chapter 6).

Example 3.3

If one were to mix together the two polydisperse PMMA polymers just mentioned in amounts of equal mass, what would the resulting polydispersity be?

Solution:

Denote the polymer with $M_n = 100,000$ as polymer 1, the polymer with $M_n = 500,000$ as polymer 2. Because of the fivefold difference in M_n or DP_n, mixing together equal masses will give us five times the number of moles of polymer 1 as of polymer 2. We can thus define a normalized generating function $G(s)$ given by:

$$G(s) = \frac{5}{6}\frac{(1 - q_1)s}{1 - q_1 s} + \frac{1}{6}\frac{(1 - q_2)s}{1 - q_2 s}$$

where $q_1 = 0.999$ and $q_2 = 0.9998$. As the generating function is normalized, the zeroth moment is unity, and the first and second moments are given by:

$$\mu_1 = \frac{5}{6}\frac{1}{1 - q_1} + \frac{1}{6}\frac{1}{1 - q_2}$$

$$\mu_2 = \frac{5}{6}\frac{(1 + q_1)}{(1 - q_1)^2} + \frac{1}{6}\frac{(1 + q_2)}{(1 - q_2)^2}$$

The polydispersity is given by $Q = (\mu_2\mu_0/\mu_1^2)$, which yields $Q = 3.6$.

One is tempted to make a generalization out of Example 3.3, saying that drift dispersion *always* leads to a further broadening of the distribution. While such a statement is the overwhelming rule, situations can be contrived in which the accumulation of polymer does the opposite.

For the other two cases we have examined, we have the following equations for the dead chain generating function. For combination only we have:

$$G(s) = \int_0^t \frac{k_{tc}}{2} H^2(1) \left(\frac{(1 - q)s}{1 - qs}\right)^2 dt' \qquad (3.5.51)$$

For production of dead chains by disproportionation, combination, and transfer, we have:

$$G(s) = \int_0^t \left\{ \left(\left(k_{td} + \frac{k_{tc}}{2}\right)H^2(1) + k_{tr,m}MH(1) + \sum_{j=1}^n k_{tr,sj}S_jH(1)\right) \left\{(1 - \xi')\frac{(1 - q)s}{1 - qs}\right. \right.$$

$$\left. \left. + \xi'\left(\frac{(1 - q)s}{1 - qs}\right)^2 \right\}dt' \right. \qquad (3.5.52)$$

Integration of these equations depends on all the particularities of the dependences of q and ξ' on time or conversion. Thus, in general we do not expect an analytic solution and thus will generally work with either the moment equations resulting from these or (if we are interested in the concentrations of a few chain lengths) the untransformed equations. Before proceeding to a brief discussion of the few analytical cases, we point out the following significant equations. Since the first moment of the distribution will be the conversion p (or proportional to it), we can write:

$$DP_n^{cumu} = \frac{p}{\int_0^p \frac{dp'}{DP_n^{inst}}} = \frac{\int_0^t R_p dt'}{\int_0^t \frac{R_p dt'}{DP_n^{inst}}} \qquad (3.5.53)$$

$$DP_n^{cumu} = \frac{1}{p} \int_0^p DP_w^{inst} dp' = \frac{\int_0^t DP_w^{inst} R_p dt'}{\int_0^t R_p dt'} \qquad (3.5.54)$$

These equations will more often than not be the most useful. Their status within or without the long-chain hypothesis depends generally on the expression used for the rate of polymerization R_p.

Analytical solution for the accumulated polymer distribution is possible only in highly idealized cases, such as that for which monomer depletion is the sole source of a changing q. We look at this case now, not because it is useful in practice (it is not), but because it shows the effects of drift dispersion and how rapidly analytic solution becomes difficult.

For the case in which no transfer reactions occur, but termination is by disproportionation and combination, under monomer drift (depletion) only, the parameter ξ is constant and the parameter q is given as a function of conversion:

$$q = \frac{1 - p}{1 - p + d} \qquad (3.5.55)$$

where the parameter d is:

$$d = \frac{k_I H(1)}{k_p M_0} \qquad (3.5.56)$$

which is a constant, since the rate of initiation is constant, and obviously is closely related to the initial value of q [$d = (1 - q_0)/q_0$]. We thus find the generating function to be:

$$G(s) = \frac{k_{tc} H(1) ds}{2k_p} \left[\frac{1}{1 + d - s} - \frac{1}{d + (1 - s)(1 - p)} \right]$$
$$+ \frac{(2k_{td} + k_{tc}) H(1)}{2k_p} \ln \left[\frac{1 + d}{1 + d - s} \frac{d + (1 - s)(1 - p)}{1 + d - p} \right] \qquad (3.5.57)$$

This equation is readily expanded to yield the following expression for P_i:

$$P_i = M_0 d \left\{ \left[\frac{(\xi/2)(1 + di) + (1 - \xi)}{i} \right] \left[\frac{1}{1 + d} \right]^i \right. $$
$$\left. - \left[\frac{(\xi/2)(1 + di/(1 - p)) + (1 - \xi)}{i} \right] \left[\frac{1 - p}{1 + d - p} \right]^i \right\} \qquad (3.5.58)$$

This can also be expressed as a normalized weight fraction $w_i = iP_i/(M_0 p)$:

$$w_i = \frac{1}{p} \left\{ \left[\frac{\xi}{2}(1 + di) + (1 - \xi) \right] \left[\frac{1}{1 + d} \right]^i \right. $$
$$\left. - \left[\frac{\xi}{2} \left(1 + \frac{di}{1 - p} \right) + (1 - \xi) \right] \left[\frac{1 - p}{1 + d - p} \right]^i \right\} \qquad (3.5.59)$$

The long-chain hypothesis has not been invoked in these equations.*

The moments of the dead chain distribution may also be derived, and one finds:

$$\mu_0 = \left(1 - \frac{\xi}{2} \right) dM_0 \ln \left(\frac{1 + d}{1 + d - p} \right) \qquad (3.5.60)$$

$$\mu_1 = pM_0 \qquad (3.5.61)$$

$$\mu_2 = pM_0 \left(1 + \xi + \frac{1 + \xi/2}{d} (2 - p) \right) \qquad (3.5.62)$$

[We had already made use of equation (3.5.61) in alteration of P_i to w_i.] Thus we have the cumulative degrees of polymerization as follows:

$$DP_n^{cumu} = \frac{p}{(1 - \xi/2) \, d \ln \left(\dfrac{1 + d}{1 + d - p} \right)} \qquad (3.5.63)$$

$$DP_w^{cumu} = 1 + \xi + \frac{1 + \xi/2}{d} (2 - p) \qquad (3.5.64)$$

These degrees of polymerization can obviously be found either from the moments derived from the expressions for $G(s)$ or from equations (3.5.63) and (3.5.64), with the appropriate expressions for DP_n^{inst} and DP_w^{inst}, given in equations (3.5.56) and (3.5.57). The reader should keep in mind that these equations are valid only if there is no transfer and the only source of drift dispersion is monomer depletion.

---

*Similar equations can be derived under the long-chain hypothesis, but this supposed "simplification" makes the math more cumbersome (e.g., a separate form is required for P_1 and P_2, and less accurate.

Example 3.4

Consider a free radical polymerization with the following data: $k_p = 900$ L/(mol · s), $k_t = 3.5 \times 10^5$ L/(mol · s), $k_d = 5 \times 10^{-5}$ s^{-1}, and $f = 0.6$. Assume that termination occurs by combination. Calculate and plot DP_n^{cumu}, DP_w^{cumu}, and Q^{cumu} for (a) a semibatch reactcor where monomer is constantly added to maintain the initial monomer concentration $M_0 = 1.5$ mol/L, and (b) a batch reactor with initial conditions given by $M_0 = 1.5$ mol/L. For the second also plot the cumulative distribution P_i. In both cases, assume that the initiator concentration is constant at $I_0 = 0.017$ mol/L.

Solution:

(a) We note that R_p and the propagation probability q appearing in equations (3.5.40) and (3.5.41) are constant. Given the data, $q = 0.99956$, resulting in $DP_n = 4521$, $DP_w = 6780$, and $Q = 1.5$ for both instantaneous and cumulative values. We do not bother to plot this!

(b) The values of DP_n, DP_w, and Q are the initial (instantaneous) values, but we must use equations (3.5.53) and (3.5.54) to get the cumulative values over the entire conversion range. Figure 3.8 shows the results for both cases. The decrease in the degrees of polymerization due to monomer depletion or drift (the only mechanism by which chain length is changing in this case) is apparent, as is the increase in polydispersity, especially as the reaction approaches completion. These observations are echoed in Figure 3.9, where we have plotted the (normalized) cumulative chain length distribution at five different conversions p. Note that the distribution broadens and its peak shifts to the left as the degree of polymerization diminishes.

The changes in the polymer product illustrated in Example 3.4 clearly pose problems to the design engineer. In some instances, the need to produce a polymer of nearly uniform properties (or at least fairly so) makes intolerable the drift caused by the depletion of monomer or initiator. Even this example is highly idealized, though, inasmuch as the rate constants themselves are

Figure 3.8.
Effect of monomer depletion on cumulative degrees of polymerization and polydispersity. (For Example 3.4.)

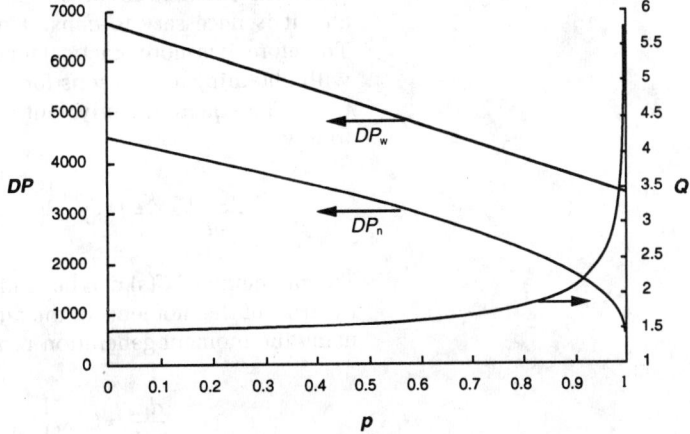

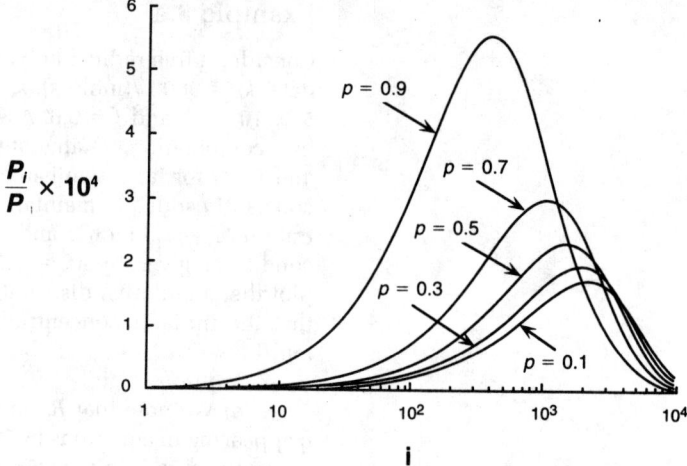

Figure 3.9.
Effect of monomer depletion on normalized chain length distribution. (For Example 3.4.) Distributions do not appear normalized because of semilog plotting.

$\dfrac{P_i}{P_1} \times 10^4$

$p = 0.9$
$p = 0.7$
$p = 0.5$
$p = 0.3$
$p = 0.1$

i

really taken to be constant, changing neither with temperature because of exothermicity nor because of any physical effects (see Section 3.6). In such cases even more drastic changes can come about.

Let us now consider our most liberal scheme, in which we have both disproportionation and combination and transfer to both monomer and agent (here restricted to a single agent). The evolution equation for the dead polymer chain distribution is written as follows:

$$\frac{dP_i}{dt} = \frac{k_{tc}}{2} R^2 (1 - q)^2 (i - 1) q^{i-2} + (k_{tr,m}M + k_{tr,s}S + k_{td}R) R(1 - q) q^{i-1} \quad (3.5.65)$$

Transfer to monomer does not allow an easy integration of this equation as was possible earlier. Analytical expressions for the chain length distribution are available when either combination or disproportionation occurs.[7] When both modes compete, as is usually the case, one must resort to numerical integrations. Even when only one termination mechanism is encountered, the expressions obtained involve terms with the exponential integral, and it is necessary to consult mathematical tables for values. Therefore, it is more practical for our purposes here to be satisfied with obtaining expressions for the moments of the distribution.

The equation equivalent to (3.5.65) in s-space is given as follows:

$$\frac{\partial G(s)}{\partial t} = \frac{k_{tc}}{2} H(s)^2 + (k_{tr,m}M + k_{tr,s}S + k_{td}R)H(s) \quad (3.5.66)$$

The moments of $G(s)$ can be readily derived from this equation in terms of the moments of the radical polymer distribution $H(s)$ using the moment generation property of the transforms. Thus:

$$\frac{d\mu_0}{dt} = k_{tc} \lambda_0 \left\{ \lambda_0 \left(\frac{1}{2} + K \right) + \beta \right\} \quad (3.5.67)$$

$$\frac{d\mu_1}{dt} = k_{tc} \lambda_1 \{\lambda_0 (1 + K) + \beta\} \qquad (3.5.68)$$

$$\frac{d\mu_2}{dt} = k_{tc} \lambda_2 \left\{ \lambda_0 (1 + K) + \beta + \frac{\lambda_1^2}{\lambda_2} \right\} \qquad (3.5.69)$$

where

$$K = \frac{k_{td}}{k_{tc}} \qquad (3.5.70)$$

$$\beta = \frac{k_{tr,m} M + k_{tr,s} S}{k_{tc}} \qquad (3.5.71)$$

The instantaneous weight- and number-average degrees of polymerization are expressed by equations (3.5.47) and (3.5.48), with q given by equation (3.5.23) and ξ given thus:

$$\xi = \frac{k_{tc} H(1)}{k_{tc} H(1) + k_{td} H(1) + k_{tr,m} M + k_{tr,s} S} \qquad (3.5.72)$$

which follows from equation (3.5.45). For $\beta = 0$ (no transfer reactions) the instantaneous polydispersity will lie between 1.5 ($K = 0$) and 2 ($k \rightarrow \infty$). In the limit of dominant transfer, $\beta \rightarrow \infty$, the polydispersity approaches 2 from below, irrespective of the termination mechanism. Cumulative values are then most conveniently found with equations (3.5.53) and (3.5.54).

3.6 Physical Effects in Free-Radical Polymerization

One of the distinctions of polymerization reactions, compared to small-molecule reactions, is the extent to which the vast changes in the state of the system affect the reactions kinetics. For example, the increasing difficulty of removing by-product during polycondensations immediately implies an effect of processing parameters on the rate of molecular weight buildup. The richest effects of this kind are found in free-radical polymerization, because of the number of concurrent reactions and because high molecular-weight polymer is formed from the beginning of the reaction.

3.6.1 Diffusional Control of Termination (Trommsdorff Effect)

Termination, as the only reaction occurring between two polymers during a linear polymerization, might be expected to be the first reaction to become diffusion-controlled. In fact, it is diffusion-controlled from the beginning of polymerization,[75] as can be seen by comparing solution polymerizations at different viscosities.[76] The physics governing how chains ends diffuse to meet in a reactive volume thus determines k_t, and, through k_t, the rate of polymerization and the molecular weights obtained. Diffusional control is thus crucial to free-radical polymerization. The most dramatic manifestation of diffusion control comes in the *Trommsdorff effect*,* wherein the rate of polymerization dramatically increases at intermediate conversions, as does the molecular weight if a chain transfer agent is not used. The effect was noted early in the history of polymer science,[77–79] and has been extensively studied,[80–83] especially for polymerizations of methyl methacrylate.

That diffusional control leads to a greater rate of polymerization may seem counterintuitive at first, but it does happen because a severe (orders of magnitude) decrease in k_t allows an increase in the concentration of radicals $H(1)$ (even though the QSSA fails here), hence an increase in the rate of polymerization. The Trommsdorff effect does not rely on an increase in temperature and is thus distinct from a thermal autoacceleration. Since, however, the heat rapidly released during the Trommsdorff effect may make it difficult to maintain isothermal conditions, the Trommsdorff effect may be coupled to a thermal autoacceleration, giving a potentially explosive situation.

A number of models of the Trommsdorff effect have been proposed,[10,81–86] and most can describe the rate of polymerization and average molecular weight data fairly well. This good agreement is in one sense unfortunate, for it means that model discrimination cannot be made on these grounds. Data on actual termination rates, rather than aggregated effects on overall rates and chain lengths, are far superior. Many models may suffice for purposes of modeling conversion and average molecular weight data, but it is better to have a model based on fundamental physics. The usual starting point for analysis of diffusion-controlled reactions is the Smoluchowski equation for the rate constant $k_{i,j}$ between species i and species j[87,88]:

$$k_{i,j} = 4\pi r(D_i + D_j) \tag{3.6.1}$$

where r is the reaction radius, and D_i and D_j are the diffusion coefficients of the species. Application of this equation is complicated by many factors (such as whether the limiting step is considered to be center-of-mass or segmental diffusion); the important point is that k_t may depend on the lengths of the chains involved in the reaction.

Diffusion of polymer chains is determined by molecular weights and by concentration, as indicated by two common observations. First, the presence of an inert diluent tends to delay the Trommsdorff effect, and a sufficient amount will erase it altogether.[78] Second, the addition of a transfer agent has the same effect.[82,83] Thus both concentration and molecular weight play roles in the course of the Trommsdorff effect, or, equivalently, in determining the value and conversion dependence of k_t.

---

*Also commonly called the Norrish–Smith effect or the gel effect. We do not use the latter term in the interest of reserving the term "gel" for physically or chemically crosslinked systems.

In the initial stages of (bulk) polymerization, the termination rate constant is altered by some subtle factors we will not delve into here.[89] At moderate conversions, when the solution becomes entangled, diffusion is more difficult, and unentangled oligomeric radicals may acquire significance in determining the termination rate because of their greater mobility.[90-94] The conversion at which the Trommsdorff effect sets in might be expected to correlate with the concentration required for entanglement, but it depends more weakly on molecular weight of the previously formed chains than would be expected ($p_{onset} \sim DP^{-0.24}$).[90] This correlation probably also points to the importance of the oligomeric radicals. In this regime, then, one expects a great dependence of the termination rate on the chain length distribution of the living chains, hence a wide variation in the values of k_t possible.[72] At higher conversions, termination becomes dominated not by movement of the chains, but by the "diffusion" of the radical ends by successive addition of monomer. Here, k_t values may be much less system dependent and related to propagation rate constants.[95-97]

The most dire consequences of the Trommsdorff effect come from the rapid exotherm that may result. The effects of the exotherm may be mitigated somewhat by polymerization in the presence of a heat sink, as in suspension and emulsion polymerizations (see Chapter 7). These do not remove the Trommsdorff effect, however. By performing a solution polymerization, one provides both a thermal reservoir and a decreased polymer concentration, thus also delaying or removing the Trommsdorff effect itself. Furthermore, by performing the polymerization near the boiling point of the solvent, one may extract the heat of polymerization through the latent heat of vaporization of the solvent. Solution polymerizations, though, have the drawback that the solvent may need to be removed, depending upon the application.

3.6.2 Diffusional Control of Propagation and Transfer Reactions (Glass Effect)

Reactions between polymer and small molecule, such as propagation and transfer, are expected to become diffusion-controlled later than termination, since the diffusion coefficient of the small molecule is comparatively large, hence dominates the rate of reaction [cf. equation (3.6.1)]. Propagation and transfer usually become diffusion-controlled when and if the system becomes glassy. Thus the glass transition temperature of the polymer being formed has a decisive effect on the polymerization at high conversion. Because of the glass effect, a system will often fail to reach complete conversion (making dead-end polymerization for the reasons given before an irrelevant issue in these cases).[75,98]

3.6.3 Diffusional Effects on Initiation

While initiator decomposition, as a first-order reaction, might not be expected to exhibit the effects described for the preceding reactions, the same cannot be said for the initiator efficiency, f.

The efficiency of an initiator depends on the physical nature of the solvent cage in which it decomposes. The lifetime of such a cage increases with increasing polymer concentration, and so the efficiency drops dramatically during the course of a polymerization, perhaps five orders of magnitude above a conversion p of 0.7.[99]

3.6.4 Volume Change in Radical Polymerization

Most free-radical polymerizations are accompanied by a volume change, and this is of course most severe for bulk polymerization, for which volume decreases of 20–30% are common. All equations thus far, however, have been written in terms of concentrations; no allowance has been made for the change of concentration because of volume change. Correcting for this omission involves dealing with numbers, rather than concentrations, of species. Thus the kinetic equations are multiplied through by the volume V and this is retained inside derivatives. A linear relation between volume and conversion, p, is then usually used[100]:

$$V = V_0 (1 + \varepsilon p) \tag{3.6.2}$$

where it is assumed that volume relaxation occurs on time scales much shorter than that of reaction. The fractional change, ε, in volume of the system between 0 and 100% conversion is a function of temperature. Such corrections generally give no appreciable changes in the kinetics of polymerization[100,101] but do affect the prediction of average molecular weights, with the result that error in the polydispersity may approach 50% at high conversions.

Example 3.5

We examine here a model[81] of the bulk free-radical polymerization of methyl methacrylate, neglecting the severe decrease in initiator efficiency and the volume change upon reaction. The model does, however, faithfully show the extreme effects of both the Trommsdorff and glass effects on rate and molecular weight. The reaction occurs under isothermal conditions at 70°C. From the following data, calculate the instantaneous and cumulative number- and weight-average molecular weights and polydispersity as a function of conversion up to $p = 0.97$. Also plot the conversion p and the cumulative molecular weights versus time to examine the kinetics.

The initiator is AIBN, for which data suggests that $k_d = 6.32 \times 10^{16} \exp^{-15,430/T}$ min^{-1} (with T in K) (slightly different from Example 3.2). Assume an initiator efficiency of 0.58, and assume that $I_0 = 0.0258$ mol/L. The initial values of the propagation and termination rate constants are given as follows: $k_{p0} = 2.95 \times 10^7 \exp^{-2191/T}$ L/(mol $\cdot$ min) and $k_{t0} = 5.88 \times 10^9 \exp^{-352.8/T}$ L/(mol $\cdot$ min).

The initial monomer concentration can be found from its density, given as $\rho = 0.973 - 1.164 \times 10^{-3} T(°C)$ g/mL.

The propagation and termination rate constants evolve in the following way:

$$k_p = k_{p0} \frac{C}{C + \theta_p k_{p0} H(1)}$$

$$k_t = k_{t0} \frac{C}{C + \theta_t k_{t0} H(1)}$$

where the parameter C is related to conversion p in the following way:

$$\log_{10}(C) = \frac{1 - p}{A + B(1 - p)}$$

and the additional parameters are given as:

$$A = 0.168 - 8.21 \times 10^6 \, [T \, (°C) - 114]^2$$
$$B = 0.03$$
$$\theta_p = 5.4814 \times 10^{-16} \, e^{13,982/T} \text{ min}$$
$$\theta_t = \frac{1.1353 \times 10^{-22} \, e^{17,420/T}}{I_0} \text{ min}$$

All temperatures are in kelvin except as noted.

Solution:

We assume that methyl methacrylate radicals terminate predominantly by disproportionation, and thus we have $\xi = 0$, and $k_t = k_{td}$. The instantaneous degrees of polymerization are given by equations (3.5.35) and (3.5.36), with the parameter q given by (3.5.17). Quasi–steady state is assumed at all times, and the long-chain hypothesis was employed in the monomer balance.

The cumulative values are calculated by the equations (3.5.53) and (3.5.54), the integration being done with Simpson's rule. A fourth-order Runge–Kutta scheme is used to integrate the monomer balance, which is necessary given the changes in the propagation rate constant and the changes in $H(1)$ occasioned by the decreases in initiator concentration and k_t.

Figure 3.10 shows the instantaneous number- and weight-average degrees of polymerization as a function of conversion. Note that during the initial stage of the polymerization (i.e., the first 20–30% conversion), monomer depletion causes a drop in the instantaneous values, as we have observed before (see Figures 3.7 and 3.8). A sharp upturn occurs, however, at approximately 30% conversion, increasing the molecular weights by an order of magnitude. This is due to the decrease in the termination rate constant, which allows more propagation steps to occur before termination (i.e., q increases). This sharp upturn is followed by a sharp downturn (by about two orders of magnitude), which is due to the decrease in the propagation rate con-

Figure 3.10.
Instantaneous number- and weight-average molecular weights for a realistic bulk polymerization of methyl methacrylate.[81]

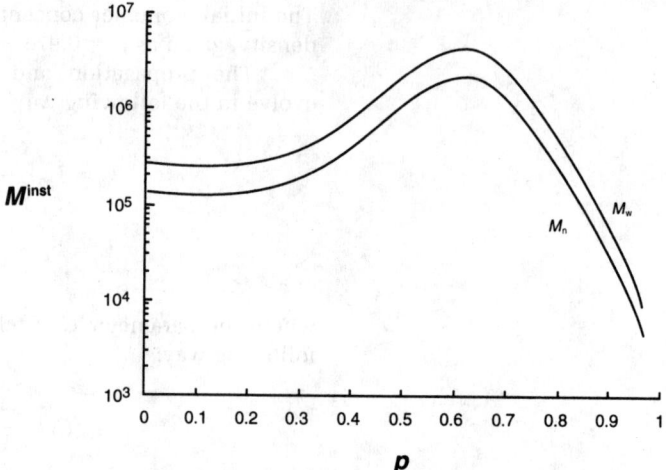

stant as the glass transition temperature of the system is reached. Note that M_n^{inst} and M_w^{inst} are parallel over the entire conversion regime on this semilog plot, meaning that the instantaneous chains have a constant polydispersity, which we know to be ~2 because of the disproportionation mechanism.

Our product, however, is not this instantaneous polymer but rather the cumulative polymer. Figure 3.11 plots M_n^{cumu} and M_w^{cumu}. The trends are the same as those shown earlier but are manifested in a less extreme way, for the accumulation tends to damp out the excursions in molecular weight. Thus, M_n^{cumu} and M_w^{cumu} increases fivefold rather than by an order of magnitude, and the increase in M_n^{cumu} is less yet. This less extreme effect observed in the cumulative values is still sufficient to yield a significant increase in the polydispersity, as seen in Figure 3.12. The Trommsdorff effect increases the polydispersity by a factor of about 3.

Figure 3.11.
Cumulative number- and weight-average molecular weights for a realistic bulk polymerization of methyl methacrylate.[81] Note linear scale when comparing to Figure 3.10.

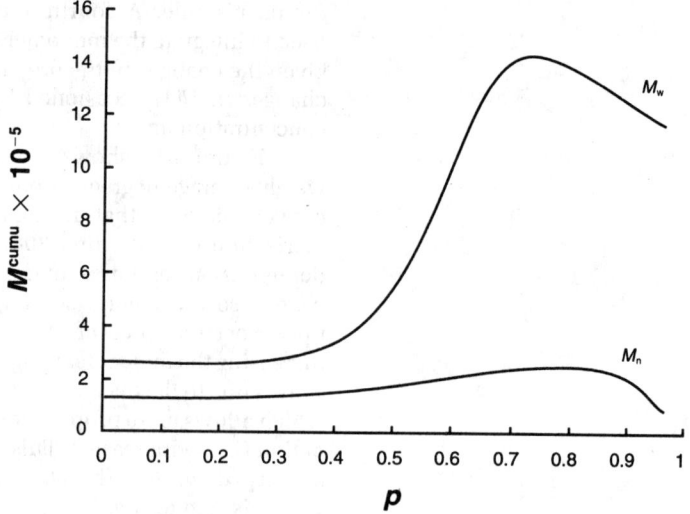

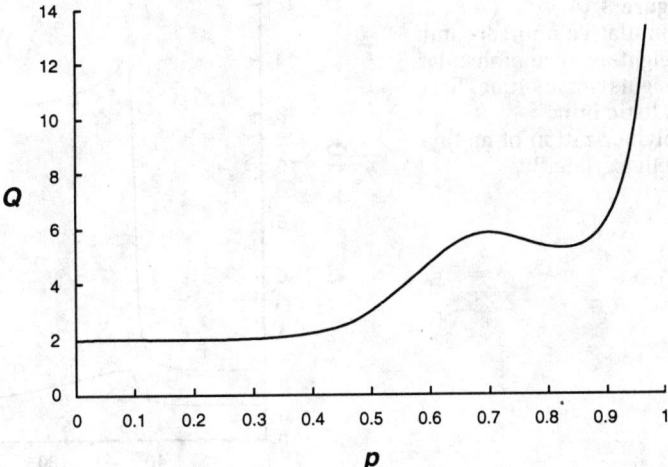

Figure 3.12.
Polydispersity (cumulative) for a realistic bulk polymerization of methyl methacrylate.[81]

Note that thereafter there occurs a slight narrowing of the distribution as chains of fairly similar distribution are added. At 90% conversion, however, the polydispersity exhibits a severe increase that is due to the addition of very short chains.

The chain length and its distribution are not the only aspects of the system to be affected. While the decrease in the termination rate constant increases the chain length of the polymer produced, it also increases the rate dramatically because of the increased number of radicals in the system. This is seen clearly in Figure 3.13. The majority of the reaction occurs in the space of a few minutes at around $t = 50$ minutes, and thus the majority of the vast increase in the cumulative molecular weights occurs in a narrow time interval as well, as shown in Figure 3.14. Note that this behavior is entirely different from that exhibited in Figure 3.6, where a monotonically decreasing rate of

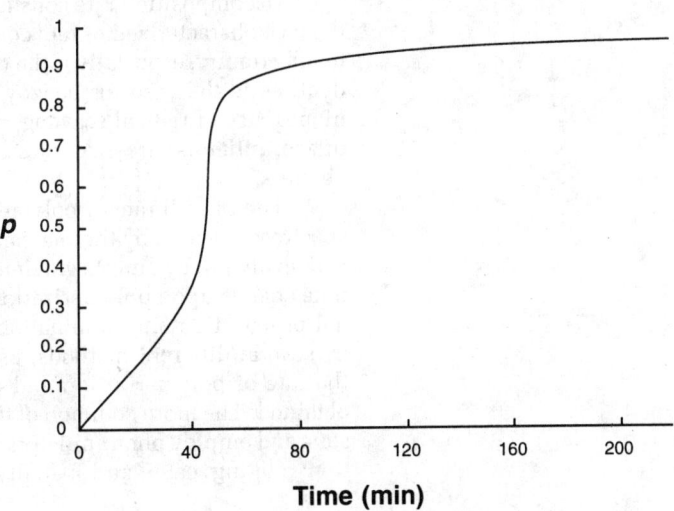

Figure 3.13.
Conversion versus time for a realistic bulk polymerization of methyl methacrylate.[81]

Figure 3.14.
Cumulative number- and weight-average molecular weights versus time, for a realistic bulk polymerization of methyl methacrylate.[81]

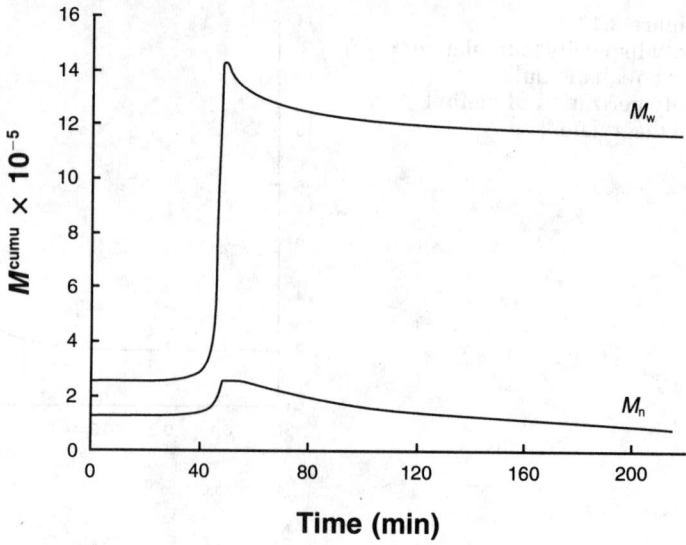

polymerization was found resulting from monomer depletion.

3.7 Parameter Estimation

Measurement of the kinetic rate constants for free-radical polymerizations, while not as difficult as for ionic polymerizations, does present challenges. Many different methods exist, many of which are based on assumptions not necessarily obeyed. Only seldom have multiple techniques been applied to a single system.[72] There is thus quite a bit of scatter in the data. What follows is a brief overview; other sources, and the references therein, are recommended.[2]

Decomposition rate constants for thermal initiators may be the best characterized of rate constants, but they are still subject to uncertainty, especially if the decomposition is solvent-dependent, as is the case for benzoyl peroxide. Spectroscopic techniques, use of radical scavengers, and N_2 evolution (in the case of azo initiators) are examples of methods that may be used to obtain k_d.

The overall rate of polymerization can of course be found spectroscopically, by thermal analysis (measuring the amount of heat evolved), by simple gravimetry, or (given a substantial volume change upon polymerization), by dilatometry. The individual propagation and termination rate coefficients can be found by several different methods, usually dependent on measuring the rate of polymerization and perhaps the molecular weights obtained. The more common of these techniques are non–steady state and employ photoinitiators receiving intermittent illumination: rotating sector[2] and spatially intermittent polymerization.[102]

In recent years these techniques have generally employed pulsed laser sources. The use of laser sources has opened the way for a different technique, laser initiation/termination, which allows k_p to be measured from DP_n.[103-106] Steady-state measurements of the radical concentration $H(1)$ through electron spin resonance (ESR) spectroscopy are difficult given the low concentration of radicals (at least before the Trommsdorff effect sets in); if feasible, however, they do allow for measurement of k_p, since both M and $H(1)$ are known.[107-109] Non–steady state analysis of the decay of radical concentration allows for measurement of k_t.[110]

Transfer constants, for transfer agents that are ideal, must be found from their effects on molecular weight rather than rate of polymerization, of course. The transfer to monomer constant, C_m, is the limit of $1/DP_n$ as initiation rate vanishes. Constants for transfer to agents are found by what is known as a Mayo plot, in which $1/DP_n$ is plotted versus S/M (the ratio of the transfer to monomer concentrations); the slope of the line is identified as C_s.[2,111]

Extensive tabulations of rate constants exist,[112,113] which provide ample testimony to the scatter present in the data. A great deal of the disagreement among parameter values is probably attributable to error in the techniques themselves, either in the analytical method or in the assumptions underlying the analysis of results. Concerted effort toward measurements of rate constants based on sound techniques (preferably several techniques) is clearly needed.[72] It should be borne in mind that rate constants such as k_t depend on system variables other than temperature (e.g., solvent, molecular weight of the chains produced) and that measurements generally give some average value related to the distributed rate constant that is dependent on the lengths of the chains participating in the termination reaction.[97]

Problems

3.1. The rate of anionic polymerization is much greater than that of free radical (often by more than 10^4 at the same temperature). Yet the propagation rate constant k_p for anionic is 10–100 times smaller. Explain.

3.2. Consider the anionic polymerization (e.g., styrene initiated by butyllithium), which is ideal except for the presence of an impurity S (such as water). Rework the problem illustrated in Section 3.3.4.2 for the case where the concentration of this impurity is not constant.
(a) Solve for the live chain concentration (λ_0), the dead chain concentration (μ_0), the monomer concentration (M), and the impurity concentration (S) as a function of time (*not* τ).
(b) Give the general equations for the live chain distribution and the dead chain distribution as functions of τ and $K = k_{tr}S/k_pM$. How are the live chains distributed? What can be said about the dead chain distribution?

3.3. Sodium naphthalene initiates anionic polymerization by an electron transfer mechanism that yields a *di*carbanion.[2,19] For example, in the case of styrene:

$$2\,\text{Na} + 2\;\bigcirc\!\!\bigcirc\; + 2\;\underset{\bigcirc}{\overset{\text{HC}=\text{CH}_2}{\big|}}\;\longrightarrow\;\text{several steps}\longrightarrow$$

$$2\;\bigcirc\!\!\bigcirc\; + 2\,\text{Na+}\quad\left[\;-:\underset{\bigcirc}{\overset{\text{H}}{\underset{|}{\overset{|}{\text{C}}}}}\text{-CH}_2\text{-CH}_2\text{-}\underset{\bigcirc}{\overset{\text{H}}{\underset{|}{\overset{|}{\text{C}}}}}:-\;\right]\quad\text{Na+}$$

Propagation then occurs at each end as usual, with a rate constant k_p. Assume that the initiation steps occur instantaneously. For ideal anionic polymerization by butyllithium, we would obtain a Poisson distribution. What do we obtain here? What are DP_n and DP_w?

3.4. Invert the generating function $G(s)$ of equation (3.3.8) to obtain the Gold distribution of equation (3.3.9).

3.5. An empirical distribution often used in the description of polymers formed by free-radical mechanisms is the Schulz–Zimm distribution:

$$P(n) = (-\ln b)^k\, n^{k-1}\,\frac{b^n}{\Gamma(k)}$$

where $\Gamma(k)$ is the gamma function, and b and k are parameters. Treating the distribution as *continuous* (i.e., not discrete) on the interval $[0, \infty]$, work (a)–(f).
(a) Derive the general equation for the ith moment μ_i.
(b) Is this distribution normalized?
(c) What are DP_n, DP_w, and the polydispersity Q?
(d) What value of k gives a polydispersity like that of a most probable distribution as $b \to 1$?
(e) Does this value of k also give a ratio of DP_z to DP_w corresponding to a most probable distribution as $p \to 1$?
(f) In a batch free-radical polymerization, the distribution broadens as time goes on, as a result of monomer depletion, initiator depletion, transfer agent depletion, diffusion limitations, and so forth. How would you account for this effect with the Schulz–Zimm distribution?

3.6. Mercaptans (thiols) are good transfer agents for free-radical polymerizations because the thiol proton is readily abstracted, hence the transfer reaction occurs with a large rate constant. Chain transfer constants C_s may often be near unity. Consider a system for which $C_s = 1$ throughout the reaction and transfer is the dominant chain-ending reaction. What is the instantaneous dead chain distribution, and what is its polydispersity? Is drift dispersion significant? Why or why not?

3.7. Consider the solution polymerization of ethyl acrylate in ethyl acetate at 60°C. The data of Rätzch and Zschach[114] suggest that at this temperature, k_p = 955 L/(mol · s) and k_t = 2.04 × 10^6 L(mol · s). The reaction is initiated with AIBN, for which k_d is 9.70 × 10^{-6} s^{-1} at 60°C and f = 0.7. A mercaptan, 1-butanethiol, may be added as a transfer agent; assume that C_s = 1.

The monomer–solvent mixture is 40% by weight monomer, and 0.006 g of initiator is added per gram of solution. The thiol, if added, is introduced at an additional 0.005 g/g solution. Take the density of ethyl acrylate to be 0.924 g/mL and that of ethyl acetate to be 0.902 g/mL, and assume no volume of mixing. Also assume the following: (1) quasi–steady state approximation, (2) long-chain hypothesis, (3) negligible volume change upon reaction, (4) rate constants truly constant throughout the reaction (e.g., no Trommsdorff effect occurs), and (5) termination by combination. Do **not** neglect monomer, initiator, and transfer agent depletion.

(a) Plot conversion versus time up to 90% conversion. Does conversion depend on whether the transfer agent was added? How significant was initiator depletion? (If initiator depletion is less than ~5% at 90% conversion, neglect initiator depletion in the rest of this problem.)

(b) In the absence of the transfer agent, calculate the number- and weight-average molecular weight (both instantaneous and cumulative) and the cumulative polydispersity and plot as a function of both of conversion and time.

(c) Repeat (b) with the transfer agent present.

(d) What is the main source of drift dispersion in the two cases? Why might one use transfer agents? In the second case, how might this dispersion be counteracted?

(e) For which polymerization (with or without transfer agent) is the neglect of the Trommsdorff effect more valid?

(f) Data in the literature suggest that k_p of methyl methacrylate at 60°C is around 660 L/mol · s). Why is this value lower than that for ethyl acrylate? (Suggest a possible reason.)

(g) The glass transition temperature of poly(methyl methacrylate) is typically around 100–110°C, while that for poly(ethyl acrylate) is −22°C. What might be the source of this difference? In what sort of application might we prefer to use poly(ethyl acrylate)?

3.8. One method of controlling temperature during a free-radical polymerization is to let the solvent evaporate (perhaps under a slight vacuum), recondense in a condenser, and return to the reactor. Consider the polymerization of Problem 3.7 and assume that the removal of latent heat is the main source of heat removal. Consider $C_p \Delta T$ changes to be minor compared to the latent heat. The heat of vaporization of ethyl acetate is 7.72 kcal/mol. The heat of polymerization of ethyl acrylate is on the order of −65 kJ/mol.

(a) Calculate the load on the condenser for Problem 3.7 as a function of conversion and time.

(b) For a 4000-gallon reactor, three-quarters full, calculate the throughput of ethyl acetate through the condenser as a function of time and conversion. Comment on how realistic these numbers are.

References

1. P. C. Hiemenz, *Polymer Chemistry*. Dekker, New York (1984).

2. G. Odian, *Principles of Polymerization*, 3rd ed. Wiley, New York (1991).

3. (a) M. Szwarc, *Carbanions, Living Polymers, and Electron-Transfer Processes*. Wiley-Interscience, New York (1968). (b) M. Szwarc and M. Van Beylen, *Ionic Polymerization and Living Polymers*. Chapman and Hall, New York (1993).

4. H. Dostal and H. Mark, *Trans. Faraday Soc.*, **32**, 54 (1936).

5. S. L. Liu and N. R. Amundson, *J. Chem. Rubber Technol.*, **34**, 995 (1961).

6. C. H. Bamford and H. Tompa, *Trans. Faraday Soc.*, **50**, 1097 (1954).

7. C. H. Bamford, W. G. Barb, A. D. Jenkins, and P. F. Onyon, *The Kinetics of Vinyl Polymerization by Radical Mechanisms*. Academic Press, New York (1958).

8. C. H. Bamford and A. D. Jenkins, *Trans. Faraday Soc.*, **56**, 907 (1960).

9. R. J. Zeman and N. R. Amundson, *Chem. Eng. Sci.*, **20**, 331, 637 (1965).

10. D. J. Coyle, T. J. Tulig, and M. Tirrell, *Ind. Eng. Chem., Fundam.*, **24**, 343 (1985).

11. T. W. Taylor, V. Gonzalez, and K. F. Jensen, Modelling and control of the molecular weight distribution in methyl methacrylate polymerization. Paper presented at the Polymer Reaction Engineering Workshop, West Berlin, October 1986.

12. N. R. Amundson and D. Luss, *J. Macromol. Sci., Rev. Macromol. Chem.*, **C2**, 145 (1968).

13. M. Tirrell, R. Galván, and R. L. Laurence, Polymerization reactors, in *Chemical Reaction and Reactor Engineering*, J. J. Carberry and A. Varma, Eds. Dekker, New York (1987).

14. W. H. Ray, *J. Macromol. Sci., Rev.*, **C8**, 1 (1972).

15. F. C. Goodrich, *J. Chem. Phys.*, **35**, 2101 (1961).

16. M. S. Falkovitz and L. A. Segel, *SIAM J. Appl. Math.*, **42**, 542 (1982).

17. (a) W. Brown and M. Szwarc, *Trans. Faraday Soc.*, **54**, 416 (1958). (b) M. Szwarc and M. Litt, *J. Phys. Chem.*, **62**, 568 (1958).

18. S. M. Ross, *Introduction to Probability Models*, 3rd ed., Academic Press, Orlando, FL (1985).

19. M. Szwarc, M. Levy, and R. Milkovich, *J. Am. Chem. Soc.*, **78**, 2656 (1956).

20. K. Ziegler, *Angew. Chem.*, **49**, 499 (1936).

21. P. J. Flory, *J. Am. Chem. Soc.*, **62**, 1561 (1940).

22. H. Dostal and H. Mark, *Z. Phys. Chem., B*, **29**, 299 (1935).

23. P. Rempp, E. Franta, and J.-E. Herz, *Adv. Polym. Sci.*, **86**, 145 (1988).

24. M. Sawamoto, *Prog. Polym. Sci.*, **16**, 111 (1991).

25. S. Katz and G. M. Saidel, *AIChE J.*, **13**, 319 (1967).

26. R. W. Lenz, *Organic Chemistry of Synthetic High Polymers.* Wiley-Interscience, New York (1967).

27. P. H. Plesch, Ed., *The Chemistry of Cationic Polymerization;* Macmillan, New York (1963).

28. M. Morton, *Anionic Polymerization: Principles and Practice.* Academic Press, New York (1983).

29. R. N. Young, R. P. Quirk, and L. J. Fetters, *Adv. Polym. Sci.*, **56**, 1 (1984).

30. M. Van Beylen, S. Bywater, G. Smets, M. Szwarc, and D. J. Worsfold, *Adv. Polym. Sci.*, **86**, 87 (1988).

31. J. Masamoto, *Prog. Polym. Sci.*, **18**, 1 (1993).

32. M. Szwarc, *Makromol. Chem., Macromol. Symp.*, **67**, 83 (1993).

33. M. Szwarc, M. Van Beylen, and D. Van Hoyweghen, *Macromolecules*, **20**, 445 (1987).

34. O. W. Webster, W. R. Hertler, D. Y. Sogah, W. B. Farnham, and T. V. RajanBabu, *J. Am. Chem. Soc.*, **105**, 5706 (1983).

35. W. J. Brittain, *Rubber Chem. Technol.*, **65**, 580 (1992).

36. K. Matyjaszewski, *Macromolecules*, **26**, 1787 (1993).

37. H. Sawada, *Thermodynamics of Polymerization*, Dekker, New York (1976).

38. R. L. Meier, *J. Chem. Soc. London*, 3656 (1950).

39. D. C. Pepper, *Q. Rev.*, **8**, 88 (1954).

40. S. Penczek and A. Duda, *Makromol. Chem., Macromol. Symp.*, **67**, 15 (1993).

41. L. H. Peebles, *Molecular Weight Distributions in Polymers.* Wiley, New York (1971).

42. (a) C.-M. Yuan and D.-Y. Yan, *Makromol. Chem.*, **187**, 2629, 2641 (1986). (b) D.-Y. Yan and C.-M. Yuan, *Makromol. Chem.*, **188**, 333 (1987). (c) C.-M. Yuan and D.-Y. Yan, *Makromol. Chem.*, **188**, 341 (1987). (d) G.-F. Cai, D.-Y. Yan, and M. Litt, *Macromolecules*, **21**, 578 (1988). (e) C.-M. Yuan and D.-Y. Yan, *Polymer*, **29**, 924 (1988).

43. V. S. Nanda and R. K. Jain, *J. Polym. Sci. A*, **2**, 4583 (1964).

44. L. Gold, *J. Chem. Phys.*, **28**, 91 (1958).

45. J. C. W. Chien, *J. Polym. Sci. A*, **1**, 425 (1963).

46. B. D. Coleman, F. Gornick, and G. Weiss, *J. Chem. Phys.*, **39**, 3233 (1963).

47. R. Chiang and J. J. Hermans, *J. Polym. Sci. A-1*, **4**, 2843 (1966).

48. S. Mochizuki, *Chem. Eng. Sci.*, **22**, 77 (1967).

49. S. Penczek, P. Kubisa, and R. Szymanski, *Makromol. Chem., Rapid Commun.*, **12**, 77 (1991).

50. G. C. Eastmond, The kinetics of free-radical polymerization of vinyl monomers in homogeneous solution, in *Comprehensive Chemical Kinetics*, Vol. 14A: *Free Radical Polymerisation*, C. H. Bamford and C. F. H. Tipper, Eds., p. 1. Elsevier, Amsterdam (1976).

51. G. Moad, E. Rizzardo, D. H. Solomon, S. R. Johns, and R. I. Willing, *Makromol. Chem., Rapid Commun.*, **5**, 793 (1984).

52. A. M. North, *The Kinetics of Free Radical Polymerization*. Pergamon Press, Oxford (1966).

53. S. P. Pappas, Photoinitiation of radical polymerization—1, in *UV Curing: Science and Technology*, S. P. Pappas, Ed. Technology Marketing Corporation, Norwalk, CT (1980).

54. H. F. Gruber, *Prog. Polym. Sci.*, **17**, 953 (1992).

55. F. A. Bovey and I. M. Kolthoff, *Chem. Rev.*, **42**, 491 (1948).

56. F. Tüdos and T. Földes-Berezsnich, *Prog. Polym. Sci.*, **14**, 717 (1989).

57. G. C. Eastmond, Chain transfer, inhibition, and retardation, in *Comprehensive Chemical Kinetics*, Vol. 14A: *Free Radical Polymerisation*, C. H. Bamford and C. F. H. Tipper, Eds. p 105. Elsevier, Amsterdam (1976).

58. H. Tompa, The calculation of mole-weight distributions from kinetic schemes, in *Comprehensive Chemical Kinetics*, Vol. 14A: *Free Radical Polymerisation*, C. H. Bamford and C. F. H. Tipper, Eds. p. 527. Elsevier, Amsterdam (1976).

59. J.-Y. Chien, *J. Am. Chem. Soc.*, **70**, 2256 (1948).

60. J. C. W. Chien, *J. Am. Chem. Soc.*, **81**, 86 (1959).

61. C. H. Bamford, *Polymer*, **6**, 63 (1965).

62. S. Katz and G. M. Saidel, *AIChE J.*, **13**, 319 (1967).

63. H. T. Davis, *Introduction to Nonlinear Differential and Integral Equations*. Dover, New York (1962).

64. A. E. Hamielec, J. W. Hodgins, and K. Tebbens, *AIChE J.*, **13**, 1087 (1967).

65. M. K. Georges, R. P. N. Veregin, P. M. Kazmaier, and G. K. Hamer, *Macromolecules*, **26**, 2987 (1993).

66. N. A. Dotson, R. Galván, and C. W. Macosko, *Macromolecules*, **21**, 2560 (1988).

67. R. J. J. Williams, *Macromolecules*, **21**, 2568 (1988).

68. A. V. Tobolsky, *J. Am. Chem. Soc.*, **80**, 5927 (1958).

69. A. V. Tobolsky, C. E. Rogers, and R. D. Brickman, *J. Am. Chem. Soc.*, **82**, 1277 (1960).

70. J. P. Van Hook and A. V. Tobolsky, *J. Am. Chem. Soc.*, **80**, 779 (1958).

71. S. W. Lansdowne, R. G. Gilbert, D. H. Napper, and D. F. Sangster, *J. Chem. Soc., Faraday Trans. I*, **76**, 1344 (1980).

72. M. Buback, L. H. Garcia-Rubio, R. G. Gilbert, D. H. Napper, J. Guillot, A. E. Hamielec, D. Hill, K. F. O'Driscoll, O. F. Olaj, J. Shen, D. Solomon, G. Moad, M. Stickler, M. Tirrell, and M. A. Winnik, *J. Polym. Sci., Polym. Lett.*, **26**, 293 (1988).

73. W. Feller, *An Introduction to Probability Theory and Its Applications*, 3rd ed. Wiley, New York (1968).

74. A. B. Scranton, J. Klier, and N. A. Peppas, *Macromolecules*, **24**, 1412 (1991).

75. I. Mita and K. Horie, *J. Macromol. Sci., Rev. Macromol. Chem Phys.*, **C27**, 91 (1987).

76. K. Yokota and M. Itoh, *J. Polym. Sci., Polym. Lett.*, **6**, 825 (1968).

77. R. G. W. Norrish and R. R. Smith, *Nature*, **150**, 336 (1942).

78. G. V. Schulz and G. Harborth, *Makromol. Chem.*, **1**, 106 (1947).

79. E. Trommsdorff, H. Köhle, and P. Lagally, *Makromol. Chem.*, **1**, 169 (1948).

80. P. Hayden and H. Melville, *J. Polym. Sci.*, **43**, 201 (1960).

81. S. T. Balke and A. E. Hamielec, *J. Appl. Polym. Sci.*, **17**, 905 (1973).

82. J. N. Cardenas and K. F. O'Driscoll, *J. Polym. Sci., Polym. Chem.*, **14**, 883 (1976).

83. J. N. Cardenas and K. F. O'Driscoll, *J. Polym. Sci., Polym. Chem.*, **15**, 1883, 2097 (1977).

84. F. L. Marten and A. E. Hamielec, *ACS Symp. Ser.*, **104**, 43 (1979).

85. T. J. Tulig and M. Tirrell, *Macromolecules*, **14**, 1501 (1981).

86. M. Tirrell and T. J. Tulig, in *Polymer Reaction Engineering*, K. H. Reichert, Ed. p. 247. Hanser, Munich (1983).

87. R. M. Noyes, *Prog. React. Kinet.*, **1**, 129 (1961).

88. A. M. North, *Q. Rev.*, **20**, 421 (1966).

89. M. S. Kent, A. Faldi, M. Tirrell, and T. P. Lodge, *Macromolecules*, **25**, 4501 (1992).

90. T. J. Tulig and M. Tirrell, *Macromolecules*, **15**, 459 (1982).

91. G. T. Russell, R. G. Gilbert, and D. H. Napper, *Macromolecules*, **25**, 2459 (1992).

92. G. T. Russell, R. G. Gilbert, and D. H. Napper, *Macromolecules*, **26**, 3538 (1993).

93. A. Faldi, M. Tirrell, and T. P. Lodge, *Macromolecules*, **27**, 4176 (1994).

94. (a) B. O'Shaughnessy, *Phys. Rev. Lett.*, **71**, 3331 (1993). (b) B. O'Shaughnessy, *Macromolecules*, **27**, 3875 (1994). (c) B. O'Shaughnessy and J. Yu, *Macromolecules*, **27**, 5067, 5079 (1994).

95. G. V. Schulz, *Z. Phys. Chem. (Munich)*, **8**, 290 (1956).

96. G. T. Russell, D. H. Napper, and R. G. Gilbert, *Macromolecules*, **21**, 2133 (1988).

97. G. T. Russell, *Macromol. Theory Simulations*, **3**, 439 (1994).

98. N. Friis and A. E. Hamielec, *ACS Symp. Ser.*, **24**, 82 (1976).

99. G. T. Russell, D. H. Napper, and R. G. Gilbert, *Macromolecules*, **21**, 2141 (1988).

100. A. W. Hui and A. E. Hamielec, *J. Appl. Polym. Sci.*, **16**, 749 (1972).

101. J. A. Biesenberger and D. H. Sebastian, *Principles of Polymerization Engineering*. Wiley, New York (1983).

102. K. F. O'Driscoll and H. K. Mahabadi, *J. Polym. Sci., Polym. Chem.*, **14**, 869 (1976).

103. O. F. Olaj, I. Bitai, and G. Gleixner, *Makromol. Chem.,* **186**, 2569 (1985).

104. M. Buback, H. Hippler, J. Schweer, and H.-P. Vögele, *Makromol. Chem., Rapid Commun.,* **7**, 261 (1986).

105. O. F. Olaj, I. Bitai, and F. Hinklemann, *Makromol. Chem.,* **188**, 1689 (1987).

106. R. A. Hutchinson, M. T. Aronson, and J. R. Richards, *Macromolecules,* **26**, 6410 (1993).

107. M. J. Ballard, R. G. Gilbert, D. H. Napper, P. J. Pomery, P. W. O'Sullivan, and J. H. O'Donnell, *Macromolecules,* **19**, 1303 (1986).

108. W. Lau, D. G. Westmoreland, and R. W. Novak, *Macromolecules,* **20**, 457 (1987).

109. R. W. Garrett, D. J. T. Hill, J. H. O'Donnell, P. J. Pomery, and C. L. Winzor, *Polym. Bull.,* **22**, 611 (1989).

110. S. Zhu, Y. Tian, A. E. Hamielec, and D. R. Eaton, *Macromolecules,* **23**, 1144 (1990).

111. R. A. Gregg and F. R. Mayo, *J. Am. Chem. Soc.,* **70**, 2373 (1948).

112. G. C. Eastmond, Kinetic data for homogeneous free-radical polymerizations of various monomers, in *Comprehensive Chemical Kinetics,* Vol. 14A: *Free Radical Polymerisation,* C. H. Bamford and C. F. H. Tipper, Eds. p. 153. Elsevier, Amsterdam (1976).

113. J. Brandrup and E. H. Immergut, Eds., *Polymer Handbook,* 3rd ed. Wiley, New York (1989).

114. M. Rätzsch and J. Zschach, *Plaste Kautsch.,* **21**, 345 (1974).

4

COPOLYMER-IZATION

4.1 Introduction

The polymerization of different monomers together to form a copolymer offers one of the simplest methods of improving (or at least changing) polymer properties, and one that has almost limitless variants. Many of the monomers discussed in the past two chapters are actually more useful when copolymerized. For example, the homopolymer of styrene is poor in its resistance to organic solvents and ability to undergo environmental degradation; these disadvantages, coupled with brittleness and low upper use temperatures, effectively disqualify it for a number of applications.[1] All these shortcomings can be alleviated by copolymerization of styrene with other monomers.

Copolymerization is often an attempt to obtain properties intermediate between those of the homopolymers. One common route, used with increasing frequency in the past 10–20 years, is to *blend* homopolymers. Since most polymers are immiscible, phase separation occurs, and the reaction engineering problem then consists of obtaining the desired size of the dispersed phase and providing stabilization of, and adhesion between, the interfaces. Polymer blends combine different monomers on the level of *microstructure;* the kinds of copolymer we discuss in this chapter combine monomers on the scale of the *backbone.* Simultaneous polymerization or some sort of sequential or multistage polymerization may be used to obtain this sort of copolymer (or, in general, multicomponent polymer). Simultaneous polymerization of two or more monomers is a powerful and commercially important process for creating polymers with "tailor-made" properties. It also creates challenges for the reaction engineer not present in homopolymerization reactions.

Copolymerizations may be either stepwise or chainwise. The reaction of 1,4-diphenylmethane diisocyanate (MDI) with a mixture of 1,4-butanediol and a hydroxy-terminated, low molecular weight polycaprolactone diol to form segmented polyurethanes is an example of the former, while the free-radical copolymerization of styrene and acrylonitrile to produce SAN polymers is an example of the latter.

A more important categorization, however, is on the basis of the resulting architecture. We know this from Section 1.3.2, which described the differences between "random" and block

163

copolymers. A *block copolymer* is one in which long sequences (or blocks) of one monomer are followed by long sequences of the other monomer:

$$\sim\sim\sim AAAAAAABBBBBBB\sim\sim\sim \qquad (4.1.1)$$

(Here we show a diblock copolymer; triblocks and higher multiblock polymers are of course possible.) Such strict block copolymers are obviously best made by sequential polymerization (e.g., by a living polymerization) or in a multistage process serving to link two telechelic polymers. *Graft copolymers* are closely related; here homopolymers of one monomer are grafted into a backbone formed by homopolymer of the other monomer:

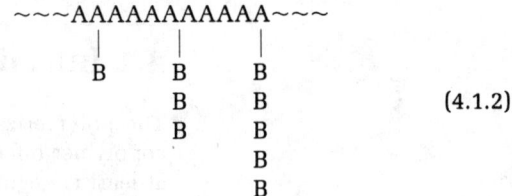

$$\qquad (4.1.2)$$

Graft copolymers are also best made by a sequence of reactions. Both of these still separate the two monomers to a large degree; there are only a few bonds between dissimilar monomers. This may result in microphase-separated structures, or such polymers may be used to stabilize the interfaces of appropriate polymer blends.

The polymers of interest in this chapter, on the other hand, have many bonds between the dissimilar monomers; an extreme example of this would be *alternating copolymers:*

$$\sim\sim\sim ABABABA\sim\sim\sim \qquad (4.1.3)$$

which may include stepwise polymers of the types $A_2 + B_2$ or AB. The usual case, however, is the statistical, or so-called "random," copolymers in which the monomers appear in irregular order:

$$\sim\sim\sim AABABBBAABBBAB\sim\sim\sim \qquad (4.1.4)$$

Alternating or blocky copolymers appear as particular limits of statistical copolymerization.

Copolymerization introduces new variables, which influence the properties of copolymers to a large extent. Chain length distribution and average molecular weights still play their important roles, and our repertoire of techniques still provides solutions to those problems (with increasingly involved mathematics). But this chapter focuses on the characteristics peculiar to copolymerizations. There are two of these: the copolymer composition (an obvious concept) and the length of sequences of each type of monomer along the chain. Both are distributed quantities; that is, there are chains of different composition, yielding a *copolymer composition distribution,* and there are sequences of different lengths, giving a *sequence length distribution.*

Both composition and sequence lengths (or *chain micro-structure*) are important determinants of product quality to be considered when designing or assessing copolymerization reactions. Copolymers of differing composition may be immiscible; differences in sequence length may affect this immiscibility,[2] or sufficiently long sequences of a monomer may microphase-separate. Both problems affect styrene–acrylonitrile; thermodynamic immiscibility between polymers of different composition diminishes mechanical properties, while long acrylonitrile sequences can react oxidatively to impart a yellowish color to the product.

While composition and sequence distribution are not clearly separable either in their origins or in their effect on final properties, it is best to introduce them separately. The origin of differences in copolymer composition is most easily examined in the context of a free-radical copolymerization (perhaps the most important case). The less obvious concept of sequence length is introduced in Section 4.3 in the context of the stepwise copolymerization $A_2 + B_2 + C_2$, then applied to free-radical copolymerization in Section 4.4.

4.2 Copolymer Composition: Free-Radical Copolymerization

Free-radical copolymerizations, like homopolymerizations, proceed by the three basic steps of initiation, propagation, and termination, and the possibilities of chain transfer and inhibition exist as well. All these steps are important for the determination of molecular weight and its distribution, as illustrated in Chapter 3, but only the most common reaction, propagation, is important in the determination of the chain composition if the chains are long. This can be verified experimentally for a given copolymerization system by changing initiators, varying initiator concentration, and adding transfer agents or inhibitors. To understand how composition is determined in a free-radical copolymerization, then, we need first to understand only the kinetics of propagation, and the different models that have been proposed for propagation.[1,3]

4.2.1 Kinetics of the Propagation Reaction

4.2.1.1 The Terminal Model

Any model proposed for the propagation reaction must fulfill at least one qualification: it must be able to predict a copolymer differing in composition from its parent comonomer solution. Monomers can be consumed at different rates, and a model must reflect that, if nothing else. It might seem sufficient for the reaction rate to depend only on the monomer being added, but in most cases the rate is found to depend on the attributes of the radical as well. The *terminal model,* in which the rate of reaction with a given type of monomer depends only on the type of monomer bearing the radical,[4-10] is the most widely used and generally succeeds where a zeroth-order model (i.e., one not depending on the identity of the radical[11]) fails.

The terminal model implies four distinct propagation reactions:

$$\sim\sim\sim M_1^{\cdot} + M_1 \xrightarrow{k_{11}} \sim\sim\sim M_1^{\cdot} \qquad (4.2.1)$$

$$\sim\sim\sim M_1^{\cdot} + M_2 \xrightarrow{k_{12}} \sim\sim\sim M_2^{\cdot} \qquad (4.2.2)$$

$$\sim\sim\sim M_2^{\cdot} + M_1 \xrightarrow{k_{21}} \sim\sim\sim M_1^{\cdot} \qquad (4.2.3)$$

$$\sim\sim\sim M_2^{\cdot} + M_2 \xrightarrow{k_{22}} \sim\sim\sim M_2^{\cdot} \qquad (4.2.4)$$

where M_1 and M_2 denote the two monomers, $M_1^{\cdot}$ and $M_2^{\cdot}$ the two radical-bearing monomers, and k_{ij} the propagation rate constant between $M_i^{\cdot}$ and M_j. (For the multicomponent polymerization of N monomers, N^2 reactions and rate constants are required.) We will refer to the concentrations of the two radicals as $R_{(1)}$ and $R_{(2)}$, respectively; these are more often in the literature referred to as P and Q, but we avoid this to eliminate notation problems and to allow for easier generalization to multicomponent polymerization. These growing copolymer chains must be double subscripted to denote the numbers of each monomer in the chain (and even this still does not uniquely define a chain, since no information about the sequences of the incorporated monomers is given.) $R_{(1)n,m}$ represents the concentration of growing chains with $n\,M_1$ units and $m\,M_2$ units, ending in an $M_1^{\cdot}$ radical. Likewise, $R_{(2)n,m}$ represents the concentration of growing chains with $n\,M_1$ units and $m\,M_2$ units, ending in an $M_1^{\cdot}$ radical.

Under the long-chain hypothesis, the rates of monomer consumption are given by the following equations:

$$-\frac{dM_1}{dt} = k_{11}M_1R_{(1)} + k_{21}M_1R_{(2)} \qquad (4.2.5)$$

$$-\frac{dM_2}{dt} = k_{12}M_2R_{(1)} + k_{22}M_2R_{(2)} \qquad (4.2.6)$$

The instantaneous copolymer composition is obtained by dividing the one equation by the other:

$$\frac{dM_1}{dM_2} = \frac{(k_{11}R_{(1)} + k_{21}R_{(2)})M_1}{(k_{12}R_{(1)} + k_{22}R_{(2)})M_2} \qquad (4.2.7)$$

The quasi–steady state approximation for a copolymerization is:

$$k_{21}R_{(2)}M_1 = k_{12}R_{(1)}M_2 \qquad (4.2.8)$$

and thus chains ending in M_1 cannot grow at the expense of chains ending in M_2. For long chains and values of k_{21} and k_{12}

not terribly different, this assumption is a good approximation. This relation gives us the ratio $R_{(1)}/R_{(2)}$, and leads to:

$$\frac{dM_1}{dM_2} = \frac{(r_1M_1 + M_2)M_1}{(M_1 + r_2M_2)M_2} \tag{4.2.9}$$

where the *reactivity ratios* r_1 and r_2 are given as follows[7]:

$$r_1 = \frac{k_{11}}{k_{12}} \tag{4.2.10}$$

$$r_2 = \frac{k_{22}}{k_{21}} \tag{4.2.11}$$

Equation (4.2.9) describes the rate of incorporation of M_1 relative to that of M_2 and as such also gives the instantaneous copolymer composition. It is usually called the Mayo–Lewis[7] equation and generally written thus:

$$F_1 = \frac{r_1f_1^2 + f_1f_2}{r_1f_1^2 + 2f_1f_2 + r_2f_2^2} \tag{4.2.12}$$

where F_1 is the mole fraction of M_1 in the copolymer and f_i is the mole fraction of M_i in the monomer mixture ($f_1 + f_2 = 1$). The terminal model is the most commonly used model of copolymerization, and so the types of behavior predicted by the Mayo–Lewis equation are expanded on shortly.

4.2.1.2 The Penultimate Model
The terminal model is not always an appropriate model for copolymerization, despite its popularity. Its inadequacies may appear in one of two ways. In the more obvious of these, the compositional behavior is not adequately described by equation (4.2.12). For example, in the copolymerization between styrene and fumaronitrile, there is a decreased reactivity of styrene-ended chains toward fumaronitrile if the chains are rich in fumaronitrile.[12] The terminal model seems to fail for this system, but a higher order model, the *penultimate model*,[13] is more successful.[14] However, penultimate effects may be invisible from a compositional viewpoint (the composition may be adequately described by the Mayo–Lewis equation), but apparent in the inability of the terminal model to provide a consistent picture of the rate of polymerization. The best example of this is styrene–methyl methacrylate, long thought to be a system conforming well to the terminal model because of the excellent description that it gives the composition. However, this system actually exhibits strong penultimate effects,[15] as do p-chlorostyrene–methyl acrylate[16,17] and styrene–acrylonitrile.[18,19]

The penultimate model[13] is one in which the identity of the last two units of the chain determines the rate of propagation. In a copolymerization, there are four different radicals, and thus eight distinct reactions and rate constants (in general, the number

of reactions now goes as N^3). We can write these reactions as
follows:

$$\sim\sim\sim M_1 M_1^{\cdot} + M_1 \xrightarrow{k_{111}} \sim\sim\sim M_1 M_1^{\cdot} \qquad (4.2.13)$$

$$\sim\sim\sim M_1 M_1^{\cdot} + M_2 \xrightarrow{k_{112}} \sim\sim\sim M_1 M_2^{\cdot} \qquad (4.2.14)$$

$$\sim\sim\sim M_1 M_2^{\cdot} + M_1 \xrightarrow{k_{121}} \sim\sim\sim M_2 M_1^{\cdot} \qquad (4.2.15)$$

$$\sim\sim\sim M_1 M_2^{\cdot} + M_2 \xrightarrow{k_{122}} \sim\sim\sim M_2 M_2^{\cdot} \qquad (4.2.16)$$

$$\sim\sim\sim M_2 M_1^{\cdot} + M_1 \xrightarrow{k_{211}} \sim\sim\sim M_1 M_1^{\cdot} \qquad (4.2.17)$$

$$\sim\sim\sim M_2 M_1^{\cdot} + M_2 \xrightarrow{k_{212}} \sim\sim\sim M_1 M_2^{\cdot} \qquad (4.2.18)$$

$$\sim\sim\sim M_2 M_2^{\cdot} + M_1 \xrightarrow{k_{221}} \sim\sim\sim M_2 M_1^{\cdot} \qquad (4.2.19)$$

$$\sim\sim\sim M_2 M_2^{\cdot} + M_2 \xrightarrow{k_{222}} \sim\sim\sim M_2 M_2^{\cdot} \qquad (4.2.20)$$

Following the procedure used for the terminal model with
the QSSA for all four types of growing radical, one can find a
relationship between F_1 and f_1. This can be put in the form of
the Mayo–Lewis equation, and it yields the following for the
instantaneous copolymer composition:

$$F_1 = \frac{\bar{r}_1 f_1^2 + f_1 f_2}{\bar{r}_1 f_1^2 + 2 f_1 f_2 + \bar{r}_2 f_2^2} \qquad (4.2.21)$$

The pseudoreactivity ratios are functions of the comonomer composition f_1 and are given by:

$$\bar{r}_1 = r_{21} \frac{r_{11} f_1 + f_2}{r_{21} f_1 + f_2} \qquad (4.2.22)$$

$$\bar{r}_2 = r_{12} \frac{f_1 + r_{22} f_2}{f_1 + r_{12} f_2} \qquad (4.2.23)$$

where

$$r_{11} = \frac{k_{111}}{k_{112}} \qquad (4.2.24)$$

$$r_{12} = \frac{k_{122}}{k_{121}} \qquad (4.2.25)$$

$$r_{21} = \frac{k_{211}}{k_{212}} \qquad (4.2.26)$$

$$r_{22} = \frac{k_{222}}{k_{221}} \qquad (4.2.27)$$

4.2.1.3 Alternate Models of Copolymerization

Numerous other models for copolymerization have been suggested over the years. Accounting for effects of increasingly longer terminal sequences is an obvious route; thus while historically copolymerization was first suggested to be independent of the identity of the radical,[11] deviations in the data made it necessary to suggest in turn the terminal model, the penultimate model, and then the antepenultimate model.[20-23] In the last, one has N^4 propagation reactions and a corresponding number of rate constants, but generally distinguishing between such high-order models is difficult because of the kind and precision of data available.

For some systems, particularly those that exhibit alternating structure like allyl acetate–maleic anhydride,[24] olefin–SO_2,[14,25] or styrene–maleic anhydride,[26] the influence of complexes (monomer–monomer, monomer–radical, or monomer–solvent) has been implicated. The *complex participation model* describes copolymerization in which monomer complexes are added to the radical ends of polymer chains.[27-29] However, it has been suggested that in various systems such as styrene–maleic anhydride, the complex participates in the propagation step but dissociates upon incorporation, and only one of the two monomers in the complex reacts; this is the *complex dissociation model*.[26,30,31] Although it is clear that these complexes exist, there are serious doubts about whether they do in fact participate in propagation.[32-35] Such doubts are based on the difficulty in distinguishing [e.g., from data on the relative amounts of the different possible sequences of three monomers (triads)] between these high-order models[36-38] and on direct evidence.[39]

It may be that the microenvironment of the radical differs from that of the overall system because of preferential solvation.[40] The relative concentrations of the two monomers thus would not be the same as the relative bulk concentrations. This "bootstrap" effect (so called because the macroradical affects its own environment) accounts for the following observation: copolymerizations in solvents in which a wide range of reactivity ratios are obtained, the copolymers exhibit the same microstructure at the same copolymer composition.[40,41] Such solvation effects may occur for those polymerizations in which complexes are known to form.[42] Bootstrap effects can come into play regardless of the mode of copolymerization (e.g., terminal or penultimate), and these effects can be quantified.[43] It has been suggested that such effects operate in many systems,[41,43] including styrene–methyl methacrylate polymerizations in solution.[43-45]

Other possible complications include matrix copolymerization with existing polymer,[46,47] equilibrium copolymerization[48-51] or copolymerizations in which depolymerization occurs[52] (perhaps more common in ionic copolymerizations; see Section 3.3), and nonequivalent energy states of otherwise identical terminal groups—the "hot radical" theory.[53,54] The subject is hardly a closed one; at the very least, penultimate effects and preferential solvation occur. In this chapter we deal with the terminal model only because of its relative simplicity and historical prominence.

4.2.2 Copolymer Composition

The Mayo–Lewis equation for the terminal model gives the relationship between the composition of the copolymer and that of the comonomer at any instant. Since it allows these two to differ, it also, in a batch reactor, allows the comonomer composition to change with time, or "drift," from its initial value. We can thus categorize specific r_1 and r_2 values by the kind of drift they impose, or how they affect composition.[1] Consider the following classification of copolymerizations in terms of reactivity ratios, with reference to Figure 4.1.

- a: $r_1 = 1$, $r_2 = 1$. In this case $F_1 = f_1$, which defines the "azeotropic" line. No composition drift occurs, so the composition of the copolymer is constant throughout the reaction and equal to the monomer feed composition. The two monomers are truly randomly distributed on the chains.

- b: $r_1 r_2 = 1$. This case, referred to as "ideal copolymerization," allows for a difference between F_1 and f_1, although the monomers are still randomly distributed on chains. The *relative* reactivity of the monomers is independent of the identity of the radical, and the rate constant k_{ij} is given by a product $k_i k_j$. If $r_1 \neq 1$, $F_1 \neq f_1$, and the composition drifts. If $r_1 > 1$, as in curve b(1), then $F_1 > f_1$, and f_1 will decrease with conversion; if $r_1 < 1$, as in curve b(2), then $F_1 < f_1$, and f_1 will increase with conversion. (Whenever F_1 lies above the azeotropic line, the composition will move to the left, or "descend" the composition curve; whenever below, the composition will move to the right, or "climb" the composition curve.)

- c: $r_1 > 1$, $r_2 \leq 1$. As for an ideal copolymerization with $r_1 > 1$, the F_1–f_1 curve lies completely above the azeotropic line, but here the monomers are no longer truly randomly distributed. Drift will cause a decrease in f_1 as a function of conversion.

Figure 4.1.
Copolymer composition versus comonomer composition for terminal model. a, $r_1 = r_2 = 1$; b, (1) $r_1 = 8$, $r_2 = 0.125$; (2) $r_1 = 0.125$, $r_2 = 8$; c, $r_1 = 10$, $r_2 = 0.9$; d, $r_1 = 0.9$, $r_2 = 10$; e, $r_1 = 0.1$, $r_2 = 0.005$; f, $r_1 = 20$, $r_2 = 10$.

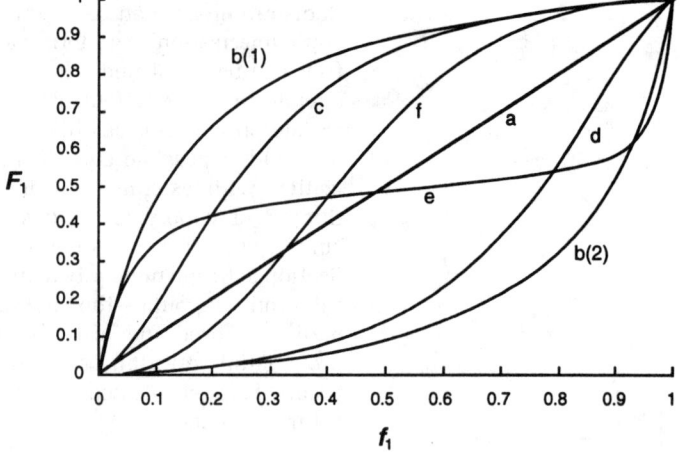

- d: $r_1 \leq 1$, $r_2 > 1$. Exactly the opposite of curve c. The F_1–f_1 curve lies below the azeotropic line and drift causes f_1 to increase with conversion.

- e: $r_1 < 1$, $r_2 < 1$. In this case, cross-propagation is preferred to homopropagation ($k_{12} > k_{11}$, $k_{21} > k_{22}$) by both kinds of radical. The polymer tends to have an alternating architecture and is strictly alternating in the limiting case, where $r_1 = 0$, $r_2 = 0$. The F_1–f_1 curve crosses the azeotropic line, and point of crossing is called, in analogy to a distillation, the azeotropic composition.[8] The azeotrope is unstable; copolymer composition will drift away from the azeotropic composition with conversion.

- f: $r_1 > 1$, $r_2 > 1$. This is the opposite of curve e and rarely is seen for free-radical copolymerization. Each growing radical prefers homopropagation to cross-propagation, resulting in a blocklike copolymer or a mixture of homopolymers, and is strictly homopolymerizing in the limiting case, where $r_1 = \infty$, $r_2 = \infty$. This case also has an azeotrope, but composition drifts toward the azeotrope with conversion.

Compositional drift can have a great effect on the properties of the final product, and thus exactly how composition drifts with conversion is of great interest. The differential equation governing the evolution of f_1 with conversion p, known as Skeist's equation,[55] can be easily derived in the following way:

$$
\frac{df_1}{dp} = \frac{d}{dp} \frac{M_1}{M_1 + M_2} = \frac{(M_1 + M_2)\dfrac{dM_1}{dp} - M_1\dfrac{d(M_1 + M_2)}{dp}}{(M_1 + M_2)^2}
$$

$$
= -\frac{1}{1 - p}\frac{dM_1}{d(M_1 + M_2)} + f_1 \frac{1}{1 - p} \tag{4.2.28}
$$

where the overall conversion, p, is given by:

$$
p = \frac{(M_1 + M_2)_0 - (M_1 + M_2)}{(M_1 + M_2)_0} \tag{4.2.29}
$$

The remaining derivative in equation (4.2.28) is simply the copolymer composition F_1. Thus, equation (4.2.28) can be rearranged to give:

$$
-\frac{dp}{1 - p} = \frac{df_1}{F_1 - f_1} \tag{4.2.30}
$$

$$
\ln(1 - p) = \int_{f_{1,0}}^{f_1} \frac{df_1}{F_1 - f_1} \tag{4.2.31}
$$

where the initial comonomer composition, $f_{1,0}$, is given as follows:

$$
f_{1,0} = \frac{(M_1)_0}{(M_1 + M_2)_0} \tag{4.2.32}
$$

Equation (4.2.31) allows the calculation of the monomer conversion as a function of final instantaneous composition, or vice versa, starting from a mixture with composition given by $f_{1,0}$. To compute the average composition up to that point, one simply uses the following:

$$\frac{F_1^{cumu}}{F_2^{cumu}} = \frac{(M_1)_0 - M_1}{(M_2)_0 - M_2} \tag{4.2.33}$$

where the superscript indicates cumulative values up to the present time or conversion. (The superscripted "inst" was omitted in the instantaneous values to correspond with common terminology.) This can be rewritten as follows:

$$\frac{F_1^{cumu}}{1 - F_1^{cumu}} = \frac{f_{1,0} - f_1(1 - p)}{f_1(1 - p) + p - f_{1,0}} \tag{4.2.34}$$

Thus far, no particular model of copolymerization has been assumed. For the terminal model, equation (4.2.31) can be integrated analytically if the reactivity ratios are constant[7,56]:

$$1 - p = \left(\frac{f_1}{f_{1,0}}\right)^{\alpha} \left(\frac{1 - f_1}{1 - f_{1,0}}\right)^{\beta} \left(\frac{f_{1,0} - \delta}{f_1 - \delta}\right)^{\gamma} \qquad r_1 \neq 1, r_2 \neq 1, r_1 + r_2 \neq 2 \tag{4.2.35}$$

where

$$\alpha = \frac{r_2}{1 - r_2} \tag{4.2.36}$$

$$\beta = \frac{r_1}{1 - r_1} \tag{4.2.37}$$

$$\gamma = \frac{1 - r_1 r_2}{(1 - r_1)(1 - r_2)} \tag{4.2.38}$$

$$\delta = \frac{1 - r_2}{2 - r_1 - r_2} \tag{4.2.39}$$

The results for the special cases follow.[56]
for $r_1 = 1, r_2 \neq 1$:

$$1 - p = \left(\frac{f_1}{f_{1,0}}\right)^{\alpha} \left(\frac{1 - f_1}{1 - f_{1,0}}\right)^{-\alpha-1} \exp\left(\frac{1}{r_2 - 1}\left(\frac{1}{1 - f_{1,0}} - \frac{1}{1 - f_1}\right)\right) \tag{4.2.40}$$

for $r_1 \neq 1, r_2 = 1$:

$$1 - p = \left(\frac{f_1}{f_{1,0}}\right)^{-\beta-1} \left(\frac{1 - f_1}{1 - f_{1,0}}\right)^{\beta} \exp\left(\frac{1}{r_1 - 1}\left(\frac{1}{f_{1,0}} - \frac{1}{f_1}\right)\right) \tag{4.2.41}$$

for $r_1 + r_2 = 2, r_1 \neq 1$:

$$1 - p = \left(\frac{f_1}{f_{1,0}}\right)^{\alpha} \left(\frac{1 - f_1}{1 - f_{1,0}}\right)^{\beta} \tag{4.2.42}$$

For $r_1 = r_2 = 1$, of course, no composition drift occurs. It should also be noted that for the case $f_{1,0} = \delta$ (which is the azeotrope composition), no drift occurs theoretically. For $r_1, r_2 < 1$, however, this azeotrope is unstable; only for $r_1, r_2 > 1$ is the azeotrope stable.

Equations (4.2.31) and (4.2.35) (and its special cases) relate the copolymer composition as a function of conversion. The relationship between composition and time, more important for the reaction engineer, is found in the following way.[57] The monomer mass balances, equations (4.2.5) and (4.2.6), can be rewritten as follows:

$$-\frac{d(\ln(M_1))}{dt} = H(1)\left\{(k_{11} - k_{21})\frac{R_{(1)}}{H(1)} + k_{21}\right\} \qquad (4.2.43)$$

$$-\frac{d(\ln(M_2))}{dt} = H(1)\left\{k_{22} + (k_{12} - k_{22})\frac{R_{(1)}}{H(1)}\right\} \qquad (4.2.44)$$

where $H(1)$ $(= R_{(1)} + R_{(2)})$ is the total growing radical chains concentration. Defining:

$$m = \frac{k_{11} - k_{21}}{k_{12} - k_{22}} \qquad (4.2.45)$$

equations (4.2.43) and (4.2.44) can be combined and integrated to yield (for nonideal copolymerization):

$$\ln\left(\frac{M_1}{(M_1)_0}\right) - m\ln\left(\frac{M_2}{(M_2)_0}\right) = -(k_{21} - mk_{22})\int_0^t H(1)dt \qquad r_1 \neq 1/r_2 \qquad (4.2.46)$$

[Recall that $H(1)$ is a function of time, although not explicitly notated as such.] Substituting equation (4.2.36) into equation (4.2.46) gives the following final result:

$$\ln\left[\left(\frac{f_1}{f_{1,0}}\right)^a\left(\frac{1 - f_1}{1 - f_{1,0}}\right)^b\left(\frac{f_{1,0} - \delta}{f_1 - \delta}\right)^c\right] = (mk_{22} - k_{21})\int_0^t H(1)dt \qquad (4.2.47)$$

under the restriction $r_1 \neq 1$, $r_2 \neq 1$, $r_1 + r_2 \neq 2$, and $r_1 \neq 1/r_2$. [Similar equations can be derived for the special cases of equations (4.2.40)–4.2.42)]. The parameters a, b, and c are defined thus:

$$a = \alpha(1 - m) + 1 \qquad (4.2.48)$$

$$b = \beta(1 - m) - m \qquad (4.2.49)$$

$$c = \gamma(1 - m) \qquad (4.2.50)$$

The utility of equation (4.2.47) depends on the possibility of evaluation of the integral. The situation is complex for free-radical copolymerization, because all the complications discussed in Chapter 3 which affect the concentration of radicals apply to copolymerizations as well. The quasi–steady state approximation, if applicable, gives the overall radical concentration $H(1)$ just as for a homopolymerization. All the foregoing can be applied to ionic copolymerization as well, if the quasi–steady state approximation holds for the two different types of ionic end group. In this case, equation (4.2.47) is quite simple because the integral merely reduces to $(I_0 t)$.

4.3 Sequence Length: $A_2 + B_2 + C_2$ Stepwise Polymerization

The second characteristic unique to copolymers is sequence length and its distribution. Copolymer composition does not uniquely determine the structure; copolymers with the same composition do not necessarily have the same sequence length distribution or average sequence length. To see this, consider the following chains:

(a) AAA BB ABBABAAAB BBBABABBAA
(b) ABABA BABAB ABA BABABABABAB
(c) ABAAAAB BAABBBBA ABBABBAAB

All three copolymers consist of 12 A units and 12 B units, thus resulting in a 50 mol % copolymer composition. The number-average sequence length $(N_x)_n$ is calculated by dividing the total number of units of monomer "X" by the number of sequences of that monomer, thus

(a) $(N_A)_n = (N_B)_n = 12/7 = 1.71$
(b) $(N_A)_n = (N_B)_n = 12/12 = 1.0$
(c) $(N_A)_n = (N_B)_n = 12/7 = 1.71$

Even though copolymers a and c have the same composition and average sequence length, they differ with regard to the sequence length distributions (and chain length distributions).

One can write kinetic expressions for sequence length distributions as for chain length distributions, but the easier and more direct route is by probabilistic arguments. In this section we use the combinatorial formalism for a stepwise copolymerization of A_2 with B_2 and C_2, for which the A functionalities react only with the B and C, and the B and C functionalities do not react among themselves.[58] The system in mind is that of a segmented polyurethane, the diisocyanate (A_2) reacting with a short or "hard" diol (B_2) and a long or "soft" diol (C_2). A good example is the common system wherein the diisocyanate if 4,4'-diphenylmethane diisocyanate:

$$O=C=N-\underset{}{\bigcirc}-\overset{\text{H}}{\underset{\text{H}}{\text{C}}}-\bigcirc-N=C=O \qquad (4.3.1)$$

the hard segment diol is 1,4-butanediol:

$$HO-(CH_2)_4-OH \qquad (4.3.2)$$

and the soft segment diol is poly(propylene oxide) diol:

$$HO\text{—}(CH_2\text{—}CH\text{-}O)_n\text{—}OH \atop \qquad\qquad | \atop \qquad\qquad CH_3 \qquad\qquad (4.3.3)$$

the molecular weight of which is generally on the order of 1000. Sequences of these two generally phase-separate on a microscopic scale (macroscopic phase separation being prevented by the covalent bonds of the polymer). One can easily imagine that this microphase separation is an important determinant of the properties of the polymer, so that the sequence length and its distribution are significant.

The polymer is of course strictly alternating with respect to diisocyanate and diol monomers; the sequences come about by the number of diols of a certain type in sequence. For example, we may look at the following:

A-AB-BA-AC-CA-AC-CA-AC-CA-AB-BA-AB-BA-AB-BA-AC-CA-AB-B
$$\underbrace{\qquad\qquad}_{soft\ segment} \quad \underbrace{\qquad\qquad}_{hard\ segment} \qquad (4.3.4)$$

The soft segment involves the more flexible diol, the hard segment the stiffer butanediol. We define p as the conversion of isocyanate (A) groups, and, following the convention of Chapter 2, the A groups or isocyanate groups are the limiting reagent; thus p ranges from 0 to 1. In this case two stoichiometries are needed: the overall stoichiometry r defined similarly to that in Chapter 2:

$$r = \frac{A_0}{B_0 + C_0} \leq 1 \qquad (4.3.5)$$

and a stoichiometry describing the fraction of hard diols:

$$r_B = \frac{B_0}{B_0 + C_0} = 1 - r_C \qquad (4.3.6)$$

In this case, contrary to the cases in the body of Chapter 2, one cannot simply switch notation between A and B depending on which reagent is limiting, inasmuch as one reagent is present in two forms (B and C). Unbalanced stoichiometry here, however, means that $r < 1$.

The fraction of A groups that have reacted with B groups is defined as p_{AB}; p_{AC} is similarly defined. We also define q_1 and q_2 as the conversion of B and C functionalities, respectively. These conversions are related by the appropriate stoichiometries as follows:

$$p_{AB} = \frac{r_B}{r} q_1 \qquad (4.3.7)$$

$$p_{AC} = \frac{r_C}{r} q_2 \qquad (4.3.8)$$

where $p_{AB} + p_{AC} = p$.

We can now find the distribution of hard (B) and soft (C) segments as follows. If we grab the end of a B sequence, the probability of reaching another B without an intervening C is the probability of the B reacting with an A (q_1) multiplied by the probability of the A reacting with a B (p_{AB} or $r_B q_1/r$), or ($r_B q_1^2/r$). Thus the probability of a sequence of one is $1 - r_B q_1^2/r$ and each additional unit gives a multiplicative term of $r_B q_1^2/r$. The distribution is thus geometric, so that the sequence distribution $S_{B,k}$ is:

$$S_{B,k} = \left(1 - \frac{r_B q_1^2}{r}\right)\left(\frac{r_B q_1^2}{r}\right)^{k-1} \tag{4.3.9}$$

Thus, the number- and weight-average sequence lengths are given as follows:

$$(N_B)_n = \frac{1}{1 - r_B q_1^2/r} \tag{4.3.10}$$

$$(N_B)_w = \frac{1 + r_B q_1^2/r}{1 - r_B q_1^2/r} \tag{4.3.11}$$

Clearly, analogous equations apply for the soft segment:

$$S_{C,k} = \left(1 - \frac{r_C q_2^2}{r}\right)\left(\frac{r_C q_2^2}{r}\right)^{k-1} \tag{4.3.12}$$

Thus, the number- and weight-average sequence lengths are given as follows:

$$(N_C)_n = \frac{1}{1 - (r_C q_2^2/r)} \tag{4.3.13}$$

$$(N_C)_w = \frac{1 + (r_C q_2^2/r)}{1 - (r_C q_2^2/r)} \tag{4.3.14}$$

Sequence distribution in many cases will be geometric, so that we can for the sequence of species "X" write:

$$S_{x,k} = (1 - P_{xx})P_{xx}^{k-1} \tag{4.3.15}$$

Thus, the number- and weight-average sequence lengths are:

$$(N_x)_n = \frac{1}{1 - P_{xx}} \tag{4.3.16}$$

$$(N_B)_w = \frac{1 + P_{xx}}{1 - P_{xx}} \tag{4.3.17}$$

where P_{xx} is the probability of reaching another "X" in sequence from an "X" (in the present case $r_B q_1^2/r$ or $r_C q_2^2/r$). These results will be important in Section 4.4.

In the foregoing model for segmented block copolymer sequence lengths, there is a liberality as to q_1 and q_2, specifically that B and C groups are not necessarily equally reactive. Regardless of the details of the reactivity, though, we note that the average sequence lengths may be quite short. If the stoichiometry is balanced ($r = 1$), as would probably be desirable, the sequence lengths at complete conversion are:

$$(N_B)_n = \frac{1}{1 - r_B} \qquad (4.3.18)$$

$$(N_C)_n = \frac{1}{1 - r_C} \qquad (4.3.19)$$

Thus, the sequence lengths are quite short (~ 2) if the two are present in equimolar quantities; if the two are unbalanced, the monomer of higher concentration will have increasingly larger sequence length while that of the other will decrease to unity. Note that:

$$\frac{(N_B)_n}{(N_C)_n} = \frac{B_0}{C_0} \qquad (4.3.20)$$

For such a polyurethane system, direct measurement of these sequence lengths is difficult. The large size of the monomers prevents direct spectroscopic measurement; selective hydrolysis, however, has been used in polyester–polyurethanes to determine the sequence length of the polyurethane blocks.[59] One can manage a correlation with the flexural modulus, the amount of hydrogen bonding (measured by IR, perhaps), and so forth. In the copolymers formed by free-radical polymerization, though, the small monomer size along the backbone of the polymer (usually two carbon–carbon bonds per monomer) allows the observation by ^{13}C NMR spectroscopy of sequences up to five monomers long (pentads). We thus return to free-radical copolymerization.

4.4 Sequence Length Distribution in Free-Radical Copolymerization

4.4.1 The Terminal Model

If we denote by P_{ii} the probability of a monomer of type i adding to a previous monomer of type i, then from equation (4.3.15), the probability of finding exactly k units of type i in series in a given chain will be given by

$$S_{i,k}^{\text{inst}} = (1 - P_{ii})P_{ii}^{k-1} \qquad (4.4.1)$$

Equation (4.4.1) gives the fraction of all type-i sequences that are k units long, as the distribution is normalized. By equations (4.3.16) and (4.3.17),

$$(N_i)_n^{\text{inst}} = \frac{1}{1 - P_{ii}} \qquad (4.4.2)$$

$$(N_i)_w^{\text{inst}} = \frac{1 + P_{ii}}{1 - P_{ii}} \qquad (4.4.3)$$

where we have appended the superscript "inst" as in Chapter 3 to indicate instantaneous values valid for polymer produced in a given increment of conversion. (The transition probability P_{ii} is conversion dependent, although this is not notated explicitly.)

So far, we have assumed only that the transition probability P_{ii} is constant along the chain (i.e., QSSA is in effect and chain lifetime is short).* This of course eliminates applicability to living copolymerizations, since the lifetime of a chain will there ideally be the lifetime of the reaction and in general we will have composition drift. These were also implicitly eliminated by use of the term "instantaneous," since that concept is foreign to living polymerization. It also will often eliminate short copolymer chains made by free-radical polymerization, for reasons discussed later. For long chains, however, constant P_{ii} is a fine assumption, and the effect of termination reactions can be neglected. Thus, a radical of type i either incorporates a monomer unit of type i or one of type j. This can be formally expressed by:

$$P_{ii} + P_{ij} = 1 \qquad (4.4.4)$$

where $i, j = 1, 2, i \neq j$; here we have restricted ourselves to a copolymerization of only two monomers [although equations (4.4.1)–(4.4.3) are valid for multicomponent polymerization]. The probability P_{ij} can be expressed as follows:

$$P_{ii} = \frac{k_{ii}M_i}{k_{ii}M_i + k_{ij}M_j} = \frac{\alpha_i}{1 + \alpha_i} \qquad (4.4.5)$$

where $\alpha_i = r_i(f_i/f_j)$. Similarly:

$$P_{ij} = \frac{1}{1 + \alpha_i} \qquad (4.4.6)$$

Thus

$$S_{i,k}^{inst} = \frac{\alpha_i^{k-1}}{(1 + \alpha_i)^k} \qquad (4.4.7)$$

and

$$(N_i)_n^{inst} = 1 + \alpha_i \qquad (4.4.8)$$

$$(N_i)_w^{inst} = 1 + 2\alpha_i \qquad (4.4.9)$$

We may also note a parallel to equation (4.3.20) in that

$$\frac{(N_1)_n^{inst}}{(N_2)_n^{inst}} = \frac{F_1}{F_2} \qquad (4.4.10)$$

---

*By arguments similar to those of Chapter 3. The idea that statistical derivation implies applicability to the relations for average sequence length under general conditions[1.60] is false, since statistical derivation implicitly assumes the QSSA.[61,62]

The difference between this relation and the preceding one for step copolymerization is that here the polymer is that instantaneously produced, whereas before, since there is no distinction between monomer and polymer, it referred to all species.

Sequence length distributions and average sequence lengths, while useful concepts, are not themselves measurable by the sensitive NMR (especially ^{13}C) techniques. This is because, at present, the longest sequences that can be distinguished are sequences of five monomers. In reality, then, model predictions can be compared only with experimental measurements of the fractions of triads or pentads having a given comonomer at the center (or some combination of those[63]). Predictions of triad fractions for the terminal model are given in Example 4.1.

Example 4.1

For a copolymerization between two monomers obeying the terminal model, derive equations for the fraction of triads centered on each of the two monomers by combinatorial reasoning. For the pair consisting of styrene (1) and methyl methacrylate (2), assume $r_1 = 0.48$, $r_2 = 0.42$, and give the initial fractions of each triad for $f_{1,0} = 0.25$, 0.5, and 0.75. Also compare the number-average sequence lengths $(N_1)_n$ and $(N_2)_n$ for each comonomer composition.

Solution:

There are six different distinguishable triads: [111], [112], and [212] for triads centered on monomer 1, and [121], [122], and [222] for triads centered on monomer 2. (We do not consider triad [211] to be distinct from [112], nor [122] from [221].) The fractions of these are simply related to the transition probabilities in the following ways:

$$F_{[iii]}^{inst} = P_{ii}^2 \tag{4.4.11}$$

$$F_{[iij]}^{inst} = 2P_{ii}P_{ij} \tag{4.4.12}$$

$$F_{[jij]}^{inst} = P_{ij}^2 \tag{4.4.13}$$

These relations are valid because the probabilities apply in either direction on the chain and because the model is first-order Markovian. Substituting in equations (4.4.5) and (4.4.6):

$$F_{[iii]}^{inst} = \left(\frac{r_i f_i / f_j}{1 + r_i f_i / f_j} \right)^2 \tag{4.4.14}$$

$$F_{[iij]}^{inst} = \frac{r_i f_i / f_j}{(1 + r_i f_i / f_j)^2} \tag{4.4.15}$$

$$F_{[jij]}^{inst} = \left(\frac{1}{1 + r_i f_i / f_j} \right)^2 \tag{4.4.16}$$

For the case of styrene copolymerized with methyl methacrylate, the results for the various triad fractions and

TABLE 4.1 / Triad Fractions for the Pair Styrene–Methyl Methacrylate

$f_{1.0}$	$F_{[111]}^{inst}$	$F_{[112]}^{inst}$	$F_{[212]}^{inst}$	$F_{[222]}^{inst}$	$F_{[221]}^{inst}$	$F_{[121]}^{inst}$	$(N_1)_n^{inst}$	$(N_2)_n^{inst}$
0.25	0.019	0.238	0.743	0.311	0.493	0.196	1.16	2.26
0.50	0.105	0.438	0.457	0.087	0.417	0.496	1.48	1.42
0.75	0.348	0.484	0.168	0.015	0.215	0.770	2.44	1.14

the number-average sequence lengths are summarized in Table 4.1. An alternating sort of structure (although not strictly alternating) tends to result from $r_1, r_2 < 1$. This keeps the sequence lengths fairly short and favors mixed triads.

4.4.2 The Penultimate Model

For the penultimate model, we can no longer use equations (4.3.15)–(4.3.17), which refer to first-order Markovian models—that is, situations in which the transition (addition of next monomer) depends only on the present state (identity of end group). Here, the penultimate state matters as well, and so we have to start over, with triply indexed probabilities P_{iii}, P_{iij}, P_{jii}, and P_{ijj}, where $i, j = 1, 2, i \neq j$. Two separate equations are needed for the probability of finding sequences n units long depending on whether n is equal or larger than 1[64,65]:

$$S_{i,1}^{inst} = 1 - P_{jii} \tag{4.4.17}$$

$$S_{i,k}^{inst} = P_{jii}(P_{iii})^{k-2}(1 - P_{iii}) \qquad k \geq 2 \tag{4.4.18}$$

This distribution is normalized, and so the preceding equations give the fractions of sequences of different length. The average sequence length is given by:

$$(N_i)_n^{inst} = 1 + \frac{P_{jii}}{1 - P_{iii}} \tag{4.4.19}$$

$$(N_i)_w^{inst} = 1 + \frac{2P_{jii}}{1 - P_{iii}} \frac{1}{1 - P_{iii} + P_{jii}} \tag{4.4.20}$$

Plainly, the penultimate results in a distribution that is not geometric. This was obvious from the distribution itself, but here we see that the "polydispersity" of the sequence length distribution $(N_i)_w^{inst}/(N_i)_n^{inst}$ may be larger than 2 for P_{iii} near unity, depending on P_{jii}. This probability determines the ease with which sequences larger than 1 may be formed. If this occurs with relative ease (P_{jii} near 1), then the distribution will be fairly geometric; if, on the other hand, sequences longer than 1 are formed with difficulty (P_{jii} near 0), the overwhelming majority of sequences will be one unit long, but there will be a small geometric "tail" to the distribution, giving rise to the large polydispersity.

As for the terminal model, these equations do not rely on the assumption of long chains but only on the assumption of constant transition probabilities along the chain. This generally restricts us to long chains anyhow, so that we can ignore the termination reactions in the formulation of the transition probabilities, which themselves will then obey the following restrictions:

$$P_{iii} + P_{iij} = 1 \qquad (4.4.21)$$

$$P_{iji} + P_{ijj} = 1 \qquad (4.4.22)$$

These transition probabilities of interest are then given as follows:

$$P_{iii} = \frac{k_{iii}M_i}{k_{iii}M_i + k_{iij}M_j} = \frac{r_{ii}f_i/f_j}{1 + r_{ii}f_i/f_j} \qquad (4.4.23)$$

$$P_{jii} = \frac{k_{jii}M_i}{k_{jii}M_i + k_{jij}M_j} = \frac{r_{ji}f_i/f_j}{1 + r_{ji}f_i/f_j} \qquad (4.4.24)$$

In terms of reactivity ratios, the number-average sequence length becomes:

$$(N_i)_n^{inst} = 1 + \frac{f_i}{f_j}\frac{f_j + r_{ii}f_i}{f_i + f_j/r_{ji}} \qquad (4.4.25)$$

Example 4.2

For a copolymerization between two monomers obeying the penultimate model, derive equations for the fraction of triads centered on each of the two monomers.

Solution:

The triad functions are not as easily reasoned from combinatorial arguments as was done in the case for terminal kinetics. However, a more general method (which could have been used for Example 4.1 as well) obtains these fractions from the sequence distributions in the following way.

The only sequence that contributes to [*jij*] triads is that of length 1. Thus the triad fraction is proportional to the fraction of *i*-type sequences of length 1:

$$F_{[jij]}^{inst} \propto S_{i,1}^{inst} = 1 - P_{jii} \qquad (4.4.26)$$

On the other hand, all sequences except those of length 1 contribute to [*iij*] triads, each sequence contributing two triads (one at either end of the sequence). Thus:

$$F_{[iij]}^{inst} \propto 2\sum_{k=2}^{\infty} S_{i,k}^{inst} = 2P_{jii} \qquad (4.4.27)$$

In similar fashion, all sequences of length 3 and greater contribute to [*iii*] triads, each sequence contributing $k - 2$ triads (since the ends contributed to [*iij*] triads). Thus:

$$F_{[iii]}^{inst} \propto \sum_{k=3}^{\infty} (k - 2)S_{i,k}^{inst} = P_{jii}\left(\frac{P_{iii}}{1 - P_{iii}}\right) \qquad (4.4.28)$$

Applying equations (4.4.23) and (4.4.24) yields the following:

$$F_{[iii]}^{inst} \propto \frac{r_{ii}r_{ji}(f_i/f_j)^2}{1 + r_{ji}(f_i/f_j)} \tag{4.4.29}$$

$$F_{[iij]}^{inst} \propto \frac{2r_{ji}(f_i/f_j)}{1 + r_{ji}(f_i/f_j)} \tag{4.4.30}$$

$$F_{[jij]}^{inst} \propto \frac{1}{1 + r_{ji}(f_i/f_j)} \tag{4.4.31}$$

Normalizing these equations yields[66]:

$$F_{[iii]}^{inst} \propto \frac{r_{ii}r_{ji}(f_i/f_j)^2}{1 + 2r_{ji}(f_i/f_j) + r_{ii}r_{ji}(f_i/f_j)^2} \tag{4.4.32}$$

$$F_{[iij]}^{inst} \propto \frac{2r_{ji}(f_i/f_j)}{1 + 2r_{ji}(f_i/f_j) + r_{ii}r_{ji}(f_i/f_j)^2} \tag{4.4.33}$$

$$F_{[jij]}^{inst} \propto \frac{1}{1 + 2r_{ji}(f_i/f_j) + r_{ii}r_{ji}(f_i/f_j)^2} \tag{4.4.34}$$

Note that for the case $r_{ii} = r_{ji}$, these relations reduce to the ones for terminal polymerization. This is thought to be nearly the case for the pair styrene–methyl methacrylate.[15] It does not necessarily imply absence of penultimate effects, but does show that in this case, penultimate effects will be invisible both in composition and in sequence length.[67] However, as has been pointed out, this invisibility holds only if $r_{ii} = r_{ji}$ exactly.[68]

4.4.3 Comparison

Information on the sequence distribution, as from triad data, provides a way to distinguish between terminal and penultimate behavior (though admittedly not in all cases). We expect the same to be true for the other models we have mentioned. The use of sequence distribution as a distinguishing tool can be seen in some of the work of Hill and co-workers.[18,31]

The bulk copolymerization of styrene and acrylonitrile at 60°C was studied by ^{13}C NMR, giving composition and sequence length distribution data.[18] Reactivity ratios for the terminal, penultimate, and complex participation and complex dissociation models were calculated from copolymer composition data; these are given in Table 4.2. Figure 4.2 shows the experimental data and models predictions for copolymer composition; all models offer good agreement, and it would be difficult to decide which one best describes the system. In comparing the predictions for certain sequence lengths (see Figures 4.3 and 4.4), however, it can be observed that the results based on the penultimate model offer the best agreement.

For the system styrene–maleic anhydride, Hill et al. calculated kinetic parameters for four models: terminal, penultimate,

TABLE 4.2 / Reactivity Ratios for Bulk Copolymerization of Acrylonitrile (1) and Styrene (2) at 60°C

Model	Value	Range
Terminal		
r_1	0.053	0.04–0.07
r_2	0.331	0.27–0.42
Penultimate		
r_{11}	0.039	0.035–0.045
r_{12}	0.634	0.57–0.73
r_{21}	0.091	0.079–0.110
r_{22}	0.229	0.21–0.26

Source: Reference 18.

complex participation, and complex dissociation.[31] Although no triad measurements were made, the predictions are sufficiently different (see Figures 4.5 and 4.6) to permit the use of such data to clearly distinguish between them (though perhaps not with the bootstrap effect[43]).

It must be emphasized that distinguishing between models by sequence length information will not always be possible. Pairs for which the penultimate effect is invisible in the composition data, such as styrene–methyl methacrylate, cannot be distinguished by sequence data.[67] For this particular system it has been shown that terminal and penultimate models, and bootstrap versions of these, all provide an adequate fit for composition and sequence length data, and that the last two fit rate data equally

Figure 4.2.
Copolymer composition curve for styrene-acrylonitrile copolymerization in bulk at 60°C. Solid line: complex model (variant 1); long dashes: complex model (variant 2); separated long dashes: penultimate model; long dash-short dash: terminal model. (Not all curves are visible.) (From reference 18, used by permission of the American Chemical Society.)

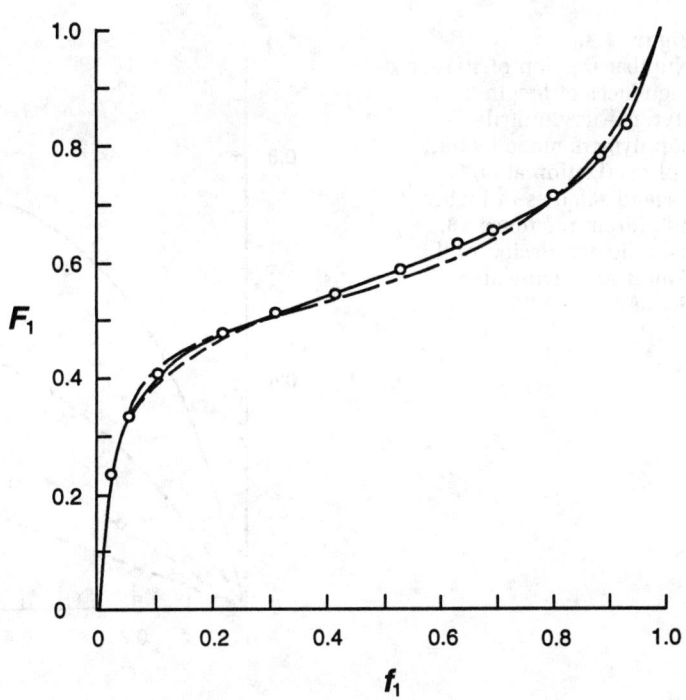

Figure 4.3.
Number fraction of styrene sequences of length 1 in styrene-acrylonitrile copolymers made by bulk polymerization at 60°C. From left to right the curves are in the order listed in the caption to Figure 4.2. (From reference 18, used by permission of the American Chemical Society.)

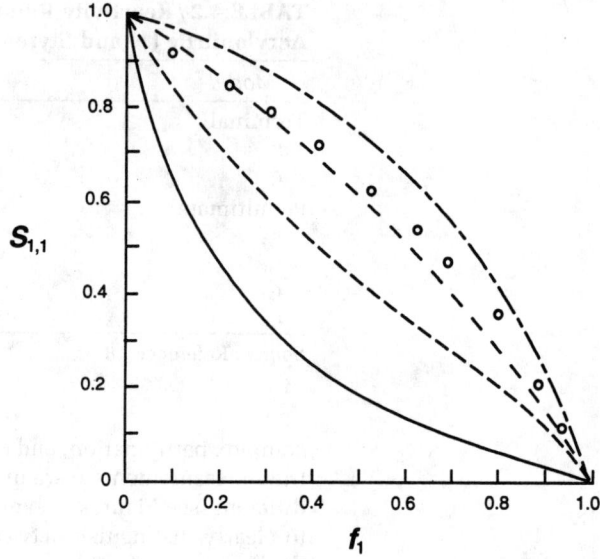

well.[45] This reflects the difficulty in distinguishing between different models with many parameters, especially when one is forced to rely, not on direct evidence in favor of a mechanism, but on aggregated effects on composition, sequence lengths, or rate.

4.4.4 A Comment on Finite Molecular Weight

Application of the formulas of Sections 4.4.1 and 4.4.2 generally is restricted to long chains, though not because the presence of

Figure 4.4.
Number fraction of styrene sequences of length 2 in styrene–acrylonitrile copolymers made by bulk polymerization at 60°C. Legend same as in Figure 4.2. (From reference 18, used by permission of the American Chemical Society.)

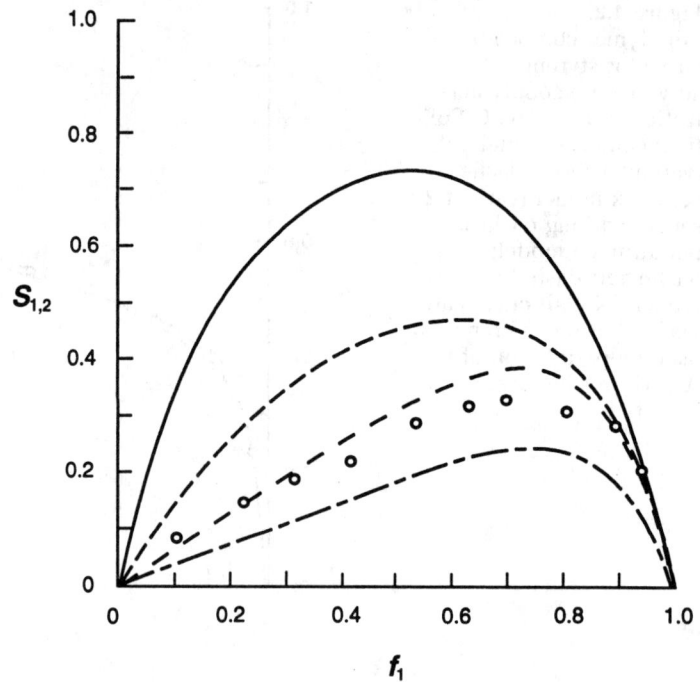

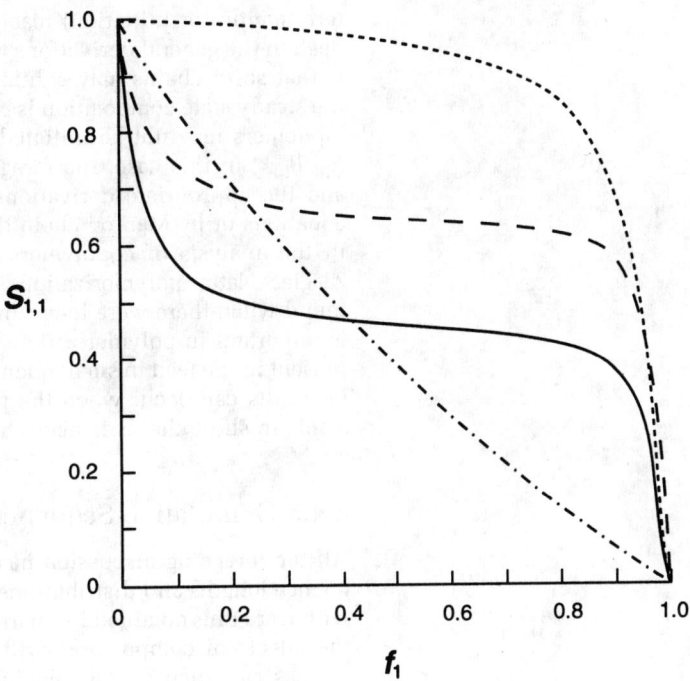

Figure 4.5.
Number fraction of styrene sequences of length 1 in styrene–maleic anhydride copolymers made by bulk polymerization at 60°C. Lines represent theoretical predictions: solid curve, complex model; long dashes, complex dissociation model; dash/dot curve; penultimate model; short dashes; terminal model. (From reference 31, used by permission of the American Chemical Society.)

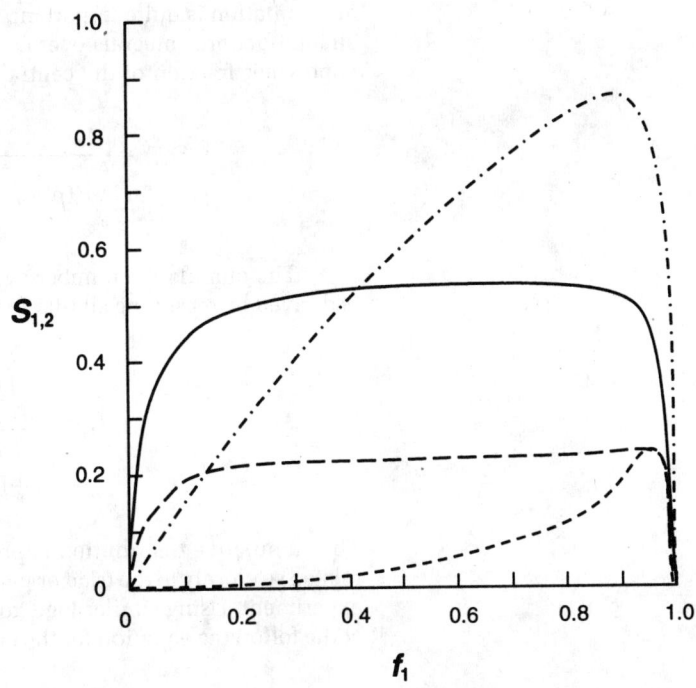

Figure 4.6.
Number fraction of styrene sequences of length 2 in styrene–maleic anhydride copolymers made by bulk polymerization at 60°C. Legend same as in Figure 4.5. (From reference 31, used by permission of the American Chemical Society.)

termination (or initiation) reactions itself poses a problem (indeed in the general derivations this is allowed for). The problem is that short chains may exhibit composition transients before the steady state composition is reached. This occurs because the monomers may not be initiated at the quasi–steady state ratio $R_{(1)}/R_{(2)}$.[69] In this case, the foregoing discussions are not valid, and the appropriate derivations can be made only by kinetic equations or by Markov chain theory. The latter has been used in the analysis of copolymers of ethylene and propylene by Ziegler–Natta polymerization.[70] Significant differences were found when there were long runs of one of the monomers. This is important in polymerizations where one of the monomers is present in the feed in small quantities. Serious misinterpretation of results can occur when the presence of compositional transients in short chains is neglected.[71]

4.4.5 Cumulative Sequence Length Distribution

All the foregoing discussion has been on the instantaneous sequence lengths and distributions; the experimental results dealt with were thus obtained in a narrow conversion interval to avoid the effects of composition drift. Indeed, only in rare circumstances can such drift be avoided in batch polymerization: if the monomers are equally reactive, if the initial monomer feed composition is at the azeotropic point δ (generally unstable), or if semibatch control strategies are used so that the monomer mixture composition is kept constant. These are rare or unrealizable situations, though, and in general for industrial production we will need to deal with the cumulative values that can be avoided in research experiments.

Most data are for triad fractions, and the proper method of accumulation is quite important. Cumulative triad functions are straightforward integrals over conversion, weighted only by the copolymer fraction of the central monomer: for example,

$$F_{[iii]}^{cumu} = \frac{1}{\int_0^p F_i(p')dp'} \int_0^p F_{[iii]}^{inst} F_i(p')dp' \qquad (4.4.35)$$

The cumulative number-average sequence length, $(N_i)_n^{cumu}$, is derived by reasoning similar to that giving DP_n^{cumu} in Chapter 3:

$$(N_i)_n^{cumu} = \frac{\int_0^p F_i(p')dp'}{\int_0^p \frac{F_i(p')dp'}{(N_i)_n^{inst}(p')}} \qquad (4.4.36)$$

The cumulative distribution is probably more useful, since it is related so closely to the triad or pentad data measured in the NMR experiment. Using similar logic to that just given, one can arrive at the following equation for the cumulative distribution:

$$S_{i,k}^{cumu}(p) = \frac{\int\limits_0^P \dfrac{S_{i,k}^{inst}(p')F_i(p')dp'}{(N_i)_n^{inst}(p')}}{\int\limits_0^P \dfrac{F_i(p')dp'}{(N_i)_n^{inst}(p')}} \qquad (4.4.37)$$

Obviously, upon taking the first moment of this distribution, one gets equation (4.4.36).*

4.5 Molecular Weight Distribution

Molecular weight and its distribution are as significant for copolymers as for homopolymers; moreover, this significance is not diminished by the presence of the additional facets of composition and sequence length. It is appropriate to see how the presence of comonomers affects chain length distribution. A review of our consideration in Section 4.2 of the different mechanisms proposed for the propagation step will of course be helpful, but we need first to discuss the termination reaction, which also has been the subject of debate.

4.5.1 Termination

As for free-radical homopolymerizations, the main point of contention concerning the termination reaction is whether it is under chemical or diffusional control. Early work in copolymerization kinetics concentrated on chemical control and on the relative importance of combination and disproportionation; generally, terminal model kinetics was assumed. Within this framework, termination can be described by six reactions, arising from the two different termination mechanisms and the three different combinations of terminating radicals present under terminal kinetics.[10,73-76]:

$$R_{(1)n,m} + R_{(1)r,s} \xrightarrow{k_{tc11}} P_{n+r,m+s} \qquad (4.5.1)$$

$$R_{(2)n,m} + R_{(2)r,s} \xrightarrow{k_{tc22}} P_{n+r,m+s} \qquad (4.5.2)$$

$$R_{(1)n,m} + R_{(2)r,s} \xrightarrow{k_{tc12}} P_{n+r,m+s} \qquad (4.5.3)$$

$$R_{(1)n,m} + R_{(1)r,s} \xrightarrow{k_{td11}} P_{n,m} + P_{r,s} \qquad (4.5.4)$$

$$R_{(2)n,m} + R_{(2)r,s} \xrightarrow{k_{td22}} P_{n,m} + P_{r,s} \qquad (4.5.5)$$

$$R_{(1)n,m} + R_{(2)r,s} \xrightarrow{k_{td12}} P_{n,m} + P_{r,s} \qquad (4.5.6)$$

---

*Equations (4.4.36) and (4.4.37) differ from those suggested in the literature.[72]

The termination rate constant is assumed to be independent of chain length. As in Chapter 3, we will denote overall termination rate constants, irrespective of mechanism, by k_t (e.g., k_{t11}).

Calculation of total rate of copolymerization—that is, the rate of disappearance of monomer ($M = M_1 + M_2$), proceeds as follows. Combining equations (4.2.5) and (4.2.6), which are based on the long-chain hypothesis, one obtains:

$$-\frac{dM}{dt} = (k_{11}M_1 + k_{12}M_2)R_{(1)} + (k_{21}M_1 + k_{22}M_2)R_{(2)} \quad (4.5.7)$$

The quasi–steady state approximation, equation (4.2.8), also implies that the rates of initiation and termination are identical as in Chapter 3, so that for initiation by thermal initiators:

$$2fk_dI = k_{t11}R_{(1)}^2 + 2k_{t12}R_{(1)}R_{(2)} + k_{t22}R_{(2)}^2 \quad (4.5.8)$$

Combining equations (4.2.8), (4.5.7), and (4.5.8) yields*

$$-\frac{dM}{dt} = \frac{(2fk_dI)^{1/2}(r_1M_1^2 + 2M_1M_2 + r_2M_2^2)}{(r_1^2\delta_1^2M_1^2 + 2\phi r_1r_2\delta_1\delta_2M_1M_2 + r_2^2\delta_2^2M_2^2)^{1/2}} \quad (4.5.9)$$

where

$$\delta_1 = \frac{k_{t11}^{1/2}}{k_{11}} \quad (4.5.10)$$

$$\delta_2 = \frac{k_{t22}^{1/2}}{k_{22}} \quad (4.5.11)$$

$$\phi = \frac{k_{t12}^{1/2}}{(k_{t11}k_{t22})^{1/2}} \quad (4.5.12)$$

In principle, the parameters δ_1 and δ_2 could be obtained from homopolymerization experiments and r_1, r_2, the reactivity ratios, from composition analysis. It is the cross-termination factor, ϕ, that causes problems both in measurement and in concept.

Even though equation (4.5.9) has been used extensively,[53,77–81] two deviations from theory are commonly seen. First, the cross-termination factor is very different from unity,[75,82] whereas low molecular weight radicals have no preference for cross-termination.[15,17] Second, ϕ seems to be composition dependent, which is not allowed for in the theory. But the most important objection is that termination is diffusion-controlled, as discussed in Chapter 3. Atherton and North[83] recognized some of the deficiencies of the chemical control treatment of copolymerization and extended some of the arguments that by then were well established for free-radical homopolymerization.[84,85] Experiments on the copolymerization of vinyl acetate and methyl methacrylate in solvent mixtures of different viscosities strongly suggested diffusion limitations.

---

*An equation for the case where one of the monomers does not homopolymerize (as might be the case for maleic anhydride) has been presented by North.[76]

If diffusion is controlling, termination is governed not by the chemical nature of the last unit but by the diffusivity of the polymer chains, implying a dependence on length and composition of the chain only. This is clearly a problem too intricate to deal with here, and so ignore the details of the dependence on the chain length and composition of the participating chains and assume dependence on the composition of the comonomer alone. Thus we have one rate constant, k_t, dependent on composition.

The quasi–steady state approximation can now be written, as for homopolymerization, as follows:

$$2fk_d I = k_t'(R_{(1)} + R_{(2)})^2 = k_t'R^2 \qquad (4.5.13)$$

Equation (4.5.9) is now rearranged to give:

$$-\frac{dM}{dt} = \frac{(2fk_d I)^{1/2}(r_1M_1^2 + 2M_1M_2 + r_2M_2^2)}{(k_t')^{1/2}(M_1/k_{12} + M_2/k_{21})} \qquad (4.5.14)$$

To better demonstrate the differences between the two approaches, the following can be defined:

$$\varepsilon_1 = \frac{(k_t')^{1/2}}{k_{11}} \qquad (4.5.15)$$

$$\varepsilon_2 = \frac{(k_t')^{1/2}}{k_{22}} \qquad (4.5.16)$$

and

$$-\frac{dM}{dt} = \frac{(2fk_d I)^{1/2}(r_1M_1^2 + 2M_1M_2 + r_2M_2^2)}{(r_1^2\varepsilon_1^2 M_1^2 + 2r_1r_2\varepsilon_1\varepsilon_2 M_1M_2 + r_2^2\varepsilon_2^2 M_2^2)^{1/2}} \qquad (4.5.17)$$

Comparing equations (4.5.9) and (4.5.17) one can see how, if a diffusive process were controlling, the cross-termination factor could be composition dependent. A fairly successful dependence of termination rate constant on composition is expressed by the Atherton–North relation[83,86]:

$$k_t' = k_{t11}F_1 + k_{t22}F_2 \qquad (4.5.18)$$

Although this does allow for fit of experimental data, it has been illustrated[15] that the large cross-termination coefficient is due to the incorrect assumption of terminal kinetics on the propagation step and that upon proper use of the penultimate rate constants, this problem is eliminated. Improper assumptions about the nature of the propagation steps thus confuse interpretation of the nature of the termination steps.

4.5.2 Detailed Description of Chain Length Distribution

In describing the chain length or molecular weight distributions in copolymerization, one is faced with two alternatives: (1) a detailed description of the growing radicals by type (i.e., $R_{(1)}$ or $R_{(2)}$ type) and length, according to each of the monomers incorporated,[10,73,80,87,88] and (2) a lumped description in which the only variable of interest is the overall chain length.[72,89-91] In the first alternative one is interested in the development of the exact quantities $R_{(1)n,m}$ and $R_{(2)n,m}$ for all n and m, in the second, the evolution of R_i, the concentration of growing radicals, irrespective of the last type of monomer incorporated, for which $n + m = i$. A similar distinction can be made for the population of dead polymer. That is, we can be interested in (1) $P_{n,m}$ or in (2) P_i $(n + m = i)$. We begin with the detailed description and then proceed to the simpler.

For this specific example we assume that terminal kinetics are operative [formulas (4.2.1)–(4.2.4)] and that termination is chemically controlled [(4.5.1)–(4.5.6)]. The formulism would not change drastically if termination were diffusion-controlled. Mass balances for growing chains are given by the following:

$$\frac{dR_{(1)1,0}}{dt} = R_{\text{init}(1)} - R_{(1)1,0}(k_{11}M_1 + k_{12}M_2 + k_{t11}R_{(1)} + k_{t12}R_{(2)}) \tag{4.5.19}$$

$$\frac{dR_{(2)0,1}}{dt} = R_{\text{init}(2)} - R_{(2)0,1}(k_{21}M_1 + k_{22}M_2 + k_{t12}R_{(1)} + k_{t22}R_{(2)}) \tag{4.5.20}$$

$$\frac{dR_{(1)n,m}}{dt} = k_{11}M_1R_{(1)n-1,m} + k_{21}M_1R_{(2)n-1,m} - R_{(1)n,m}(k_{11}M_1 + k_{12}M_2 + k_{t11}R_{(1)} + k_{t12}R_{(2)}) \tag{4.5.21}$$

$$\frac{dR_{(2)n,m}}{dt} = k_{22}M_2R_{(2)n,m-1} + k_{12}M_2R_{(1)n,m-1} - R_{(2)n,m}(k_{21}M_1 + k_{22}M_2 + k_{t12}R_{(1)} + k_{t22}R_{(2)}) \tag{4.5.22}$$

where $R_{\text{init}(i)}$ is the rate of initiation of monomer i. Applying QSSA allows solution for the homopolymers:

$$R_{(1)n,0} = \frac{R_{\text{init}(1)}}{k_{11}M_1} \alpha_1^n \tag{4.5.23}$$

$$R_{(2)0,m} = \frac{R_{\text{init}(2)}}{k_{22}M_2} \alpha_2^m \tag{4.5.24}$$

which are of course geometric distributions. The probabilities of homopropagation (αs) are:

$$\alpha_1 = \frac{k_{11}M_1}{k_{11}M_1 + k_{12}M_2 + k_{t11}R_{(1)} + k_{t12}R_{(2)}} \tag{4.5.25}$$

$$\alpha_2 = \frac{k_{22}M_2}{k_{21}M_1 + k_{22}M_2 + k_{t12}R_{(1)} + k_{t22}R_{(2)}} \tag{4.5.26}$$

The concentrations of radicals of either type is given as follows:

$$R_{(1)} = \left(\frac{R_{\text{init}(1)} + R_{\text{init}(2)}}{k_{t11} + 2k_{t12}\beta + k_{t22}\beta^2} \right)^{1/2} \tag{4.5.27}$$

$$R_{(2)} = \beta R_{(1)} \tag{4.5.28}$$

where the ratio β is given according to the QSSA between radical types, equation (4.2.8):

$$\beta = \frac{k_{12}M_2}{k_{21}M_1} \tag{4.5.29}$$

Notice that the equations for growing radicals are of the same type as in homopolymerization $(R_{init}/k_t)^{1/2}$, where R_{init} is the total initiation rate, $2fk_dI$. Similar expressions would be found if termination were diffusion-controlled.[57]

Using QSSA in equations (4.5.21) and (4.5.22), one can write:

$$R_{(1)n,m} = \alpha_1 R_{(1)n-1,m} + \frac{k_{21}}{k_{11}} \alpha_1 R_{(2)n-1,m} \tag{4.5.30}$$

$$R_{(2)n,m} = \alpha_2 R_{(2)n,m-1} + \frac{k_{12}}{k_{22}} \alpha_2 R_{(1)n,m-1} \tag{4.5.31}$$

which resemble equations for homopolymerization, in terms of propagation probabilities, modified to account for cross-propagation. The solutions to these equations are[80]:

$$R_{(1)n,m} = \alpha_1^n \alpha_2^m \left[\sum_{j=1}^{n} \left(C_{11} \frac{n-j}{j} + C_{12} \right) \binom{n-1}{j-1} \binom{m-1}{j-1} \left(\frac{1}{r_1 r_2} \right)^j \right] \tag{4.5.32}$$

$$R_{(2)n,m} = \alpha_1^n \alpha_2^m \left[\sum_{j=1}^{n} \left(C_{21} + C_{22} \frac{m-j}{j} \right) \binom{n-1}{j-1} \binom{m-1}{j-1} \left(\frac{1}{r_1 r_2} \right)^j \right] \tag{4.5.33}$$

where

$$C_{ij} = \frac{R_{init(j)}}{k_{ij}M_j} \tag{4.5.34}$$

In both equations, it is observed that the first term is the contribution of chains started with an M_1 while the second term is the contribution of chains started with an M_2, as can be seen from the definitions of the C_{ij}. More general results including transfer to agent and monomer are expressed by equations (4.5.32) and (4.5.33) with the parameters α_i and C_{ij} modified accordingly.[80]

Given these analytical solutions, one can obtain the dead polymer by:

$$\frac{P_{n,m}}{dt} = R_{(1)n,m}(k_{td11}R_{(1)} + k_{td12}R_{(2)}) + R_{(2)n,m}(k_{td12}R_{(1)} + k_{td22}R_{(2)})$$

$$+ \frac{1}{2} \sum_{i=0}^{n} \sum_{j=0}^{m} \left[k_{tc11}R_{(1)i,j}R_{(1)n-i,m-j} + 2k_{tc12}R_{(1)i,j}R_{(2)n-i,m-j} + k_{tc22}R_{(2)i,j}R_{(2)n-i,m-j} \right] \tag{4.5.35}$$

or the corresponding generating function or moment equations. Equation (4.5.35) makes it clear that even a description of this detail does not capture all the features of the copolymer. For example, a random copolymer is indistinguishable from a block copolymer made by combination.

4.5.3 Less Detailed Description

The treatment of chain length distribution in free-radical copolymerization can be handled by means similar to those developed for homopolymerization, but the mathematics rapidly becomes complicated. Moreover, what one more often measures is the total chain length P_i or molecular weight distribution W_i

$$P_i = \sum_{n=0}^{i} P_{n,i-n} \tag{4.5.36}$$

$$W_i = \sum_{n=0}^{i} (nw_1 + (i-n)w_2)P_{n,i-n} \tag{4.5.37}$$

where w_1 and w_2 are the molecular weights of monomers M_1 and M_2, respectively.

This lumping is often quite valid because most chains, if sufficiently long, have virtually the same composition. Under a continuous variable approximation, the weight fraction of radicals of length r (a continuous variable) and composition deviation y is given as follows[73]:

$$w(r,y)dr\, dy = \frac{re^{-r/DP_n}}{DP_n^2} \left(\frac{r}{2\pi F_1 F_2 K}\right)^{1/2} \exp\left(\frac{-ry^2}{2F_1 F_2 K}\right) dr\, dy \tag{4.5.38}$$

where $y = F_1' - F_1 = F_2 - F_2'$, F_i' is the mole fraction of M_i in the particular chain, F_1 is the average mole fraction of M_i in the polymer, DP_n refers to the living chains, and the constant K is given as follows:

$$K = \frac{r_1 M_1^2 + 2r_1 r_2 M_1 M_2 + r_2 M_2^2}{r_1 M_1^2 + 2M_1 M_2 + r_2 M_2^2} \tag{4.5.39}$$

The composition distribution is Gaussian, and it becomes narrower with increasing chain length and for F_1 and F_2 not terribly different.

The molecular weight distribution of the living chains, irrespective of composition, corresponds to an exponential chain length distribution, which is merely the continuous analogue to the discrete geometric distribution:

$$w(r)dr = \frac{re^{-r/DP_n}}{DP_n^2} dr \tag{4.5.40}$$

So, within the long-chain approximation, we derive relations identical to those that would apply to homopolymerizations.

Treating copolymerization as a homopolymerization leads to perhaps the best method in the class of these lumped descriptions, the "pseudokinetic rate constant" method.[89,90] The resulting live chain distribution is similar to that just given:

$$w(r)dr = (\tau + \beta)\left(\tau + \frac{1}{2}\beta(\tau + \beta)r\right)re^{-r(\tau+\beta)}dr \tag{4.5.41}$$

The parameters τ and β (the latter differing from that in the analysis of the preceding section) are defined as follows:

$$\tau = \frac{k_{td}R}{k_pM} + \frac{k_{tr,m}}{k_p} + \frac{k_{tr,s}S}{k_pM} \qquad (4.5.42)$$

$$\beta = \frac{k_{tc}R}{k_pM} \qquad (4.5.43)$$

The parameter τ is thus related to the probability of a chain-ending step that leaves the chains uncombined, and β to the probability of chain ending by combination. In formulating the problem as a homopolymerization problem, the "pseudokinetic rate constants" are functions of composition:

$$k_p = k_{11}\,\phi_1 f_1 + k_{12}\,\phi_1 f_2 + k_{21}\,\phi_2 f_1 + k_{22}\,\phi_2 f_2 \qquad (4.5.44)$$

$$k_{tr,m} = k_{tr12}\,\phi_1 f_1 + k_{tr12}\,\phi_1 f_2 + k_{tr21}\,\phi_2 f_1 + k_{tr22}\,\phi_2 f_2 \qquad (4.5.45)$$

$$k_{tr,s} = k_{tr1,s}\phi_1 + k_{tr2,s}\phi_2 \qquad (4.5.46)$$

$$k_{td} = k_{td,11}\phi_1^2 + 2k_{td12,}\,\phi_1\phi_2 + k_{td,22}\,\phi_2^2 \qquad (4.5.47)$$

$$k_{tc} = k_{tc,11}\phi_1^2 + 2k_{tc,12}\,\phi_1\phi_2 + k_{tc,22}\,\phi_2^2 \qquad (4.5.48)$$

The radical concentration is given by the standard homopolymerization relation coming from QSSA $((2fk_dI)^{1/2}/k_t)$, and the fraction of radicals of type 1, ϕ_1, is given as follows:

$$\phi_1 = \frac{R_{(1)}}{R_{(1)} + R_{(2)}} = \frac{(k_{21} + k_{tr21})f_1}{(k_{21} + k_{tr21})f_1 + (k_{12} + k_{tr11})f_2} \qquad (4.5.49)$$

These pseudokinetic rate constants are thus not at all constant, but are dependent on composition. The foregoing definitions of these constants can be derived from the kinetic equations if one assumes that the concentration of radicals of the two types is independent of chain length. This is a good approximation for long chains, and an exact one if the two monomers are initiated with a ratio ϕ_1/ϕ_2 (see Section 4.4.4). The great advantage of this framework is the ease of extension to other multicomponent polymerizations and to other kinetic schemes, such as penultimate kinetics.[91] Thus this is a very powerful method, and probably the easiest to implement. The only caveat is its limit to long chains, for if the chains are too short, there may be compositional transients and the composition of each chain may not be of average composition [see equation (4.5.38)]. In the case of short chains the equations[80] of Section 4.5.2 or equivalent Markov chain calculations[92] must be used. One can also treat the problem from a recursive viewpoint,[93] but this is also limited to long chains and provides no substantial advantages over the present method at the cost of more difficult mathematics.

The formulas for M_n and M_w come from the continuous distribution, where once again a long-chain assumption has been made. Instantaneous averages are thus:

$$M_n^{inst} = \frac{\overline{M}}{\tau + \beta/2} \qquad (4.5.50)$$

$$M_w^{inst} = \frac{\overline{M}(2\tau + 3\beta)}{(\tau + \beta)^2} \qquad (4.5.51)$$

where $\overline{M}$ is the effective molecular weight of a repeat unit in the instantaneously formed copolymer:

$$\overline{M} = w_1 F_1 + w_2 F_2 \qquad (4.5.52)$$

Cumulative values are given by:

$$w(r)^{cumu} = \frac{\int_0^t \overline{M} R_p w(r,t') dt'}{\int_0^t \overline{M} R_p dt'} = \frac{\int_0^p \overline{M} w(r,p') dp'}{\int_0^p \overline{M}\, dp'} \qquad (4.5.53)$$

$$M_n^{cumu} = \frac{\int_0^t \overline{M} R_p dt'}{\int_0^t \overline{M}(R_p/M_n) dt'} = \frac{\int_0^p \overline{M}\, dp'}{\int_0^p \left(\frac{\overline{M}}{M_n}\right) dp'} \qquad (4.5.54)$$

$$M_w^{cumu} = \frac{\int_0^t \overline{M} R_p M_w dt'}{\int_0^t \overline{M} R_p dt'} = \frac{\int_0^p \overline{M} M_w dp'}{\int_0^p \overline{M}\, dp'} \qquad (4.5.55)$$

These equations are of course analogous to the general equations for the cumulative values for homopolymers found in Chapter 3.

4.5.4 Moment Representation

For some copolymerization systems, analytical solutions for the distributions of the polymeric product may not be possible. One such case, for example, is the heterogeneous copolymerization of α-olefins using Ziegler–Natta catalysts.[94] Under those circumstances one may resort to the moment generating functions or directly use the balance equations for $R_{(1)n,m}$ and $R_{(2)n,m}$ to develop differential equations for the moments of the copolymer chain length distribution. However, care must be taken in using the right expressions for the moments, because for copolymerization the simple relations between moments of the chain length distribution and moments of the molecular weight distribution do not exist.

It is clear that the number-average degree of polymerization is written as:

$$DP_n = \frac{\sum\limits_{n=0}^{\infty} \sum\limits_{m=0}^{\infty} (n+m)P_{n,m}}{\sum\limits_{n=0}^{\infty} \sum\limits_{m=0}^{\infty} P_{n,m}} \qquad (4.5.56)$$

However, to obtain the weight-average degree of polymerization one needs clear definitions of terms. One may mean that the weighting is simply by the number of monomers, so that

$$DP_w = \frac{\sum\limits_{n=0}^{\infty} \sum\limits_{m=0}^{\infty} (n+m)^2 P_{n,m}}{\sum\limits_{n=0}^{\infty} \sum\limits_{m=0}^{\infty} (n+m)P_{n,m}} \qquad (4.5.57)$$

One, however, may mean "weight" in the literal sense—weighted by the mass of the chain. If there is no composition distribution, one merely has a "characteristic" molecular weight of the repeat unit, as in equation (4.5.52). If not, it is better to write the following:

$$DP_w = \frac{\sum\limits_{n=0}^{\infty} \sum\limits_{m=0}^{\infty} (n+m)(nw_1 + mw_2)P_{n,m}}{\sum\limits_{n=0}^{\infty} \sum\limits_{m=0}^{\infty} (nw_1 + mw_2)P_{n,m}} \qquad (4.5.58)$$

Some of this confusion is eliminated if one works directly with the molecular weight distribution. In this case the moments are defined by:

$$\mu'_k = \sum\limits_{n=0}^{\infty} \sum\limits_{m=0}^{\infty} (nw_1 + mw_2)^k P_{n,m} \qquad (5.4.59)$$

such that

$$M_n = \frac{\mu'_1}{\mu'_0} = \frac{\sum\limits_{n=0}^{\infty} \sum\limits_{m=0}^{\infty} (nw_1 + mw_2)P_{n,m}}{\sum\limits_{n=0}^{\infty} \sum\limits_{m=0}^{\infty} P_{n,m}} \qquad (5.4.60)$$

$$M_w = \frac{\mu'_2}{\mu'_1} = \frac{\sum\limits_{n=0}^{\infty} \sum\limits_{m=0}^{\infty} (nw_1 + mw_2)^2 P_{n,m}}{\sum\limits_{n=0}^{\infty} \sum\limits_{m=0}^{\infty} (nw_1 + mw_2)P_{n,m}} \qquad (5.4.61)$$

4.6 Parameter Estimation and Characterization Techniques

As we have progressed from the discussion of step polymerizations to the discussion of chainwise polymerizations to the foregoing discussion of copolymerization, we have witnessed a great proliferation in the number of rate constants. These rate constants are rarely if ever well known. Even though the mathematical models are difficult and complex, they are in one sense easy compared to the task of actually measuring the rate constants on which those theories rest.

The most important constants for copolymerization are the reactivity ratios, which solely determine the composition and sequence length, and their evolution with conversion. Traditionally, the different models proposed for the propagation reaction have been tested for their ability to predict copolymer composition. This has been largely due to the complications in obtaining reliable measures of other properties such as sequence distribution and molecular weight distribution. With the advent of more sophisticated techniques such as high resolution NMR and combinations of preparative gel permeation and thin-layer chromatography, better assessment of the models has become possible. Nonetheless, it is best to first elaborate on the methods of estimating reactivity ratios.

4.6.1 Reactivity Ratios

If the terminal model or a higher order model is assumed, it is clear that the relation between F_1 and f_1 is nonlinear. It thus seems reasonable to rearrange the equation to permit a linear relationship in the reactivity ratios to be obtained. Such a rearrangement will always confuse the dependent (F_1) and independent variable (f_1), as in the Fineman–Ross method[95]:

$$G = r_1 H - r_2 \qquad (4.6.1)$$

where

$$G = \frac{f_1}{f_2}\left(1 - \frac{F_1}{F_2}\right) \qquad (4.6.2)$$

$$H = \left(\frac{f_1}{f_2}\right)^2 \frac{F_1}{F_2} \qquad (4.6.3)$$

Reactivity ratios can then be evaluated, graphically or by least squares, from the intercept and slope. Other linearizations are possible.[7,96] The Fineman–Ross method has the undesirable feature of unequal weighting of the data points, the data at low and high f_1 being weighted more heavily, with the clearly unacceptable result that the reactivity ratios depend on the indexing (i.e., which monomer is called monomer "1").[97]

Kelen and Tüdos[98] refined the Fineman–Ross method to give equal weighting to all observations by introducing an arbitrary constant α, yielding the following equation:

$$\eta = \left[r_1 + \frac{r_2}{\alpha}\right]\xi - \frac{r_2}{\alpha} \qquad (4.6.4)$$

where

$$\eta = \frac{G}{\alpha + F} \tag{4.6.5}$$

$$\xi = \frac{F}{\alpha + F} \tag{4.6.6}$$

Still, because the method does represent a linearization, there are statistical limitations,[99] and it is best to simply perform the nonlinear fit, since it is the most statistically sound.[97] If this is true for the terminal model with its relatively small number of constants, nonlinear search methods are even more imperative for all other models, given the greater number of parameters.[29,35]

The second point is that all such representations, equations (4.6.1) and (4.6.4) as well as the original nonlinear equation, relate instantaneous values, rather than cumulative ones. The polymer product is always cumulative, and so the usual solution has been to rely on the measurement of compositions at low conversions, which is experimentally very difficult and wasteful. Thus in the past 10 years it has become increasingly common to find use of the integrated form of the copolymer composition equation.[99,100]

The third point, perhaps the most important, is the experimental error incurred in the measurement of the various variables, in particular the dependent variable f_1. For some time, deviations from the terminal model observed in the F_1–f_1 curves have been attributed to experimental error, based on the confidence region for the calculated reactivity ratios.[97,101] This thinking has prompted the use of techniques such as "error in variables" method (EVM).[100,102–104]

4.6.2 Characterization Techniques

The inherent complexity of copolymerization systems has stimulated research in developing suitable techniques that can yield information on the different distributions or, at least, some of their moments. It should be recognized that since a complete description would be a three-dimensional plot of chain length, sequence length, and composition, the task is not at all trivial. An excellent review of this has been presented by Garcia-Rubio.[72a] The synopsis that closes this chapter follows closely his work.

4.6.2.1 Copolymer Composition

The appropriate technique for the determination of copolymer composition depends on the particular copolymer. Thus one may use elemental analysis if the comonomers are sufficiently different in composition: styrene–methyl methacrylate copolymers or styrene–acrylonitrile copolymers are well suited to this technique, but it would be useless on a copolymer of methyl acrylate and vinyl acetate. Mass spectrometry may also be used if one expects to find different fragments from the different monomers (e.g., styrene–methyl methacrylate). Any other sort of chemical analysis, such as titration, may be used if appropriate monomers (e.g., methacrylic acid) are present. The suitability of spectroscopic analysis by ultraviolet, infrared, or NMR depends on peak overlap and so forth. Refractive index increment, *dn/dc*, may be related to the increments of the homopolymers by the weight fractions of the two.

4.6.2.2 Molecular Weight

The techniques for molecular weight determination are, basically, the same as for homopolymerization. Osmometry and end group analysis provide information on the M_n; light scattering yields M_w (although the refractive index increment *dn/dc* will depend on the copolymer composition); sedimentation may provide M_z; and viscometry provides M_v.

The analysis of the overall molecular weight distribution is a little more complicated than in homopolymerization because of the effect of composition in the hydrodynamic properties of the copolymer when analyzed by size exclusion chromatography. However, cross-fractionation techniques have been successfully developed, to separate by composition and by molecular weight. For example, the joint chain length and composition distribution of low-conversion styrene–methyl methacrylate copolymers has been determined by this method[105] to conform to the predictions[73] set forth in Section 4.5.3. Combining size exclusion chromatography with thin-layer chromatography, similar information on that system has been obtained for high-conversion copolymers; the results are explained in terms of the Trommsdorff effect and the drift in comonomer composition.[106,107]

4.6.2.3 Sequence Length

The determination of triad fractions (and tacticity in both homopolymers and copolymers) has benefited immensely from high resolution NMR. This technique requires the assignment of the different peaks to nuclei in particular configurations; model compounds can be used for this purpose, but more important are the sophisticated NMR techniques that substantiate or prove the proper assignment of peaks. From this information, as discussed in Section 4.4, it sometimes is possible to discriminate between models. Reports on the analyses of a large number of copolymers using this technique are available.[108,109]

Problems

4.1 Prove equation (4.2.21), the equivalent to the Mayo–Lewis equation for systems obeying penultimate kinetics, given the definitions of the apparent reactivity ratios, equations (4.2.22)–(4.2.27).

4.2 Derive the equations for $(N_i)_n^{cumu}$ [equation (4.4.36)] and $S_{i,k}^{cumu}$ [equation (4.4.37)]. From the latter derivation, propose a simple way to calculate cumulative chain length distributions for free-radical homopolymerizations.

4.3 Compare the copolymerization of styrene with acrylonitrile with that of styrene and methyl methacrylate. For a system consisting of styrene (1) and acrylonitrile (2), assume that $r_1 = 0.331$ and $r_2 = 0.053$. For a styrene–methyl methacrylate system, assume $r_1 = 0.48$, and $r_2 = 0.42$. For each pair:
(a) Plot the instantaneous copolymer composition F_1 versus conversion for $f_{1,0} = 0.1, 0.3, 0.5, 0.7, 0.9$.
(b) Determine the azeotrope composition.
(c) Plot $(N_1)_n^{inst}$, $(N_2)_n^{inst}$, $(N_1)_n^{cumu}$, $(N_2)_n^{cumu}$, p_1, and p_2 as a function of overall conversion.

4.4 Under typical conditions for the copolymerization of styrene (2) and butadiene (1) to make SBR (styrene–butadiene rubber), the reactivity ratios are $r_1 = 1.4$, $r_2 = 0.8$. Ignore reaction of the second double bond in butadiene.
(a) Plot F_1 versus $f_{1,0}$.
(b) Plot the instantaneous copolymer composition F_1 versus overall conversion p for $f_{1,0} = 0.1, 0.3, 0.5, 0.7$, and 0.9 (to $p = 1$).
(c) What is the azeotrope composition?
(d) For $f_{1,0} = 0.7$, make the following plots:
$(N_1)_n^{inst}$, $(N_1)_w^{inst}$, $(N_1)_n^{cumu}$, and $(N_1)_w^{cumu}$ versus p
$(N_2)_2^{inst}$, $(N_2)_2^{inst}$, $(N_2)_n^{cumu}$, and $(N_2)_w^{cumu}$ versus p
p_1 and p_2 versus p
F_2 and F_2^{cumu} versus p
(e) In emulsion polymerization of styrene–butadiene with $f_{1,0} = 0.7$, the reaction is stopped at $p = 0.75$, and the resulting polymer has ~25% styrene content.[110] Do the foregoing calculations support this? What problems might result from too high a styrene content?

4.5 (a) Under what conditions will the analog of the Mayo–Lewis equation for the penultimate model reduce to that for the terminal model? Does this condition necessarily imply absence of penultimate effects?
(b) In the ^{1}H NMR spectra of styrene–methyl methacrylate copolymers, the methoxy resonance, which in the PMMA homopolymer appears at ~3.6 ppm, appears as many peaks in the range 2.10–3.70 ppm. Assume that all the methoxy resonance in the range 3.40–3.70 ppm is contributed by methoxy units centered in triads like MMM (M = MMA), while the remaining portion is contributed by methoxy units centered in triads like MMS or SMS (S = styrene). Given the areas $A(2.10$–$3.40)$ and $A(3.40$–$3.70)$, what is the average sequence length for MMA? What assumptions did you make in this calculation? Should the penultimate effect peculiar to the MMA–styrene system affect this result?

4.6 (a) Consider the anionic copolymerization of two monomers with $r_1 r_2 = 1$ (which is not too rare). What is the resulting chain length distribution? Assume that initiation is instantaneous and that no termination or transfer occurs. Is the sequence length distribution geometric?
(b) Consider the free-radical copolymerization of two mono-

mers with $r_1 r_2 = 1$ (which is pretty rare). What is the resulting live chain length distribution?

4.7 For the system $A_2 + B_2 + C_2$ described in Section 4.3, derive, by recursive arguments, equations for M_n and M_w.

4.8 For the two-component system styrene–methyl methacrylate, assume the following rate constants (and ignore the fact that this system exhibits penultimate effects):

$k_{11} = 176$ L/(mol · s)

$k_{21} = 734$ L/(mol · s)

$k_{tc11} = 7.20 \times 10^7$ L/(mol · s)

$k_{tc12} = 4.76 \times 10^8$ L/(mol · s)

$k_{tc22} = 1.86 \times 10^7$ L/(mol · s)

$2 f k_d I_0 = 8.00 \times 10^{-9}$ mol/(L · s)

$k_{12} = 352$ L/(mol · s)

$k_{22} = 367$ L/(mol · s)

Assume no initiator depletion, no diffusional limitations, no gel effect, no volume change, and so forth. The only drift occurs thus in overall monomer concentration and in comonomer composition. We polymerize with three different initial conditions:

$(M_1)_0 = 2.226$ mol/L

$(M_2)_0 = 6.678$ mol/L

$(M_1)_0 = 4.359$ mol/L

$(M_2)_0 = 4.359$ mol/L

$(M_1)_0 = 6.403$ mol/L

$(M_2)_0 = 2.134$ mol/L

Use the pseudokinetic rate constant method to predict the M_n^{cumu} and M_w^{cumu} to 80% conversion these three conditions, plotting them as a function of conversion. Was it necessary to integrate any differential equations (e.g., by Runge–Kutta)?

References

1. G. Odian, *Principles of Polymerization*, 3rd ed. Wiley, New York (1991).

2. For example, see C. Zhikuan, S. Ruona, and F. E. Karasz, *Macromolecules*, **25**, 6113 (1992).

3. T. Fukuda, K. Kubo, and Y.-D. Ma, *Prog. Polym. Sci.*, **17**, 875 (1992).

4. H. Dostal, *Monatsh. Chem.*, **69**, 424 (1936).

5. R. G. W. Norrish and E. F. Brookman, *Proc. R. Soc. London*, **A171**, 147 (1939).

6. E. Jenckel, *Z. Phys. Chem.*, **A190**, 24 (1942).

7. F. R. Mayo and F. M. Lewis, *J. Am. Chem. Soc.*, **66**, 1594 (1944).

8. F. T. Wall, *J. Am. Chem. Soc.*, **66**, 2050 (1944).

9. T. Alfrey and G. Goldfinger, *J. Chem. Phys.*, **12**, 205 (1944).

10. R. Simha and H. Branson, *J. Chem. Phys.*, **12**, 253 (1944).

11. F. T. Wall, *J. Am. Chem. Soc.*, **63**, 1862 (1941).

12. R. G. Fordyce and G. E. Ham, *J. Am. Chem. Soc.*, **73**, 1186 (1951).

13. E. Merz, T. Alfrey, and G. Goldfinger, *J. Polym. Sci.*, **1**, 75 (1946).

14. (a) W. G. Barb, *J. Polym. Sci.*, **10**, 49 (1953). (b) W. G. Barb, *J. Polym. Sci.*, **11**, 117 (1953).

15. (a) T. Fukuda, Y.-D. Ma, and H. Inagaki, *Polym. Bull*, **10**, 288 (1983). (b) T. Fukuda, Y.-D. Ma, and H. Inagaki, *Macromolecules*, **18**, 17 (1985).

16. T. Fukuda, Y.-D. Ma, and H. Inagaki, *Polym. J.*, **14**, 705 (1982).

17. Y.-D. Ma, T. Fukuda, and H. Inagaki, *Macromolecules*, **18**, 26 (1985).

18. D. J. T. Hill, J. H. O'Donnell, and P. W. O'Sullivan, *Macromolecules*, **15**, 960 (1982).

19. D. J. T. Hill, A. P. Lang, J. H. O'Donnell, and P. W. O'Sullivan, *Eur. Polym. J.*, **25**, 911 (1989).

20. F. P. Price, *J. Chem. Phys.*, **36**, 209 (1962).

21. K. Ito and Y. Yamashita, *J. Polym. Sci. A*, **3**, 2165 (1965).

22. C. W. Pyun, *J. Polym. Sci., A-2*, **8**, 1111 (1970).

23. C. X. Liu, X. Z. Kong, S. X. Zhang, and H. Z. Wang, *J. Macromol. Sci.-Chem.*, **A28**, 1063 (1991).

24. P. D. Bartlett and K. Nozaki, *J. Am. Chem. Soc.*, **68**, 1495 (1946).

25. F. S. Dainton and K. J. Ivin, *Proc. R. Soc. London*, **212**, 96, 207 (1952).

26. E. Tsuchida and T. Tomono, *Makromol. Chem.*, **141**, 265 (1971).

27. (a) J. A. Seiner and M. Litt, *Macromolecules*, **4**, 308 (1971). (b) M. Litt, *Macromolecules*, **4**, 312 (1971). (c) M. Litt and J. A. Seiner, *Macromolecules*, **4**, 314, 316 (1971).

28. C. U. Pittman and T. D. Rounsefell, *Macromolecules*, **8**, 46 (1975).

29. R. E. Cais, R. G. Farmer, D. J. T. Hill, and J. H. O'Donnell, *Macromolecules*, **12**, 835 (1979).

30. P. Karad and C. Schneider, *J. Polym. Sci., Polym. Chem.*, **16**, 1137 (1978).

31. D. J. T. Hill, J. H. O'Donnell, and P. W. O'Sullivan, *Macromolecules*, **16**, 1295 (1983).

32. K. Dodgson and J. R. Ebdon, *Eur. Polym. J.*, **13**, 791 (1977).

33. K. Dodgson and J. R. Ebdon, *Makromol. Chem.*, **180**, 1251 (1979).

34. R. G. Farmer, D. J. T. Hill, and J. H. O'Donnell, *J. Macromol. Sci.-Chem.*, **A14**, 51 (1980).

35. D. J. T. Hill, J. H. O'Donnell, and P. W. O'Sullivan, *Macromolecules*, **18**, 9 (1985).

36. J. R. Ebdon, *Makromol. Chem., Macromol. Symp.*, **10/11**, 441 (1987).

37. A. S. Brown, K. Fujimori, and I. Craven, *Makromol. Chem.*, **189**, 1893 (1988).

38. A. S. Brown, K. Fujimori, and I. E. Craven, *J. Polym. Sci., Polym. Chem.*, **27**, 3315 (1989).

39. G. S. Prementine, S. A. Jones, and D. A. Tirrell, *Macromolecules*, **22**, 770 (1989).

40. H. J. Harwood, *Makromol. Chem., Macromol. Symp.*, **10/11**, 331 (1987).

41. Yu. D. Semchikov, V. V. Izvolenskii, L. A. Smirnova, N. A. Kopylova, and T. G. Sveshnikova, *Polym. Sci.*, **35**, 594 (1993) (*Vysokomol. Soed. A*, **35**, 495).

42. M. Z. Elsabee, M. W. Sabaa, H. F. Naguib, and K. Furuhata, *J. Macromol. Sci.-Chem.*, **A24**, 1207 (1987).

43. B. Klumperman and K. F. O'Driscoll, *Polymer*, **34**, 1032 (1993).

44. T. P. Davis, *Polym. Commun.*, **31**, 442 (1990).

45. I. A. Maxwell, A. M. Aerdts, and A. L. German, *Macromolecules*, **26**, 1956 (1993).

46. Y. Y. Tan and G. Challa, *Makromol. Chem., Macromol. Symp.*, **10/11**, 215 (1987).

47. S. Polowinski, *J. Polym. Sci., Polym. Chem.*, **22**, 2887 (1984).

48. M. Szwarc and C. L. Perrin, *Macromolecules*, **18**, 528 (1985).

49. R. Szymanski, *Makromol. Chem.*, **187**, 1109 (1986).

50. (a) R. Szymanski, *Macromolecules*, **19**, 3003 (1986). (b) M. Szwarc, *Macromolecules*, **19**, 3003 (1986).

51. R. Szymanski, *Makromol. Chem.*, **188**, 2605 (1987).

52. S. Iwatsuki, T. Itoh, T. Higuchi, and K. Enomoto, *Macromolecules*, **21**, 1571 (1988).

53. F. Tüdos, T. Kelen, and T. Földes-Berezhnikh, *J. Polym. Sci., Symp.*, **50**, 109 (1975).

54. I. Czajlik, T. Földes-Berezsnich, F. Tüdos, and E. Madár-Vértes, *Eur. Polym. J.*, **19**, 147 (1983).

55. I. Skeist, *J. Am. Chem. Soc.*, **68**, 1781 (1946).

56. V. E. Meyer and G. G. Lowry, *J. Polym. Sci. A*, **3**, 2843 (1965).

57. K. F. O'Driscoll and R. Knorr, *Macromolecules*, **1**, 367 (1968).

58. F. Lopez-Serrano, J. M. Castro, C. W. Macosko, and M. Tirrell, *Polymer*, **21**, 263 (1980).

59. R. S. Dias, M. F. Barreiro, and M. R. N. Cosa, in *Fourth International Workshop on Polymer Reaction Engineering, DECHEMA Monographs*, Vol. 127, p. 199. VCH Publishers, New York (1992).

60. M. Farina, *Makromol. Chem.*, **191**, 2795 (1990).

61. M. H. Thiele, *J. Polym. Sci., Polym. Chem.*, **21**, 633, 1558 (1983).

62. L. H. Garcia-Rubio, J. F. MacGregor, and A. E. Hamielec, *ACS Symp. Ser.*, **197**, 87 (1982).

63. For example, see A. M. Aerdts, J. W. de Haan, and A. L. German, *Macromolecules*, **26**, 1965 (1993).

64. R. L. Miller and L. E. Nielsen, *J. Polym. Sci.*, **46**, 303 (1960).

65. D. A. Tirrell, Copolymerization, in *Encyclopedia of Polymer Science and Engineering*, Vol. 4, 2nd ed., H. F. Mark, N. M. Bikales, C. G. Overberger, and G. Merges, Eds. Wiley, New York (1986).

66. J. L. Koenig, *Chemical Microstructure of Polymer Chains*. Wiley, New York (1980).

67. K. F. O'Driscoll and T. P. Davis, *J. Polym. Sci. Lett.*, **27**, 417 (1989).

68. G. Moad, D. H. Solomon, T. H. Spurling, and R. A. Stone, *Macromolecules*, **22**, 1145 (1989).

69. F. P. Price, in *Markov Chains and Monte Carlo Calculations in Polymer Science*, G. G. Lowry, Ed. Dekker, New York (1970).

70. R. Galván and M. Tirrell, *J. Polym. Sci., Polym. Chem.*, **24**, 803 (1986).

71. (a) S. Sakakibara and K. Ito, *Polym. Commun.*, **29**, 339 (1988). (b) K. F. O'Driscoll and T. P. Davis, *Polym. Commun.*, **30**, 317 (1989).

72. (a) L. H. Garcia-Rubio, Ph.D. thesis, McMaster University (1981). (b) L. H. Garcia-Rubio, M. G. Lord, J. F. MacGregor, and A. E. Hamielec, *Polymer*, **26**, 2001 (1985).

73. W. H. Stockmayer, *J. Chem. Phys.*, **13**, 199 (1945).

74. H. W. Melville, B. Noble, and W. F. Watson, *J. Polym. Sci.*, **2**, 229 (1947).

75. C. Walling, *J. Am. Chem. Soc.*, **71**, 1930 (1949).

76. A. M. North, *The Kinetics of Free Radical Polymerization*. Pergamon Press, Oxford (1966).

77. H. W. Melville and L. Valentine, *Proc. Soc. London*, **A200**, 337 (1950).

78. K. Ito, *J. Polym. Sci. A-1*, **8**, 2819 (1970).

79. K. Ito, *J. Polym. Sci. A-1*, **9**, 867 (1971).

80. W. H. Ray, T. L. Douglas, and E. W. Godsalve, *Macromolecules*, **4**, 166 (1971).

81. C.-C. Lin, W.-Y. Chiu, and C. T. Wang, *J. Appl. Polym. Sci.*, **23**, 1203 (1979).

82. G. M. Burnett and H. R. Gersmann, *J. Polym. Sci.*, **28**, 655 (1958).

83. J. N. Atherton and A. M. North, *Trans. Faraday Soc.*, **58**, 2049 (1962).

84. S. W. Benson and A. M. North, *J. Am. Chem. Soc.*, **81**, 1339 (1959).

85. A. M. North and G. A. Reed, *Trans. Faraday Soc.*, **57**, 859 (1961).

86. O. Procházka and P. Kratochvíl, *J. Polym. Sci., Polym. Chem.*, **21**, 3269 (1983).

87. W. H. Ray, *Macromolecules*, **4**, 162 (1971).

88. D. S. Achilias and C. Kiparissides, *J. Macromol. Sci.-Rev. Macromol. Chem. Phys.*, **C32**, 183 (1992).

89. A. E. Hamielec and J. F. MacGregor, in *Polymer Reaction Engineering*, K. H. Reichert and W. Geiseler, Eds. p. 21. Hanser, Munich (1983).

90. H.Tobita and A. E. Hamielec, *Makromol. Chem., Macromol. Symp.*, **20/21**, 501 (1988).

91. H.Tobita and A. E. Hamielec, *Polymer*, **32**, 2641 (1991).

92. D. R. Miller, *Makromol. Chem., Macromol. Symp.*, **30**, 57 (1989).

93. N. A. Dotson, Ph.D. thesis, University of Minnesota (1991).

94. (a) R. Galván and M. Tirrell, *Comp. Chem. Eng.*, **10**, 77 (1986). (b) R. Galván and M. Tirrell, *Chem. Eng. Sci.*, **41**, 2385 (1986).

95. M. Fineman and S. D. Ross, *J. Polym. Sci.*, **5**, 259 (1950).

96. D. Braun, W. Brendlein, and G. Mott, *Eur. Polym. J.*, **9**, 1007 (1973).

97. P. W. Tidwell and G. Mortimer, *J. Polym. Sci. A*, **3**, 369 (1965).

98. T. Kelen and F. Tüdos, *J. Macrom. Sci.-Chem.*, **A9**, 1 (1975).

99. K. F. O'Driscoll and P. M. Reilly, *Makromol. Chem., Macromol. Symp.*, **10/11**, 355 (1987).

100. H. Patino-Leal, P. M. Reilly, and K. F. O'Driscoll, *J. Polym. Sci., Polym. Lett.*, **18**, 219 (1980).

101. P. W. Tidwell and G. A. Mortimer, *J. Macrom. Sci.-Rev. Macrom. Chem.*, **C4**, 281 (1970).

102. M. J. Box, *Technometrics*, **12**, 219 (1970).

103. B. Yamada, M. Itahashi, and T. Otsu, *J. Polym. Sci., Polym. Chem.*, **16**, 1719 (1978).

104. M. Dube, R. A. Sanayei, A. Penlidis, K. F. O'Driscoll, and P. M. Reilly, *J. Polym. Sci., Polym. Chem.*, **29**, 703 (1991).

105. S. Teramachi and Y. Kato, *Macromolecules*, **4**, 54 (1971).

106. S. Teramachi, A. Hasegawa, M. Akatsuka, A. Yamashita, and N. Takemoto, *Macromolecules*, **11**, 1206 (1978).

107. S. Teramachi, A. Hasegawa, and S. Yoshida, *Macromolecules*, **16**, 542 (1983).

108. F. A. Bovey, *High Resolution NMR of Macromolecules*. Academic Press, New York (1972).

109. J. C. Randall, *Polymer Sequence Determination—Carbon-13 NMR Method*. Academic Press, New York (1977).

110. F. Rodriguez, *Principles of Polymer Systems*, 2nd ed. McGraw-Hill, New York (1982).

5

NONLINEAR POLYMERIZATION

5.1 Introduction: Chemistry of Branching

If bulk properties of polymers can be greatly affected by such subtle attributes as chain microstructure, it is reasonable to assume that structural characteristics on larger scales can have profound effects as well. Thus far in the text we have dealt almost exclusively with linear polymers, but properties should be greatly altered if chain architecture is allowed to be nonlinear (i.e., if branching and network formation occur). We begin with a brief list of typical chemistries or processes that can lead to branching or network formation, either by design or intrinsically.

5.1.1 Long- and Short-Chain Branching

5.1.1.1 Polyethylene

The most important commodity plastic is polyethylene, and that knowledge alone should warrant our concern in its structure, In fact, the different kinds of polyethylene (low-density, high-density, and linear low-density) are distinguished precisely by the degree and kind of branching. Polymerization of ethylene by free-radical polymerization in high-pressure reactors is marked by *long-chain branching* resulting from transfer to polymer. This transfer reaction occurs through abstraction of a secondary proton,[1] as shown in Figure 5.1. In this process a live chain is killed, and a dead chain resurrected to form a polymer with a trifunctional branch point.

When the same basic chemistry occurs through intramolecular backbiting reactions, as shown in Figure 5.2, the result is short-chain branching. The existence of branches of these kinds has been confirmed by numerous studies, especially by ^{13}C NMR.[1]

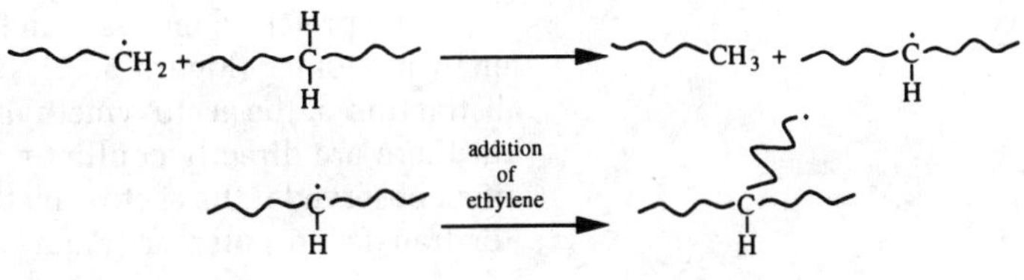

Figure 5.1.
Long-chain branching in the free-radical polymerization of ethylene.[1]

Figure 5.2.
Short-chain branching in the free-radical polymerization of ethylene, showing formation of (a) *n*-hexyl, *n*-amyl, and *n*-butyl branches and (b) ethyl branches (circled).[1]

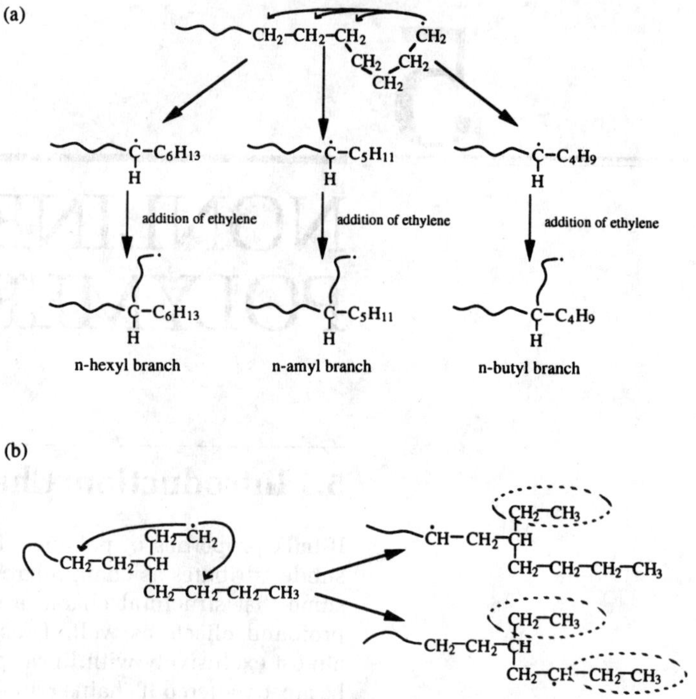

(a)

n-hexyl branch   n-amyl branch   n-butyl branch

(b)

The reactions are favored sterically; the *n*-butyl branches are especially prevalent (occurring on the average once every 100 monomers), as are the ethyl branches, which probably form after the *n*-butyl branches as shown in Figure 5.2b.

Long- and short-chain branching affects the properties of polyethylene greatly, primarily through crystallinity. Because polyethylene produced by the high-pressure technology is highly branched, the degree of crystallinity, and thus the density, is lower. Polyethylene produced by coordination catalysts (see Chapter 7), which forbid such reactions, is highly crystalline and thus of high density. The two products are different and have different end uses.

5.1.1.2 Poly(vinyl acetate)

If the polymerization of vinyl acetate is run to high conversions, the product is a highly branched polymer with long branches, which suggests that transfer to polymer occurs. The high reactivity of the vinyl acetate radical makes both transfer to polymer and transfer to monomer very likely; that is, both C_m and C_p, the constants for transfer to monomer and to polymer, are relatively large. The similarity between the two reactions is shown in Figure 5.3, which shows the proposed sites and resulting structures from transfer to polymer and to monomer.

The preferred mechanism for transfer is unclear. For transfer to monomer (Figure 5.3b), considerable evidence supports abstraction of the acetoxymethyl proton as the preferred route,[2,3] but there are directly conflicting reports on the kinetic isotope effect observed if the acetoxymethyl group is triply deuterated.[3,4] For transfer to polymer (Figure 5.3a), the evidence is likewise

Figure 5.3.
Transfer mechanisms (a) to polymer and (b) to monomer in the polymerization of vinyl acetate.[1]

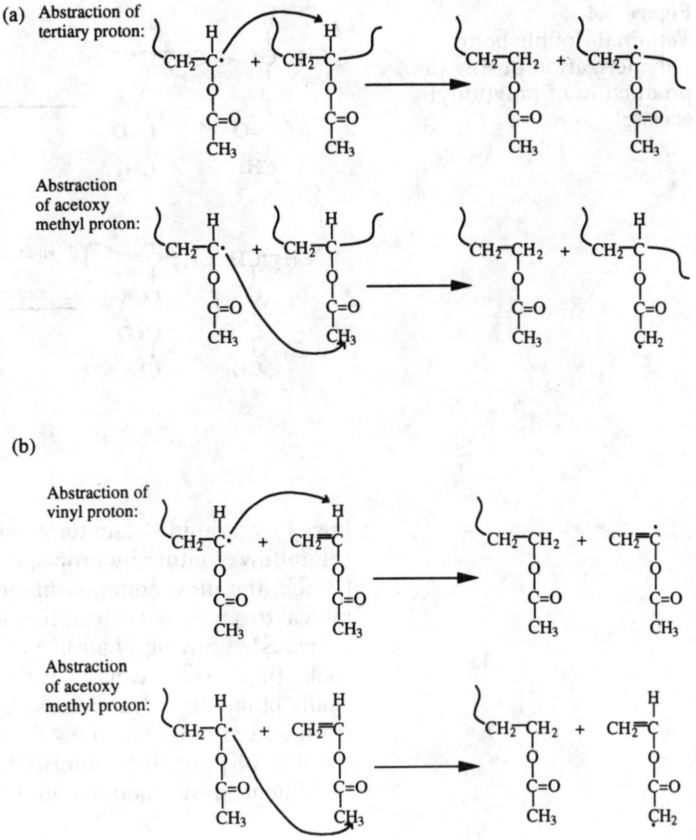

(a) Abstraction of tertiary proton:

Abstraction of acetoxy methyl proton:

(b)

Abstraction of vinyl proton:

Abstraction of acetoxy methyl proton:

divided, but if it is true that transfer of the tertiary proton occurs two to four times more easily than transfer of the acetoxy proton,[5] doubt is cast on other work relying on hydrolysis of the ester linkage to obtain the "primary chains."[6] Preference of the tertiary proton in the transfer to polymer is not necessarily inconsistent with preference of the acetoxymethyl proton in transfer to monomer. Short-chain branching may also occur, although the types of branch formed are not as well established.[2]

Transfer to polymer, regardless of the site of transfer, results in a trifunctional branch point, like that of the long-chain branching in polyethylene. In addition, though, note that transfer to monomer leaves at the beginning of the new chain a terminal double bond that can later react to form a trifunctional branch, as shown in Figure 5.4. (In the same vein, the terminal double bond left by disproportionation may also polymerize to form a trifunctional branch point.) Thus the high reactivity of the vinyl acetate radical encourages branching through transfer to both polymer and monomer.

5.1.1.3 Poly(vinyl chloride)

Another significant commodity plastic is poly(vinyl chloride), the polymerization of which is much more complex than that of vinyl acetate.[1] Transfer to monomer apparently occurs through a complex mechanism comprising head-to-head addition, fol-

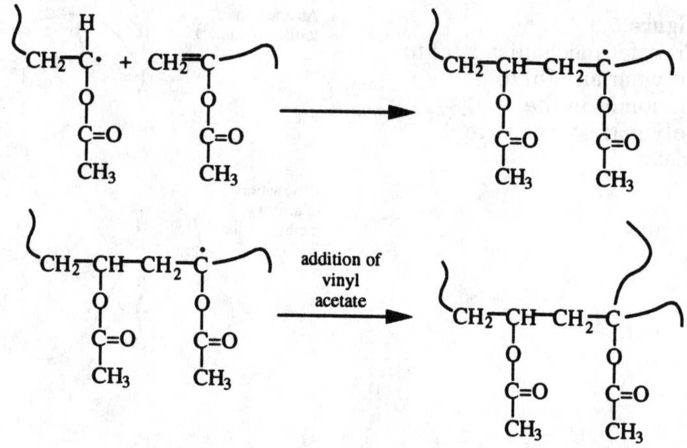

Figure 5.4.
Terminal double-bond polymerization during the production of poly(vinyl acetate).

lowed by a rapid, 1,2-chlorine shift to yield a more stable radical, followed either by propagation (resulting in a chloromethyl branch, the most common branch) or by transfer of a chlorine radical to a monomer (resulting in an allylic double bond on the *previously* growing chain).[7-9] Short-chain branching occurs by backbiting mechanisms as well as by chlorine shifts.[7-12] Long-chain branching also occurs. The complicated structures resulting from these reactions have a great impact on the thermal stability of poly(vinyl chloride), and the concentration thereof is influenced by reaction conditions.[9,13,14]

5.1.2 Graft Copolymers

We have already mentioned graft copolymers in the context of copolymers (Chapter 4), but since they are nonlinear, they more properly belong in this chapter. We could devise a number of synthetic routes leading to graft copolymers: by growing a second chain from a premade chain bearing an active group, by reacting two premade dissimilar polymers bearing mutually reactive groups, by copolymerization with "macromonomers" (polymers end-functionalized with reactive double bonds), and so forth. We do not extensively catalog the numerous possibilities here; rather, we give a few good examples.

5.1.2.1 Acrylonitrile–Butadiene–Styrene (ABS) Resins

ABS resins are not strict terpolymers made by simultaneous terpolymerization. Instead, they are copolymers of styrene and acrylonitrile grafted onto butadiene.[15,16] In this way the desired heterogeneous structure is guaranteed. There are many different methods of producing ABS resins, some grafting in a solution of the acrylonitrile and styrene monomers, and others grafting onto polybutadiene latices.[16]

Figure 5.5.
Grafting of nylon polymer
onto maleic anhydride
functionalized rubber.[15]

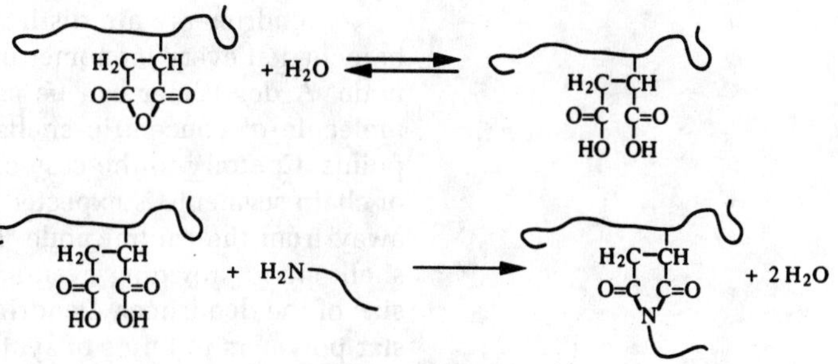

5.1.2.2 Reactive Blending of Nylon and Ethylene–Propylene Rubber

In a blend of two immiscible polymers, stability of the material, which prevents macroscopic phase separation, can be provided by a blending in a compatible block copolymer that will tend to reside at the interface, or by producing a block copolymer (or a more intricate architecture) in situ by reacting the two polymers at the interface. In the case of ethylene–propylene copolymer rubber (EPR) and nylon, we form, as in the case of ABS resins, rubbery and glassy (or semicrystalline) phases.

With the appropriate functionalization of the two polymers, any of the stepwise chemistries of Chapter 2 could be used to accomplish the reactive blending. The particular route is determined by ease of functionalization, stability before blending, and so forth. One method is to use a maleic anhydride functionalized rubber and an amine-terminated nylon, which react as is shown in Figure 5.5.[17] This route has been used for blends of polypropylene with nylon 6,[18] of polyethylene with nylon 6/6,[19] and of poly(ethylene-*co*-propylene) with nylon 6.[20]

5.1.3 Star Polymers and Dendrimers

Polymers with a star architecture, as shown in Figure 5.6, are desirable for many reasons; one end use is as an additive to increase viscosity—as is done in the case of synthetic polyisoprene, which lacks the branches of natural rubber (natural polyisoprene). There are numerous routes to star polymers, and these have varying degrees of uniformity of the number of branches and the length thereof. We mention only examples here. Either termination of anionic polymerization with a stoichiometric amount of multifunctional chain killer or anionic polymerization with a multifunctional initiator would yield great uniformity in both the number of arms and their length. A broader distribution in the number of arms could be obtained with nonstoichiometric amounts of chain killer. One might also allow for a much more random process, and a larger connecting node, by forming a functionalized core from, say, divinyl benzene, either following or preceding the linear anionic polymerization. In a free-radical polymerization, one might use a multifunctional transfer agent or a multifunctional initiator.

Figure 5.6.
Star polymer.

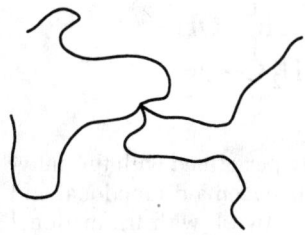

Dendrimers[21] are distinguished from star polymers being branched at every monomer unit, rather than just at the central node. A dendrimer can be envisioned as a roughly spherical molecule of concentric shells that contain successive branch points. Contrary to the case of star polymers, then, the density of chain segments is expected to increase, rather than decrease, away from the central node. Because of the branching at each shell, steric crowding eventually occurs, limiting the ultimate size of the dendrimer. Dendrimers are thus quite different from star polymers in terms of synthesis, properties, and end uses.

5.1.4 Crosslinking (Vulcanization)

The vulcanization or crosslinking of rubber is probably familiar to most students, and it has been a viable technology since 1839 when Goodyear started making tires from natural rubber reacted with sulfur. The resulting crosslinked rubber is desirable because of its superior physical strength and its insolubility in any non-degradative solvent.

The chemistry of sulfur crosslinking is complicated, but appears to be ionic.[1,22] The process is generally quite inefficient if left to itself, between 40 and 50 sulfur atoms being consumed per crosslink. Because of this, vulcanization is often carried out in the presence of organosulfur compounds, which accelerate the process. It is also possible to abandon the route of crosslinking with sulfur and use some other mechanism, such as crosslinking by peroxides or radiation, both of which result in a free-radical mechanism.

5.1.5 Nonlinear Polymerizations

Thus far in this chapter we have given examples of nonlinear structure produced either by side reactions (long- and short-chain branching) or by deliberate modification of existing polymer chains (grafting and vulcanization). However, any of the chemistries discussed in Chapters 2 and 3 can be used to produce branched or network polymers directly, provided the monomers are *multifunctional*. In this way, the polymerization itself is nonlinear and produces branched or network polymers.

For example, one can produce a crosslinked polyester through the reaction of phthalic acid and glycerin:

$$
\underset{\text{(phthalic acid)}}{\text{HO}\overset{|}{\underset{O=C}{C}}\quad\overset{OH}{\underset{C=O}{C}}}\quad + \quad \begin{array}{l} H_2C-OH \\ HC-OH \\ H_2C-OH \end{array} \tag{5.1.1}
$$

This condensation reaction (generally performed with the anhydride to eliminate the condensate[1]) between a difunctional and a trifunctional monomer leads to a network with trifunctional branch points. Perhaps a more common example is a crosslinked polyurethane, which can be made from a triol of some sort and a diisocyanate, such as 1,6-hexane diisocyanate or toluene diisocyanate:

$$HO \qquad\qquad OH$$

$$+ \; O=C=N-(CH_2)_6-N=C=O \qquad (5.1.2)$$

$$OH$$

Branch points are not restricted to being trifunctional, but of course can have higher functionality.

Chainwise mechanisms can also be used to form nonlinear polymers. The most common chainwise route, not surprisingly, is free-radical (co)polymerization between, for example, a mono-unsaturated and a diunsaturated monomer. The copolymer of styrene and divinylbenzene, [see (5.1.3)]

$$HC=CH_2 \qquad HC=CH_2$$

$$+ \qquad\qquad (5.1.3)$$

$$HC=CH_2$$

is commonly used for chromatographic packing and as supports for ion-exchange resins. After the copolymerization of these into a chain, the remaining double bond on the divinylbenzene can be incorporated into a second chain. This resembles vulcanization condensed into one step: chains are produced as they are cross-linked, rather than these two processes occurring sequentially.

Nonlinear polymerization can of course involve previously produced polymers, generally oligomers. The "curing" or cross-linking of an epoxy resin (Section 2.3.5) with amine is a good example of this. The reaction is as follows:

$$\begin{array}{cc} O & OH \\ / \backslash & | \\ R_{(1)}-CH_2-CH + H_2N-R_{(2)} \rightarrow R_{(1)}-C-CH_2-NH-R_{(2)} \end{array} \qquad (5.1.4)$$

The remaining secondary amine can also react in similar fashion, although with lower reactivity. Thus primary amines are bifunctional, and so the usual curing agents, such as diethylene triamine, are of functionality greater than 3, leading to network formation. This is a stepwise route, but epoxy resins may also be cured by chainwise (ring-opening) routes. In similar fashion, the unsaturated polyester resins mentioned at the end of Section 2.3.2 are commonly "cured" by free-radical copolymerization with styrene to form crosslinked networks.

Some chemical mechanisms are used mainly, or can only be used, for nonlinear polymerization, hence were neglected in Chapters 2 and 3. The reaction between phenols and aldehydes, as shown in Figure 5.7, to form phenolic resins is significant both commercially (see Table 1.1) and historically, since it pro-

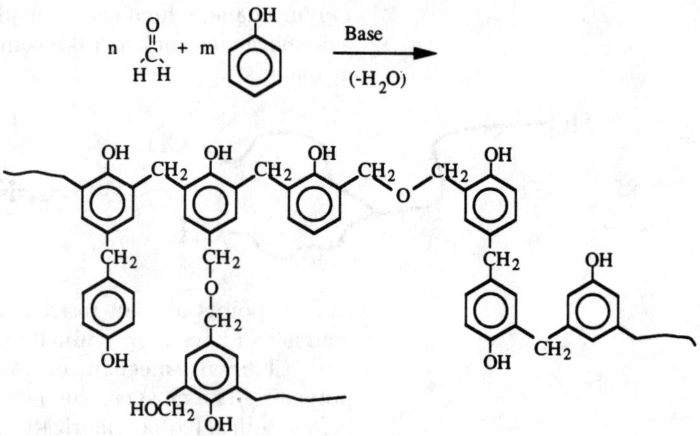

Figure 5.7.
Figure 5.7.
Reaction of phenol and
formaldehyde to form
phenolic resins. Product
may contain a mixture of
singly, doubly, or triply
reacted phenol and
methylene or etheric
linkages, as shown.

duced the first fully synthetic polymer (Bakelite). Aldehydes
other than the formaldehyde shown in Figure 5.7 can be used,
as can substituted phenols, such as p-cresol (which leads to a
linear polymer). Although phenolic resins enjoy a wide variety
of uses, their most significant application is as adhesive wood
products (e.g., particle board). The related amino plastics, which
involve the reaction of formaldehyde with urea or melamine,
are similar in synthesis and applications. A second example, of
increasing importance in recent years, consists of the inorganic
ceramic networks formed by the sol–gel process, which avoid
high temperature processes used for making traditional ceramics
and glass. Formation of these polysilicates may be catalyzed by
acid or base, although the resulting morphology differs greatly
between these two.[23] Again, the complete reaction scheme is
complicated,[23,24] but it involves first the hydrolysis (perhaps partial) of a monomer such as tetraethoxysilane [$Si(OCH_2CH_3)_4$] followed by condensation, giving a by-product of water or alcohol.

Nonlinear polymerization and vulcanization will be our
focus for the majority of the chapter, although the effects of long-
chain branching are investigated in Section 5.8. We first turn to
the paradigm of step homopolymerization of A_f monomers, to
illustrate the basic features of a nonlinear polymerization, gelation, and network formation.

5.2 Paradigm of the Pregel Regime: A_f Homopolymerization

The traditional paradigm for nonlinear polymerization is homopolymerization of monomers bearing f identical functional
groups, A_f, a case more useful in theory than in practice. The
usual example, polyetherification of pentaerythritol ($f = 4$), does
not conform to the assumptions of equal and independent reactivity (i.e., no unequal reactivity or substitution effect), which we
assume here. As we have seen, the case of $A_f + B_g$ step polymerization is more common, but it introduces mathematical difficulties
that obscure the salient features of nonlinear polymerization.

The natural successor to the paradigm of Chapter 2, AB_{f-1} poly-merization, is not useful because it cannot form a network,[25] although the inability of AB_2 polymerization to form a network is used to prepare hyperbranched polymers.[21]

5.2.1 Derivation from Kinetic Equations

Writing down the kinetic equations for this reaction is subtle, so we proceed step by step. The number of unreacted functional groups remaining on an i-mer is related to the number of links in the i-mer. If no cycles form, then there are $i - 1$ links, just as in a linear polymerization. Two functional groups disappeared per link, and since there were originally fi functional groups on the monomers that were incorporated into this i-mer,

$$\text{number of functional groups on an } i\text{-mer} = fi - 2(i - 1) = (f - 2)i + 2 \quad (5.2.1)$$

The reaction rate between an i-mer and a j-mer is then simply given as follows:

$$\text{rate of reaction between an } i\text{-mer and a } j\text{-mer} = k[\text{number of functional groups on an } i\text{-mer}]P_i\,[\text{number of functional groups on a } j\text{-mer}]P_j \quad (5.2.2)$$

which becomes, in light of equation (5.2.1):

$$\text{rate of reaction between an } i\text{-mer and a } j\text{-mer} = k[(f - 2)i + 2]P_i\,[(f - 2)j + 2]P_j \quad (5.2.3)$$

The final step is to incorporate this knowledge into the evaluation equation for an i-mer:

$$\frac{dP_i}{dt} = \{\text{production of } i\text{-mer from reaction of a } j\text{-mer with an } (i - j)\text{-mer}\} - \quad (5.2.4)$$

$$\{\text{disappearance of } i\text{-mer by reaction with all } j\text{-mers}\}$$

or:

$$\frac{dP_i}{dt} = \frac{1}{2}k \sum_{j=1}^{i-1} [(f - 2)j + 2][(f - 2)(i - j) + 2]P_j P_{i-j}$$

$$- k[(f - 2)i + 2]P_i \sum_{j=1}^{\infty} [(f - 2)j + 2]P_j \quad (5.2.5)$$

The remarkable thing about this equation is that, for $f > 2$, the summand or kernel depends on the size of the participating polymers (i and j) as well as their concentrations. Indeed, these equations show us that for sufficiently large i and j, the rate of reaction between the two is proportional to $(ij)^1$. For the linear polymers of Chapter 2, the reaction rate is independent of i and j [i.e., goes as $(ij)^0$] because each chain bears only two functional groups. But for A_f homopolymerization, the number of functional groups increases with the size of the molecule i. Thus, the larger a molecule is, the more rapidly it reacts, to become yet larger, and so on. The growth of a molecule accelerates, or cascades, in this nonlinear polymerization.

This accelerating growth should not be interpreted to mean that the rate of polymerization itself increases; nothing of the kind happens. The rate of disappearance of A groups is given by:

$$\frac{dA}{dt} = -kA^2 \tag{5.2.6}$$

which has the solution:

$$\frac{A}{A_0} = \frac{1}{1 + kA_0 t} \tag{5.2.7}$$

or that the conversion p is given as follows:

$$p = \frac{kfP_{1,0}t}{1 + kfP_{1,0}t} \tag{5.2.8}$$

where the initial monomer concentration $P_{1,0}$ has been substituted for A_0. All this is no different from the case of an A_2 homopolymerization; the cascading is in the size distribution only.

With the appropriate transform techniques,[26] one can, with difficulty, obtain the size distribution, which is given by:

$$P_i = \frac{[(f-1)i]!}{[(f-2)i+2]!i!} \frac{(1-p)^2}{p} \left(p(1-p)^{f-2} \right)^i fP_{1,0} \tag{5.2.9}$$

It is not difficult, on the other hand, to show that this solution satisfies equation (5.2.5).[27] One can also statistically derive this distribution, which we will do.

The moments of the distribution or the average molecular weights are more easily found by direct moment integration than from the distribution. The results are:

$$\frac{d\mu_0}{dt} = -\frac{1}{2}k[(f-2)\mu_1 + 2\mu_0]^2 = -\frac{1}{2}kA^2 \tag{5.2.10}$$

$$\frac{d\mu_1}{dt} = 0 \tag{5.2.11}$$

$$\frac{d\mu_2}{dt} = k[(f-2)\mu_2 + 2\mu_1]^2 \tag{5.2.12}$$

The first two equations are intuitive. The first states that the rate of decrease of the number of molecules is half that of A functional group disappearance, which we have already noted, since two functional groups react for each link. The second relation is a statement of conservation of monomeric units (not mass, since a by-product such as water may be removed). If these equations are solved with the initial condition that only monomer exists at $t = 0$, so that $\mu_i = P_{1,0}$ for all i, then:

$$\mu_0 = P_{1,0}\left(1 - \frac{f}{2}p\right) \tag{5.2.13}$$

$$\mu_1 = P_{1,0} \tag{5.2.14}$$

Figure 5.8.
Evolution of DP_w and DP_n for A_3 step homopolymerization.

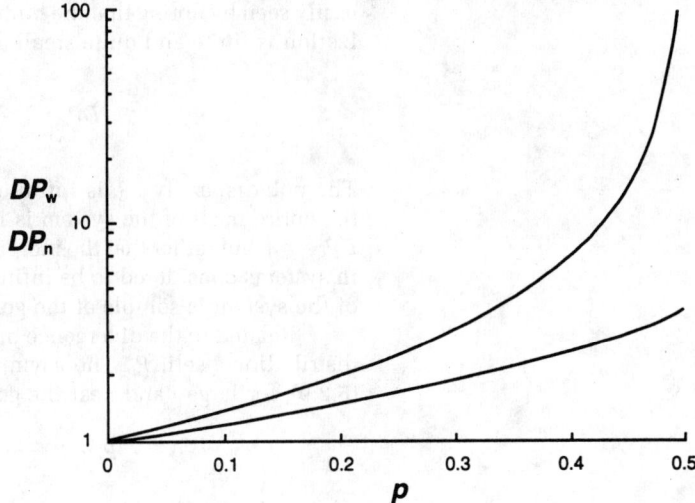

This leaves us with the number- and weight-average degrees of polymerization as follows:

$$DP_n = \frac{1}{1 - (f/2)p} \tag{5.2.16}$$

$$DP_w = \frac{1 + p}{1 - (f - 1)p} \tag{5.2.17}$$

In equation (5.2.17) we note the most important feature we will encounter in this chapter: DP_w diverges at a conversion less than unity, specifically at a critical conversion p_c given by:

$$p_c = \frac{1}{f - 1} \tag{5.2.18}$$

This defines the *gel point*. The higher the functionality of the monomer, the smaller the conversion necessary for gelation.

The divergence of DP_w^* shown in Figure 5.8 comes from the cascading growth in size leading to the formation of a (single) molecule of macroscopic proportions, an incipient gel the size of which is limited only by the size of the container in which it resides. It is "incipient" because despite the drastic effects on physical properties, which we will describe shortly, it is not very substantial. It contains little of the mass of the system, which is

---

The apparent (but not real) divergence of DP_n at p = 2/f is discussed later in the chapter. For the time being, note that 2/f > 1/(f − 1) for f > 2.

$$\mu_2 = P_{1,0} \frac{1 + p}{1 - (f - 1)p} \tag{5.2.15}$$

easily seen by noting that the number-average degree of polymerization is finite and quite small at the gel point:

$$DP_n(p_c) = 2\frac{f-1}{f-2} \qquad (5.2.19)$$

The polydispersity, Q, is infinite. Gelation does not mean that the entire mass of the system is in the gel (which would imply $DP_n = \infty$), but rather that the incipient gel has zero weight fraction in systems considered to be infinitely large. Nearly all the mass of the system is soluble at the gel point.

Related to the divergence of DP_w is the behavior of the size distribution itself, P_i. The asymptotic form of the distribution (5.2.9), for large i and near the gel point, is rather simple:

$$P_i \propto i^{-5/2} \exp(-c\varepsilon^2 i) \qquad (5.2.20)$$

where ε is a reduced distance to the gel point:

$$\varepsilon = \frac{p_c - p}{p_c} \qquad (5.2.21)$$

This can be put into the general form:

$$P_i \propto i^{-\tau} \exp(-c\varepsilon^{1/\sigma} i) \qquad (5.2.22)$$

where $\tau = 5/2$ and $\sigma = 0.5$. The form of the distribution is entirely different from, say, the geometric distribution. The geometric distribution essentially decays exponentially with size, while this distribution is power law in size but has an exponential cutoff factor depending on the (square of the) distance from the gel point. This exponential factor keeps the molecular weights finite until the gel point, at which point the exponential decay vanishes and we are left with a power law. The exponent of this power

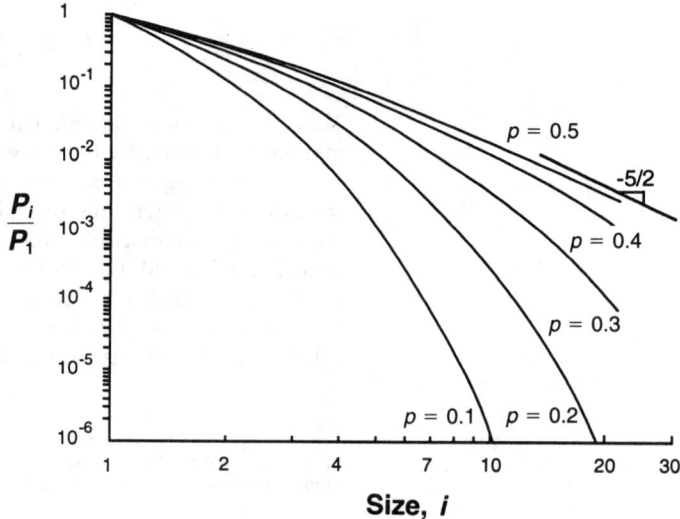

Figure 5.9.
Size distribution in A_3 step homopolymerization, up to $i = 22$.

law determines which moments diverge; in this case, all moments greater than $\mu_{3/2}$ diverge at the gel point, and so DP_w diverges. At the gel point, then, the size distribution becomes purely power law with no exponential character, as well as being the point at which DP_w diverges. In Figure 5.9 we show the exact size distribution (up to $i = 22$) for various conversions, illustrating the power law behavior as modified by the exponential cutoff, which shifts to progressively higher values of i as $p \to p_c$.

Both these definitions assign to gelation a particular point in time or conversion, but in practice neither definitions is used because of limitations of the analytical techniques. At sufficiently high molecular weights, M_w cannot be determined by light scattering; indeed, this would be impossible even if preparative cutoffs (e.g., from filtering) could be avoided.[28] Size exclusion chromatography also suffers from similar preparative cutoffs and from the more intricate relation between molecular weight and hydrodynamic volume which exists in randomly branched polymers.

Thus the more common ways of determining the gel point are more mechanical. As illustrated in Figure 5.10, the viscosity of the material diverges, and a modulus begins to appear.[29] Either viscosity or modulus may be used to measure the gel point, but will obviously give a small range rather than a precise point because of instrumental limitations. Figure 5.10 also indicates that a gel fraction, the portion of the material that is insoluble, appears only after the gel point; solubility in a suitable (good) solvent may thus also be used to determine the gel point. A more recent method is to measure the loss and storage moduli G' and G'' and determine the time at which both exhibit a power-law dependence on frequency, hence are parallel.[30] Many other less precise techniques, such as bubble rise, also are used.

5.2.2 Statistical Derivations

Statistical methods provide a somewhat simpler approach to finding the DP_n, DP_w, and p_c, and one of the simplest entails the

Figure 5.10.
Physical characteristics of a gelling system: divergence of the viscosity η at the gel point, disappearance of soluble fraction w_{sol} after gelation, and the growth of modulus E after gelation.

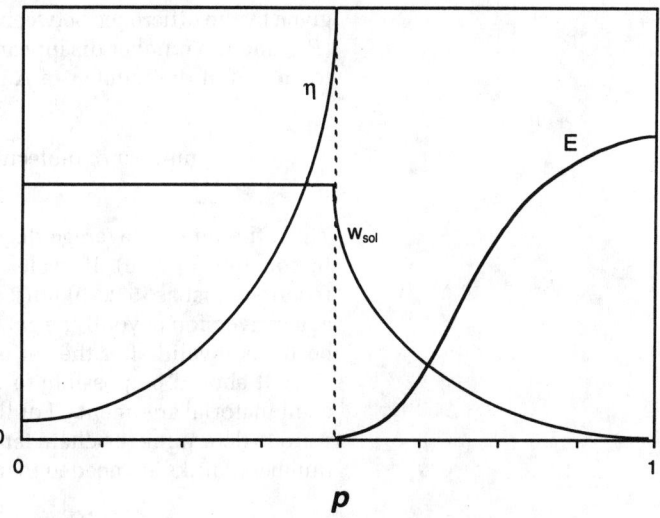

recursive approach.[31] For example, DP_w is the average number of monomers on a randomly chosen monomer. Clearly:

$$DP_w = 1 + fE(N_A^{out}) \qquad (5.2.23)$$

The expected number of groups looking out of an A group is:

$$E(N_A^{out}) = pE(N_A^{in}) \qquad (5.2.24)$$

Similarly:

$$E(N_A^{in}) = 1 + (f - 1)E(N_A^{out}) \qquad (5.2.25)$$

Solving these equations yields:

$$E(N_A^{out}) = \frac{p}{1 - (f - 1)p} \qquad (5.2.26)$$

Thus DP_w is given as in equation (5.2.17).

For linear polymers, we determined DP_n by a straightforward recursive argument, which began by grabbing a molecule by its end and determining the average length of the molecule. This worked because the molecules had the same number of ends (two) and so had an equal chance of being grabbed. For star polymers monodisperse in the number of arms, this technique would still work, but not for randomly branched polymers, which have a distribution of the number of "ends." Such an incorrect derivation gives $1/[1 - (f - 1)p]$, very similar to DP_w, which is not surprising in the light of our earlier finding that the number of functional groups or ends eventually goes as the size.

To find DP_n, then, we resort to a stoichiometric argument that is always valid in the absence of cycles. First, we write:

$$DP_n = \frac{\text{number of monomers}}{\text{number of molecules}} \qquad (5.2.27)$$

The number of monomers is $P_{1,0}$. The number of molecules is given by the difference between the original number of molecules ($P_{1,0}$) and the number disappeared, which we have already noted to be half of the number of A groups disappeared. Hence:

$$\text{number of molecules} = P_{1,0}\left(=1 - \frac{f}{2}p\right) \qquad (5.2.28)$$

Thus, the number-average degree of polymerization is as given in equation (5.2.16). If cycles occur, then equation (5.2.16) is incorrect, just as (5.2.10) must be wrong. The divergence of DP_n at a conversion beyond the gel point is fictional, as the equation becomes invalid after the gel point (see Section 5.4).

It should be possible to derive the size distribution from combinatorial arguments. For linear polymers, it was straightforward to determine the chain length distribution because a certain number of links are needed to make an i-mer: $i - 1$. The (normalized) distribution was then obvious: $(1 - p)p^{i-1}$. For A_f homopolymerization, the existence of an acyclic or treelike i-mer requires $i - 1$ links and $(f - 2)i + 1$ unreacted groups, not including the end we have chosen (the easiest way to see this is to draw the molecule out in linear configuration, since this quantity is independent of configuration). Now, we realize that we need a term ω_i which accounts for the number of possible routes to an i-mer; however, even the term

$$\omega_i p^{i-1} (1 - p)^{(f-2)i+1} \tag{5.2.29}$$

is not the probability of a chain of size i. Rather, it is the probability of finding an A group on a molecule of size i; the normalized first moment of this distribution would give us the wrongly calculated DP_n mentioned earlier. However, since each i-mer has $(f - 2)i + 2$ unreacted A groups, it is clear that

$$\omega_i p^{i-1} (1 - p)^{(f-2)i+1} = \frac{[(f - 2)i + 2]P_i}{fP_{1,0}(1 - p)} \tag{5.2.30}$$

Rearranging:

$$P_i = \frac{f\omega_i}{[(f - 2)i + 2]} p^{i-1} (1 - p)^{(f-2)i+2} P_{1,0} \tag{5.2.31}$$

The real difficulty lies in calculating ω_i. To do this, we work with a system that for these purposes is equivalent to our present system, following Flory[25] (a different approach is given by Stockmayer[27]). As for linear step polymerizations, the probability that a group has reacted is p, not p^2, because it is necessary that a reacted group have a partner. We can distinguish these groups for our purposes here, and call them B groups. This means that there are $(f - 1)i$ A groups remaining (not counting our chosen "end") from which to choose the reacted groups. The total number of ways that this choice can be made is given by

$$\binom{(f - 1)i}{i - 1} = \frac{((f - 1)i)!}{((f - 2)i + 1)!(i - 1)!} \tag{5.2.32}$$

Different possible routes to an i-mer (which are not configurations in the strict sense of Chapter 1), however, are made by $i - 1$ different pairings, so that the term needs to be multiplied by $(i - 1)!$ and then further divided by $i!$ to account for the indistinguishability of monomers. Therefore:

$$\omega_i = \frac{((f - 1)i)!}{((f - 2)i + 1)!i!} \tag{5.2.33}$$

Substituting this into equation (5.2.31) yields equation (5.2.9).

This alternative derivation of the size distribution shows that the complicated form of the distribution comes from the number of isomers available for an i-mer. This important mathematical point reflects also the physical differences that can exist between different configurations or isomers for a particular value of i. For example, there is not a direct correspondence between i and hydrodynamic volume, as there was for a linear polymerization, because of the different configurations available (note that for $f = 2$, $\omega_i = 1$). This makes size exclusion chromatography on branched systems more difficult to interpret.

Several different literature sources on different approaches to this paradigmatic case and its variations exist. The theory of branching processes[32] has been applied to examine A_f step homopolymerization.[33] Solutions have been obtained for arbitrary initial conditions (i.e., when the initial condition is other than pure monomer),[34] for polydisperse monomer mass,[35] and for unequal reactivity.[36] The alternate paradigm of A_gB_{f-g} has also been examined.[37]

5.3 Pregel Description of Other Chemical Systems

Although the paradigm of Section 5.2 provides a general understanding of nonlinear polymerizations and gelation, we expect other parameters to enter into the analysis with more realistic chemical systems, as stoichiometric ratio r did for $A_2 + B_2$ polymerization in Chapter 2. We proceed to describe a few different situations now, restricting ourselves to M_w or DP_w and the resulting p_c.

5.3.1 $A_f + B_g$ Step Polymerization

Here we examine the polymerization of A_f and B_g monomers, A functionalities only reacting with B functionalities, and vice versa, such as reaction between diisocyanate and triol. Stoichiometry is allowed to be unbalanced, r being defined here as in Section 2.4:

$$r = \frac{A}{B} \leq 1 \tag{5.3.1}$$

Also as before, the monomers can be interchanged since both are given arbitrary functionality.

The derivation of M_w for this system closely parallels that of Section 2.4.4; we select a unit of mass at random and determine the average mass of the chosen molecule. Thus:

$$M_w = w_{Af}[M_{Af} + fE(W_A^{out})] + (1 - w_{Af})[M_{Bg} + gE(W_B^{out})] \tag{5.3.2}$$

where w_{Af} is the weight fraction of A_f monomers. We can write the following recursive relations, skipping the intermediate step of dealing with "in" expected quantities:

$$E(W_A^{out}) = p[M_{Bg} + (g - 1)E(W_B^{out})] \tag{5.3.3}$$

$$E(W_B^{out}) = rp[M_{Af} + (f - 1)E(W_A^{out})] \tag{5.3.4}$$

These two equations can be solved to yield:

$$E(W_A^{out}) = \frac{pM_{Bg} + rp^2(g-1)M_{Af}}{1 - rp^2(f-1)(g-1)} \qquad (5.3.5)$$

$$E(W_B^{out}) = \frac{rpM_{Af} + rp^2(f-1)M_{Bg}}{1 - rp^2(f-1)(g-1)} \qquad (5.3.6)$$

The weight fraction, w_{Af}, of equation (5.3.2) can be written:

$$w_{Af} = \frac{rgM_{Af}}{rgM_{Af} + fM_{Bg}} \qquad (5.3.7)$$

M_w is thus given as follows:

$$M_w = \frac{rgM_{Af}^2 + fM_{Bg}^2}{rgM_{Af} + fM_{Bg}} + \frac{fgrp\{rp(g-1)M_{Af}^2 + 2pM_{Af}M_{Bg} + p(f-1)M_{Bg}^2}{(rgM_{Af} + fM_{Bg})[1 - rp^2(f-1)(g-1)]} \qquad (5.3.8)$$

which reduces to equation (2.4.49) for $f = g = 2$. The gel point is obviously given by the second part of the denominator of the second term being zero, or:

$$p_c = \frac{1}{[r(f-1)(g-1)]^{1/2}} \qquad (5.3.9)$$

As we found that imbalanced stoichiometry in $A_2 + B_2$ polymerization enforced a ceiling on the attainable molecular weight, here there is an imbalance r_c below which gelation will not occur:

$$r_c = \frac{1}{(f-1)(g-1)} \qquad (5.3.10)$$

The polymerization is obviously more and more "forgiving" with respect to stoichiometric imbalance as functionality increases, since the gel point shifts to successively lower conversions.

We have given here only the simplest of examples of the more common stepwise copolymerizations. The more general case of the polymerization of a mixture of A-functionalized monomers of different functionalities reacting with a mixture of B-functionalized monomers of different functionalities has been treated combinatorially[38] and by the recursive approach.[31,35] Cases involving unequal reactivity have been treated statistically as well.[36,39,40] Cases in which substitution effect and cyclization are allowed to occur are mentioned in Section 5.7.

5.3.2 Crosslinking of Linear Primary Chains

Consider a vulcanization of linear "primary" chains by some arbitrary mechanism.[41] The chains have an arbitrary distribution in chain length, but every monomer has one potentially reactive site; thus there is no joint distribution in number of sites, for the number of sites is equal to the degree of polymerization. We assume that the crosslinks contribute negligible mass.

To calculate M_w of the crosslinked polymer, we randomly grab a unit of mass, which immediately produces a weight-average primary chain plus whatever is attached to it:

$$M_w = M_w^0 + DP_w^0 E(W_A^{out}) \qquad (5.3.11)$$

The superscripted zeros indicate the averages appropriate to the primary chains before any crosslinking occurs. The expected weight looking out of monomer or functional group, $E(W_A^{out})$, depends on the branching probability α and is given by the recursion:

$$E(W_A^{out}) = \alpha\{M_w^0 + (DP_w^0 - 1)E(W_A^{out})\} \qquad (5.3.12)$$

Solving this and inserting into equation (5.3.11) yields:

$$M_w = M_w^0 \frac{1 + \alpha}{1 - \alpha(DP_w^0 - 1)} \qquad (5.3.13)$$

Note the similarity between this relation and equation (5.2.17) for A_f homopolymerization. The gel point here is quite low for long primary chains:

$$\alpha_c = \frac{1}{DP_w^0 - 1} \qquad (5.3.14)$$

The use of a generalized branching probability α allows us to look at several separate cases. For example, this derivation is valid for homopolymerization of A_f monomers polydisperse in size and functionality (but such that the two are always proportional). Here, the crosslink probability is simply $\alpha = p$. It also describes the copolymerization of polydisperse B_f monomers with much smaller A_2 monomers. Here, the crosslink probability would be $\alpha = rp^2$. An example of this case might be the crosslinking of poly(vinyl alcohol) with adipic acid.

There are numerous approaches to this case in the literature, most of them statistical.[33,41-46] The most general treatments[45] deal with crosslinking of primary chains of arbitrary length and site distribution. This approach has been extended to the particular case of the crosslinking of chains formed by free-radical copolymerization, the statistical properties of the primary chains by Markov chain theory (necessary for short chains with compositional transients).[46]

5.3.3 Free-Radical Crosslinking Copolymerization

Consider a very simple case of free-radical polymerization of a monounsaturated or vinyl monomer with a multiply unsaturated monomer. We could do an exhaustive treatment of this by the same recursive arguments as before, but instead note that to a first approximation this falls into the category of the crosslinking of linear polymer, described earlier. The difference here is that the polymer chains are being formed concurrent with the crosslinking process; furthermore, the two proceed by the same chemical mechanism (i.e., propagation). The main effect of this, then, is to redefine the crosslink probability α as follows[47]:

$$\alpha = a_4 p \qquad (5.3.15)$$

where p is the conversion of double bonds and a_4 is the fraction of double bonds in the monomer contributed by the crosslinker (double bonds being considered to be bifunctional); the 4 refers to the functionality of the crosslinker. The assumption on which this is based is that all double bonds are equally reactive and suffer no substitution effects. Again, note that depending on the amount of the crosslinker, the conversion necessary for gelation can be quite low.

This situation has been studied extensively, first as a special case of the crosslinking of long linear chains.[41] The theory of branching processes has been applied,[33,48,49] as has the recursive method, which allows liberality with respect to allowing for drift in the instantaneous primary chain length distribution[47] and composition drift in unequally reactive systems.[50] Other statistical methods have also been applied,[51–55] but probably the most useful method (at least prior to gelation) is that based on the pseudokinetic rate constant method presented in Chapter 4.[56,57] This method more easily accommodates nonidealities such as unequal reactivity.

Example 5.1

Express the gel point for the crosslinking free-radical copolymerization in terms of the parameters q (the probability of propagation) and ξ (the probability that a chain-ending step is by combination).

Solution:

If the pregel regime is short enough to permit both q and ξ to be considered constant, then equation (3.5.48) for DP_w^{inst} also applies to DP_w^{cumu}, identified with DP_w^0 of equation (5.3.14). Applying equation (5.3.15) gives the gel point as follows[47,52]:

$$p_c = \frac{1-q}{a_4(2q+\xi)} \qquad (5.3.16)$$

If the pregel period is long, so that drift occurs, then DP_w^0 is identified with DP_w^{cumu} calculated from equation (3.5.54). Thus, at the gel point:

$$DP_w^0 = DP_w^{cumu} = \frac{1}{p_c}\int_0^{p_c} DP_w^{inst}\, dp' = \frac{1}{p_c}\int_0^{p_c} \frac{1+q+\xi}{1-q}\, dp' \qquad (5.3.17)$$

Thus, the gel point is defined by the following equation[47]:

$$1 = a_4 \int_0^{p_c} \frac{2q+\xi}{1-q}\, dp' \qquad (5.3.18)$$

which is not explicit in p_c. Equation (5.3.18) reduces to (5.3.16) for constant q and ξ.

5.4 Paradigm of the Postgel Regime: A_f Homopolymerization

Up to the gel point, we have dealt with molecules that are easy to visualize, specifically treelike structures with no loops, as shown in Figure 5.11a. These may be architecturally intricate, but being acyclic, they do not achieve the intricacy that is possible when loops are allowed. After the gel point, though, the situation becomes different. First, we have already noted that DP_w is infinite while DP_n is finite. This means that there is an infinite molecule in the system, but that it is not the only molecule in the system. As Figure 5.11b shows, the system thus comprises two parts: a soluble (or *sol*) fraction, and an insoluble (or *gel*) fraction. One might think of these two as distinguishable "phases"; the two are molecularly mixed (or at least may be), however, even if they can be separated from one another by Soxhlet extraction.

As may be guessed, and as we will shortly see, the gel is qualitatively a different object from the rest of the system. For

Figure 5.11.
Treelike structure permitted prior to gelation (a) compared with mixture of sol and gel after gelation (b). In (b), the thick lines indicate the gel fraction and the thin lines the sol. Cycles are permitted in the gel; the sol remains treelike.

(a)

(b)

that matter, then, the postgel regime is qualitatively different from the pregel, demanding an approach to the analysis of structure unlike that appropriate before the gel point. The postgel regime is "statistically heterogeneous."[28] One very practical result is that solution of kinetic equations is not as powerful, and the statistical method can be used to predict quantities beyond the reach of deterministic approaches. It is here, then, that the statistical approach comes into its own, because nothing else allows for probing of the internal structure of the gel. The amount or fraction of the system that is gelled, however, can be found by either method[26,58-60]; we will attend first to this determination.

5.4.1 Weight Fraction Solubles and the Probability of a Finite Structure

The weight fraction of the sol, w_{sol}, is the probability that a randomly grabbed unit of mass is part of a finite structure. For A_f homopolymerization, if we ignore the mass of the condensate, this is the probability that if we were to look out of each functional group of a monomer, we would see a finite structure. Since the groups react independently, this is simply given by the *product* of the f individual probabilities, or:

$$w_{sol} = (P(F_A^{out}))^f \tag{5.4.1}$$

where we have introduced a new quantity not addressed in the pregel description, the probability of a finite structure looking out of an A group. This is easily solved for recursively.[61]

If a functional group has not reacted, the probability of a finite structure is unity; if it has reacted, this is uncertain: it depends on the probability that the structure is finite looking into an A group. Thus:

$$P(F_A^{out}) = (1 - p)(1) + pP(F_A^{in}) \tag{5.4.2}$$

The probability of a finite structure looking into an A group is simply the probability that looking out of the remaining $f - 1$ groups does not show us an infinite structure. Thus:

$$P(F_A^{in}) = (P(F_A^{out}))^{f-1} \tag{5.4.3}$$

Thus we have closed a recursion just as we did with expected weights, looking out of a group when deriving M_w, but here we obtain a polynomial, rather than linear, equation:

$$P(F_A^{out}) = (1 - p) + p(F_A^{out}))^{f-1} \tag{5.4.4}$$

The solution $P(F_A^{out}) = 1$ is always a root to this equation, and it is clear that this is the appropriate solution prior to the gel point, when no infinite molecule exists within the system. This cannot be the solution after the gel point, and so it is neglected and in this case can be removed. The resulting

$(f - 2)$-order polynomial equation is first-order for $f = 3$, and the solution is:

$$P(F_A^{out}) = \frac{1 - p}{p} \tag{5.4.5}$$

a function that monotonically decreases from infinity to zero and intersects the constant pregel solution at $p = 0.5 = p_c$. For $f = 4$:

$$P(F_A^{out}) = \left(\frac{1}{p} - \frac{3}{4}\right)^{1/2} - \frac{1}{2} \tag{5.4.6}$$

where a negative root has been discarded. Again, this solution lies in the physically meaningful range of $0 \leq P(F_A^{out}) \leq 1$ after the gel point, $p_c = 1/3$. It can be shown that there will only be one root between zero and unity after the gel point, and this root must be equal to unity at the gel point. Calculation of the $P(F_A^{out})$ thus allows an alternative derivation of the gel point.

$P(F_A^{out})$ allows calculation of w_{sol}, our objective; for example, for $f = 3$, w_{sol} is:

$$w_{sol} = \left(\frac{1 - p}{p}\right)^3, \quad p \geq 1/2 \tag{5.4.7}$$

Neither w_{sol} nor $P(F_A^{out})$ is discontinuous at the gel point, although their respective first derivatives are. As pointed out before, the gel fraction $w_{gel} = 1 - w_{sol}$ contains practically none of the mass of the system at the gel point, consistent with a finite DP_n.

The sol fraction persists up to complete conversion, meaning that up until complete conversion, there exists in the system more than one molecule. This could have been seen without this derivation, for the probability of unreacted monomer, which is necessarily part of the sol fraction, is $(1 - p)^f$. The divergence of DP_n at $p = 2/f$, which implies that all mass is in a single molecule, must be incorrect, implying that a basic assumption in the calculation of DP_n is violated. The source of this is easiest to see in the statistical derivation. We have assumed equal reactivity, independent reactivity, and no cyclization. The preceding derivation clearly maintains the first two, but it is not clear whether the third is held. In fact, it is not. After the gel point, cycles are allowed (see Figure 5.11b), although they are confined to the gel fraction. The gel is allowed by virtue of its infinite size to react with itself in accordance with a standard mass action law.*

Another approach to seeing this important point is to note that while assigning $p_c = 2/f$ is wrong, it is true to say that

---

*The conclusion that the equation for DP_n must be invalid after the gel point p_c also renders invalid the oft-quoted Carothers equation[1,62] for the gel point. The Carothers equation states that the gel point is the conversion at which DP_n diverges. Apparently this equation is still quoted because it and the correct criterion usually bracket what occurs in real systems,[1,63] but these cases must represent a coincidence due to violations of the principles on which the correct theory is based, rather than adherence to an equation which is fundamentally wrong.

in the incipient gel, the conversion is $2/f$. If cycles are strictly prohibited (i.e., the gel is not allowed to react with itself), then the conversion in the gel is limited to $2/f$, and indeed the overall conversion as well as limited to $2/f$. At this point, all material would be in the gel, and the equation for DP_n would have been correct over the entire conversion regime $0 \le p \le 2/f$. This is essentially the Stockmayer picture of the postgel regime.[26,27] It is inconsistent with the notions of equal and independent reactivity, for functionalities within the gel are allowed fewer potential partners for reaction than those in the sol (a kind of unequal reactivity); moreover, the reactivities of groups toward one another depend on their residence within or without the gel (an "infinite-shell" substitution effect). The Flory picture of the postgel regime is internally consistent, even if its implications were unclear initially.[27] Equal and independent reactivity are maintained throughout the reaction, with the consequence that cycles are not allowed in the sol but are allowed in the gel and are quite important there.[64]

5.4.2 Internal Properties of the Gel

The probability of a finite structure, $P(F_A^{out})$, is the quantity of central importance in the postgel regime.[†] Thus far, however, we have only used $P(F_A^{out})$ to calculate the w_{sol}, which could have been calculated from kinetic equations, had we been clever enough.[28,58–60] But the details of the internal structure of the gel *cannot* be derived from kinetic equations, and so it is here that the statistical methods come into their own, because (aside from simulations) to describe the internal structure of the gel, statistical arguments have to be made.

5.4.2.1 Weight Fraction of Pendant Material
Pendant material is polymer in the gel that can fully relax after a deformation has been applied—that is, material with only one "arm" to the network. The weight fraction of this pendant material, $w_{pendant}$, depends on reactions occurring possibly infinitely far away in contour length, and so writing down a kinetic equation for $w_{pendant}$ seems difficult. From a statistical viewpoint, on the other hand, we can quite simply write:

$$w_{pendant} = \frac{f!}{(f-1)!1!} (P(F_A^{out}))^{f-1} (1 - P(F_A^{out})) = f(P(F_A^{out}))^{f-1} (1 - P(F_A^{out})) \qquad (5.4.8)$$

The combinatorial term, hereafter abbreviated as follows:

$$\binom{f}{i} = \frac{f!}{(f-i)!i!} \qquad (5.4.9)$$

[†]*It is related to the extinction probability v in the theory of branching processes.[28,42] The name comes from the study of the disappearance of family names, which was the origin of the study of branching processes.[30]*

simply gives the number of distinct ways i objects can be taken from a set of f objects. The "objects" are functionalities on a monomer, and i is the number of them leading to the infinite structure. For $i = 1$, the material being pendant requires that one arm be to the infinite network, and there are f possible ways of doing this.

For $f = 3$, then, we can write the answer explicitly:

$$W_{pendant} = 3\left(\frac{1-p}{p}\right)^2\left(\frac{2p-1}{p}\right) \tag{5.4.10}$$

5.4.2.2 Weight Fraction of Elastic Material

Our system comprises sol fraction and gel fraction, and the latter may be subdivided into pendant and elastic material. Clearly, then, this last can be found by the equation:

$$W_{elastic} = 1 - W_{sol} - W_{pendant} \tag{5.4.11}$$

One could also derive it from the following:

$$W_{elastic} = \sum_{i=2}^{f}\binom{f}{i}(P(F_A^{out}))^{f-i}(1 - P(F_A^{out}))^i \tag{5.4.12}$$

in other words, all the material that has two or more links to the network. For $f = 3$:

$$W_{elastic} = \left(\frac{2p-1}{p}\right)^2\left(\frac{2-p}{p}\right) \tag{5.4.13}$$

In Figure 5.12, we see how the quantities discussed evolve with conversion. All material is soluble prior to the gel point, but after it the sol fraction drops sharply, yielding to mostly pendant material, so that the gel is very weak. The pendant material peaks very early, yielding in turn to elastic material, which at last claims all the material in the system. The absence

Figure 5.12.
Evolution of sol, pendant, and elastic fractions for A_3 step homopolymerization.

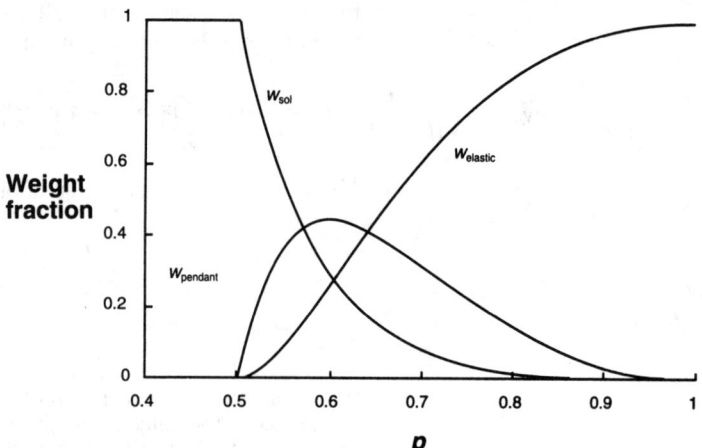

of any pendant defects at complete conversion is not necessarily a universal feature of gelling systems, even if ideal. The presence of network defects in the completely reacted system may either be expected (as in an $A_f + B_g$ step polymerization with unbalanced stoichiometry or a crosslinking free-radical copolymerization), or it may be the result of violations of the restrictions of ideality.

5.4.2.3 Elastic Junctions

A junction is a monomer of functionality $f > 2$, and it is elastically active if three or more of its arms lead to the network. Such a monomer has i arms to network with probability $P(X_{i,f})$. $P(X_{0,f})$ corresponds to the sol fraction, and $P(X_{1,f})$ to the pendant fraction, and in the case at hand are equal (although this is usually not so). The elastic fraction corresponds to $P(X_{i,f})$ for $i > 2$, but the mass of elastic material does not describe the resulting polymer as well as the number of elastically active chains, which is the number of chains between elastically active crosslinks.

Hence, it is important to distinguish between elastically effective crosslinks with different numbers of arms to the network, for they will have different numbers of elastically active chains connected to them. This probability is obviously given as follows:

$$P(X_{i,f}) = \binom{f}{i} (P(F_A^{out}))^{f-i} (1 - P(F_A^{out}))^i \qquad (5.4.14)$$

Since for the case of A_3 polymerization there is only one kind of elastically active crosslink, this is not too interesting. For the case of A_4 polymerization, however, there are elastically active crosslinks with three and four arms to the network. The evolution of these is seen in Figure 5.13. After the gel point ($p_c = 1/3$) the sol fraction decreases rapidly, yielding in turn to pendant material $P(X_{1,4})$, junctions connecting elastic junctions $P(X_{2,4})$, elastic junctions with three arms to the network and finally to elastic junctions with all four arms to the network, which had been steadily building up throughout the reaction.

Figure 5.13.
Evolution of $P(X_{m,4})$ for A_4 step homopolymerization.

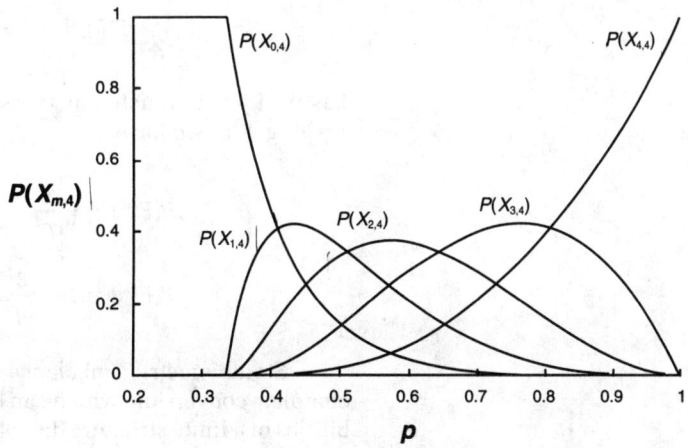

5.5 Postgel Descriptions of Other Systems

5.5.1 $A_f + B_g$ Step Polymerization

We return to the system of Section 5.3.1. Two coupled polynomials arise in this case:

$$P(F_A^{out}) = (1 - p) + p(PF_B^{out}))^{g-1} \tag{5.5.1}$$

$$P(F_B^{out}) = (1 - rp) + rp(P(F_A^{out}))^{f-1} \tag{5.5.2}$$

This in general gives us an $\{(f - 1) \times (g - 1)\}$-order polynomial to solve, although this can be reduced by one owing to the pregel solution. Thus we find analytical solutions in four different cases in which only one of the two monomers acts as a crosslinker. These differ in functionality of the crosslinker and with respect to whether the crosslinker or the bifunctional monomer is in excess. For trifunctional crosslinkers in stoichiometric excess, $f = 2, g = 3$:

$$P(F_A^{out}) = \frac{1 - 2rp^2 + r^2p^3}{r^2p^3} \tag{5.5.3}$$

$$P(F_B^{out}) = \frac{1 - rp^2}{rp^2} \tag{5.5.4}$$

For tetrafunctional crosslinkers in stoichiometric excess, $f = 2$, $g = 4$:

$$P(F_A^{out}) = 1 - \frac{3}{2rp} + \frac{1}{rp}\left(\frac{1}{rp^2} - \frac{3}{4}\right)^{1/2} \tag{5.5.5}$$

$$P(F_B^{out}) = \left(\frac{1}{rp^2} - \frac{3}{4}\right)^{1/2} - \frac{1}{2} \tag{5.5.6}$$

For trifunctional crosslinkers in stoichiometric lack, $f = 3, g = 2$:

$$P(F_A^{out}) = \frac{1 - rp^2}{rp^2} \tag{5.5.7}$$

$$P(F_B^{out}) = \frac{1 - 2rp^2 + rp^3}{rp^3} \tag{5.5.8}$$

Lastly, for tetrafunctional crosslinkers in stoichiometric lack, $f = 4, g = 3$, we have:

$$P(F_A^{out}) = \left(\frac{1}{rp^2} - \frac{3}{4}\right)^{1/2} - \frac{1}{2} \tag{5.5.9}$$

$$P(F_B^{out}) = 1 - \frac{3}{2p} + \frac{1}{p}\left(\frac{1}{rp^2} - \frac{3}{4}\right)^{1/2} \tag{5.5.10}$$

Stoichiometric imbalance allows for network defects at complete conversion, which can be seen by noting that the probabilities of a finite structure do not fall to zero at complete conversion. While network defects will in general be present because real polymerizations do not follow the ideal restrictions on which the theory is based, it is clear that stoichiometric imbalance promotes network defects. For example, a sol fraction will persist to complete conversion. This is sensible: since complete conversion here means complete conversion of the limiting species, the species in excess may still exist in monomer, for example, not to mention larger soluble structures. The sol fraction and pendant fraction can be calculated from the following equations:

$$w_{sol} = w_{Af}(PF_A^{out}))^f + (1 - w_{Af})(P(F_B^{out}))^g \qquad (5.5.11)$$

$$w_{pendant} = fw_{Af}(P(F_A^{out}))^{f-1} (1 - P(F_A^{out})) + g(1 - w_{Af})(P(F_B^{out}))^{g-1} (1 - P(F_B^{out})) \qquad (5.5.12)$$

Again, the weight fraction of elastic material is the fraction remaining from the sol and pendant fractions. Elastic junctions arising from the A_f monomer are found from the following:

$$P(X_{i,f}) = \binom{f}{i} (P(F_A^{out}))^{f-i} (1 - P(F_A^{out}))^i \qquad (5.5.13)$$

and similarly for the B_g monomer.

5.5.2 Crosslinking of Linear Primary Chains

Unlike the pregel M_w discussed in Section 5.3.2, where only the average length of the primary chains entered in, here we must know the entire distribution of primary chain lengths.[45] This is easily seen in the calculation of $P(F_A^{out})$:

$$P(F_A^{out}) = 1 - \alpha + \alpha P(F_A^{in}) \qquad (5.5.14)$$

The probability of a finite chain looking into an A group is a function of the length or equivalently the functionality of the chain (since we consider every unit to be functionalized). So, we can write:

$$P(F_A^{in}) = \frac{\sum_{j=1}^{\infty} P(F_A^{out})^{j-1} i P_i}{\sum_{i=1}^{\infty} i P_i} \qquad (5.5.15)$$

Thus one sees a weighting by the normalized weight distribution, which is expected because one is more likely to end up on a longer chain. Equation (5.5.15) looks complicated, but it can be put in compact notation by relating it to the generating function, $G(s)$, of the primary chain distribution:

$$P(F_A^{in}) \frac{\left(\dfrac{\partial G(s)}{\partial s}\right)_{s = P(F_A^{out})}}{\left(\dfrac{\partial G(s)}{\partial s}\right)_{s = 1}} \qquad (5.5.16)$$

Thus the recursive equation is given by:

$$P(F_A^{out}) = 1 - \alpha + \alpha \frac{\left(\dfrac{\partial G(s)}{\partial s}\right)_{s = P(F_A^{out})}}{\left(\dfrac{\partial G(s)}{\partial s}\right)_{s = 1}} \qquad (5.5.17)$$

The exact nature of this equation (polynomial of perhaps infinite order) depends on the particular primary chain length distribution.

For monodisperse primary chains of length f of concentration P, the generating function $G(s)$ is given as follows:

$$G(s) = s^f P \qquad (5.5.18)$$

The recursion thus becomes:

$$P(F_A^{out}) = (1 - \alpha) + \alpha (P(F_A^{out}))^{f-1} \qquad (5.5.19)$$

just as in equation (5.4.4). Thus, this applies to A_f homopolymerization where the branching probability $\alpha = p$. It also applies, approximately, to the crosslinking of chains made by anionic polymerization. For chains of true Poisson distribution with parameter τ, we have the generating function:

$$G(s) = P s e^{-(1-s)\tau} \qquad (5.5.20)$$

which results in the following recursive equation:

$$P(F_A^{out}) = (1 - \alpha) + \alpha \exp\left[-(1 - P(F_A^{out}))\tau\right] \frac{1 + P(F_A^{out})\tau}{1 + \tau} \qquad (5.5.21)$$

This equation must be solved by numerical methods, although the solution to (5.5.19) would provide a close approximation for large τ.

From $P(F_A^{out})$ one can calculate postgel properties as in the other cases. The soluble fraction is given by the following equation, where w_{Ai} is the weight fractions of chains i units long:

$$W_{sol} = \sum_{i=1}^{\infty} w_{Ai} P(F_A^{out})^i = \frac{\left(\dfrac{\partial G(s)}{\partial \ln(s)}\right)_{s = P(F_A^{out})}}{\left(\dfrac{\partial G(s)}{\partial \ln(s)}\right)_{s = 1}} \qquad (5.5.22)$$

Calculation of the fraction of pendant material is more complicated because one wants to calculate not only the weight fraction of entire primary chains which are pendant, but that of parts of the primary chains which are pendant. The calculation of elastic junctions may be quite different when one is dealing with long chains, for it is likely that the crosslink or junction is actually the (unspecified) crosslinking agent, which will most often act as a tetrafunctional junction, if the crosslink is not located near the end of a primary chain. Because of all these complications, we do not delve into these issues here.

Example 5.2

Find $P(F_A^{out})$ for the case of chains of geometric distribution of parameter q.

Solution:

The generating function for the primary chains in this case is:

$$G(s) = P\frac{(1 - q)s}{1 - qs} \tag{5.5.23}$$

The recursive equation then becomes:

$$P(F_A^{out}) = (1 - \alpha) + \alpha \frac{(1 - q)^2}{(1 - qP(F_A^{out}))^2} \tag{5.5.24}$$

This equation can be solved analytically to yield:

$$P(F_A^{out}) = \frac{1}{q} - \frac{\alpha}{2} - \left[\alpha\left(\frac{\alpha}{4} + \frac{1}{q} - 1\right)\right]^{1/2} \tag{5.5.25}$$

This could also be applied to the case of Section 5.3.3 (free-radical crosslinking copolymerization), but only in the rare case in which drift in chain length is absent. This application would correspond to results arrived at before* by different methods.[47,61] In the realistic cases in which drifts, unequal reactivity, substitution effects, and cyclization are allowed, other sophisticated approaches have been developed which are much more useful.[56,65-67]

5.6 Structure–Property Relations

The peculiar nature of the postgel state is more than a mathematical curiosity; it has real consequences with regard to characterization and properties. The insolubility of the gel fraction is one such property, and it allows w_{sol} to be found by extraction, as well as further characterization of the sol to yield the DP_n and DP_w[68] and the determination of the compositions and conversions in the sol and gel by standard techniques. The internal characteristics of the gel, such as the number of elastically effective junctions, are more troublesome. This is because these quantities depend on long-range connectivity to which none of these approaches are sensitive. A kinetic approach will not yield internal properties of the gel for the same reason. The mathematical problem corresponds to a difficulty in characterization. This is seen as a problem because for linear polymers, many of the structural

---

The $P(F_A^{out})$ given here corresponds to $(a_4P(E) + (1 - a_4))$ in the notation of reference 47. P(E) here is the probability that the remaining double bond does not lead to an infinite structure. The differences arises from a different designation of what the functional group A is.

characteristics could be measured directly, and so we could think in the following way:

$$\text{formation} \rightarrow \text{structure} \qquad (5.6.1)$$

A particular mechanism or route of formation results in a certain structure. This in turn gives certain properties that are important in processing or end use, but unimportant in characterization as long as well-established analytical methods on firm ground are available. For polymer networks, many of the important structural details cannot be measured directly, and so one must resort to thinking in the following way:

$$\text{formation} \rightarrow \text{structure} \rightarrow \text{properties} \qquad (5.6.2)$$

The *structure–property relations* cannot be ignored, for nothing else allows us to probe the internal structure of the gel. We shall look at two standard and established techniques for this—rubber elasticity and swelling—and briefly outline how they are affected by network structure.

5.6.1 Rubber Elasticity

Crosslinked rubber, such as in rubber bands, has the interesting characteristic of a high extensibility combined with complete recovery. There are three requirements for rubber elasticity: long chains, segmental mobility, and permanence of structure.[25] The long polymeric chains do not have to be carbon-based; they may be made of silicon or even sulfur, but in any case they must be long enough to permit a large number of conformations. In addition, the segments of these chains need to be mobile, to ensure that these numerous conformations are not only possible but kinetically attainable. Thus glassy, highly crystalline, or associated polymers are excluded, for the strong and long-lived interactions prohibit the chains from exploring the multitude of configurations. Nonetheless, while length and mobility do provide high extensibility, some permanence of structure is necessary to ensure the complete mechanical recovery, which is the other notable feature of rubber. We could certainly get high degrees of extensibility from a melt of linear polymer, but complete recovery will not occur unless the strain is released very quickly and the molecular weight is high (i.e., it is highly entangled). We thus need crosslinks, and the covalent crosslinks dealt with in the preceding sections will do, although a small amount of crystallinity or a microphase-separated structure may provide a thermoreversible crosslink.

These criteria limit the applicability of rubber elasticity in the characterization of polymer networks. Chemical crosslinks necessarily provide permanence of structure but do not ensure that the first two criteria will be met. In dense networks such as phenolics, high extensibility is lacking. Likewise, a styrene–divinylbenzene network may be below the glass transition temperature of the material, in which case mobility of segments does not exist. On the other hand, polyurethane networks made from a long flexible triol and crosslinked rubber would qualify.

The question is how to relate stress and strain in a system that meets the foregoing requirements. The starting point is to realize that the free energy change of the rubber upon deformation should be overwhelmingly entropic. That is, stretching the chains does not change the energy of the conformation or of the interactions between chains so much as it reduces the number of configurations available. If one assumes that the network comprises v moles of elastically active chains, described by Gaussian statistics, joined at f-functional crosslinks, then one has the following equation for the change in Gibbs free energy upon elongation, ΔF:

$$\frac{\Delta F}{RT} = \frac{v}{2}\{\alpha_x^2 + \alpha_y^2 + \alpha_z^2 - 3\} - \frac{2}{f} v \ln(\alpha_x\alpha_y\alpha_z) \qquad (5.6.3)$$

The elongation ratios along the three axes, indicated by α, appear both in the first term, which deals with the reduction in the number of conformations, and in the second term, which is a bulk term requiring the crosslinked chains to actually be joined (this term will vanish in the case of interest). Since there is no enthalpic change, the ΔF is related to the entropy change, ΔS, by:

$$\Delta F = -\frac{\Delta S}{T} \qquad (5.6.4)$$

Thus:

$$-\frac{\Delta S}{R} = \frac{v}{2}\{\alpha_x^2 + \alpha_y^2 + \alpha_z^2 - 3\} - \frac{2}{f} \ln(\alpha_x\alpha_y\alpha_z) \qquad (5.6.5)$$

If we consider elongation at constant volume, we have:

$$\alpha_x = \alpha = \frac{L}{L_0} \qquad (5.6.6)$$

$$\alpha_y = \alpha_z = \alpha^{-1/2} \qquad (5.6.7)$$

In this case, the entropic change is:

$$-\frac{\Delta S}{R} = \frac{v}{2}\{\alpha^2 + 2\alpha - 3\} \qquad (5.6.8)$$

The retractive force f is given by:

$$f = -T\left(\frac{\partial S}{\partial L}\right)_{T,V} \qquad (5.6.9)$$

The stress τ, that is, this retractive force per unit area, is therefore:

$$\tau = RT\frac{v}{V}\left\{\alpha - \frac{1}{\alpha^2}\right\} \qquad (5.6.10)$$

The elongation modulus E is finally given as follows:

$$E = \left(\frac{\partial \tau}{\partial \alpha}\right)_{\alpha=1} = RT\frac{v}{V} \qquad (5.6.11)$$

The number of elastically active chains ν is obviously related to the number of elastically active junctions. If we return to $P(X_{i,j})$, we note that if $i \geq 3$ the monomer bears i different network strands. However, each strand is always shared with another crosslink, so that

$$\frac{\nu}{V} = \sum_{i=3}^{f} \frac{i}{2} P(X_{i,j}) A_f \qquad (5.6.12)$$

We have thus related our elastically active crosslink density to the bulk property of modulus.

Equation (5.6.11) is not the only theoretically justifiable equation. Indeed, the above derivation is based on the assumptions that the crosslinks or junctions move affinely with the deformation and that they are not allowed to fluctuate freely about their mean positions. If junctions are allowed to fluctuate about their positions freely, one obtains the relation:

$$E = RT\frac{\nu - \mu}{V} \qquad (5.6.13)$$

where μ is the number of elastically active junctions. This fluctuation may of course be restricted somewhat; similarly physical entanglements may be trapped between crosslinks, hence not able to relax although not covalently bonded. We may thus write:

$$E = RT\frac{\nu - h\mu}{V} + T_e E^0 \qquad (5.6.14)$$

where h is a junction fluctuation parameter ranging between 0 and 1, T_e is the probability of a trapped entanglement, and E^0 is the plateau modulus of the uncrosslinked material of network strands. Which relation best describes reality has been the subject of great debate, as illustrated by recent reviews of rubber elasticity theory.[69–72] The important thing to understand is that although we have a structure–property relation, it is not as solidly grounded as we would like.

5.6.2 Swelling

As has been mentioned before, the polymer network structure is insoluble owing to its macroscopic three-dimensional structure. Nonetheless, networks can swell up with solvent if the free energy change is favorable. Here, unlike the preceding case, we will have enthalpic terms, since we are replacing polymer–polymer interactions with polymer–solvent interactions.

The main assumption in the derivation of the swelling equation[25] is that the free energy change of elastic deformation and of mixing are separable: that is,

$$\Delta F = \Delta F_{\text{mixing}} + \Delta F_{\text{elastic}} \qquad (5.6.15)$$

Under conditions of chemical equilibrium, the chemical potential difference of the solvent (species 1) is zero:

$$\mu_1 - \mu_1^0 = \left(\frac{\partial \Delta F}{\partial n_1}\right)_{T,P} = 0 \tag{5.6.16}$$

The (entropic) elastic part of the free energy is given by equation (5.6.3), and for the isotropic deformation expected from swelling, one obtains:

$$\Delta F_{elastic} = RT \frac{3\nu}{2V}\left[\alpha^2 - 1 - \frac{4}{f}\ln\alpha\right] \tag{5.6.17}$$

$$\left(\frac{\partial \Delta F_{elastic}}{\partial n_1}\right)_{T,P} = RT\frac{\nu v_1}{V_0}\left\{v_2^{1/3} - \frac{2}{f}v_2\right\} \tag{5.6.18}$$

where v_1 is the partial molar volume of the solvent and v_2 is the volume fraction of the polymer (species 2) in the swollen sample.

The free energy of mixing, on the other hand, is given by Flory–Huggins theory[25]:

$$\Delta F_{mixing} = RT\left\{n_1\ln(1 - v_2) + \chi_1 n_1 v_2^2\right\} \tag{5.6.19}$$

where χ_1 is the solvent–polymer interaction parameter. Upon derivation:

$$\left(\frac{\partial \Delta F_{mixing}}{\partial n_1}\right)_{T,P} = RT\left\{\ln(1 - v_2) + v_2 + \chi_1 v_2^2\right\} \tag{5.6.20}$$

Thus, we find that the chemical potential is given as follows:

$$\mu_1 - \mu_1^0 = RT\left\{\ln(1 - v_2) + v_2 + \chi_1 v_2^2 + \frac{\nu v_1}{V_0}\left(v_2^{1/3} - \frac{2}{f}v_2\right)\right\} = 0 \tag{5.6.21}$$

Finding the zero of this equation provides an alternative method for finding ν from the measured degree of swelling q_m ($= 1/v_2$), given that the interaction parameter χ_1 is known. Again, this is only the most basic of derivations.[73] The theory for swelling is on less sure grounds than that for rubber elasticity, since any objections to rubber elasticity apply to swelling as well, and there may be additional objections, such as the separability of the mixing and elastic terms.[74]

5.7 Critique of Gelation Theory

As we have mentioned before, the physical state of our polymerizing system upon and after gelation is qualitatively different from that before gelation, or that in a linear polymerizing system. After gelation one can think of the system as having two phases, although here the distinguishing traits are not density or enthalpy or magnetization, but rather connectivity—specifically, whether the branched structure is infinite. The gel point is thus analogous to a critical point above which one phase exists (gas, or a solid with no net magnetization) and below which two phases exist (gas and liquid, or magnetized and unmagnetized). Gelation is thus termed a critical phenomenon,[75-77] and what is known about critical phenomena in general applies to gelation as well.

Thus far in this book, we have assumed that the rate of reaction of any species with another is based on the average concentrations of the two, or likewise, in statistical approaches, that functional groups are statistically equivalent and share the same probability of reaction. Such an assumption is technically referred to as a "mean field" assumption; that is, every molecule is assumed to experience an environment identical to the average bulk environment. This convention has served us well for homogeneous systems thus far; generally, however, the so-called mean field theories are not applicable in the vicinity of a critical point. Near a critical point, there are large fluctuations of properties (e.g., density) from point to point (possibly the most familiar example is critical opalescence). Calculation of the state of a molecule based on an average environment is then clearly wrong, because no molecule in the system (or few) experience an average environment. Although this reasoning applies to an equilibrium situation and the properties involved are density, enthalpy, and so forth, the ideas carry directly over to a polymerizing system where the property of interest is connectivity. Near the gel point there must be large fluctuations in connectivity, and reacting molecules in their neighborhood do not "see" the average concentrations P_i. This in turn must affect the evolution and shape of the distribution P_i itself.

These effects must become progressively more pronounced, the closer to the gel point one is. The effects are seen on either side of the gel point, either in the pregel regime or in the postgel regime, and so we redefine the rescaled conversion ε, which appeared in equation (5.2.21):

$$\varepsilon = \frac{|p_c - p|}{p_c} \tag{5.7.1}$$

so that either case can be examined. We are interested in the limit as $\varepsilon \to 0$. The predictions that suffer are those relating material properties to this reduced distance from the critical point, ε. These divergent or vanishing quantities generally scale with ε raised to some power, a critical exponent. For example, we can note that in the pregel regime:

$$DP_w \propto \varepsilon^{-\gamma} \tag{5.7.2}$$

From everything that we have done so far in this chapter, $\gamma = 1$ [cf. equations (5.2.17), (5.3.8), and (5.3.13)]; this is the mean field,

Figure 5.14.
Interior section of a Bethe lattice of coordination number (or functionality) $f = 3$. The progressive crowding at each generation out from the central node in this two-dimensional rendering occurs in any finite-dimensional space.

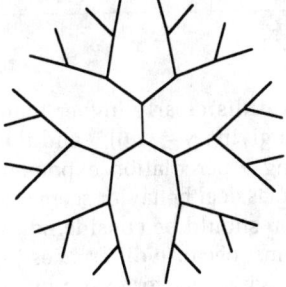

or "classical," exponent. Fluctuations alter the value of γ, so that $\gamma = 1$ will not hold sufficiently close to the gel point.

What can one propose in place of this classical theory? Perhaps the best way to answer this question is to reformulate the classical theory somewhat. Consider the Bethe lattice shown in Figure 5.14, which has a coordination number or functionality of 3. The random placement of bonds between nodes on this lattice corresponds directly to the random A_3 step homopolymerization; that on an f-functional lattice, to A_f polymerization. One can obviously construct more intricate lattices for $A_f + B_g$ polymerization and so forth. This process of placing bonds on a lattice is termed the percolation problem,[78] because there comes a point at which a fluid could "percolate" through the network formed because a continuous path exists. This "percolation threshold" is analogous to the gel point.

The "crowding" that occurs at the periphery of the Bethe lattice shown in Figure 5.14 is artifact of rendering the graph on a two-dimensional page; the same crowding, however, would also be seen if it were drawn in three-dimensional space, or any higher dimensional space. Attempting to place or "embed" a Bethe lattice in a finite-dimensional space always leads to crowding, because the treelike structure of this lattice cannot be embedded in space. The lack of "space embeddedness" is what yields its mean field behavior and keeps cycles from forming in finite species. It gives a simple picture, but one that does not correspond to the true situation. Apparently what is needed, then, is to replace the Bethe lattice with a different, truly three-dimensional lattice, which can allow for the spatial effects of fluctuations and so forth.[75,79] The so-called critical properties of the gelling system are then altered. For example, equation (5.7.2) is still obeyed, albeit with a different value for γ, $\gamma = 1.74$. Such a value cannot be predicted from classical theories, whether statistical or kinetic.* In Table 5.1 we survey and compare the classical and percolation exponents for polymer network formation.

The literature examining these critical exponents is quite vast, even though restricted to the past 15 years. For the sake of brevity, we here review only the literature for the exponent γ. Values of γ have been measured as 1.71 for a polyurethane system,[81] 1.8 for a polyester system,[82] and 1.7 for an epoxy system,[83] all of which support the percolation viewpoint. On the other hand, exponents closer to the classical predictions have been found for an epoxy system ($\gamma = 0.97$),[84] or between the two predictions, tending toward percolation values as dilution increases.[85,86] Exponents much higher than those expected from percolation have been measured for sol–gel systems ($\gamma = 2.7$).[87] For crosslinking free-radical copolymerization, the system methyl methacrylate–ethylene glycol dimethacrylate has yielded

---

Kinetic equations can be written which can yield nonclassical and even percolation exponents.[80] This comes about through the reformulation of the kernel of the polymerization equation so that the rate of reaction between an i-mer and a j-mer eventually goes as (ij)^x rather than simply as ij for large i and j.

TABLE 5.1 / Comparison of Classical and Percolation Predictions for Critical Exponents

Property	Dependence	Exponent	Classical	Percolation
DP_w	$\sim\varepsilon^{-\gamma}$	γ	1	1.74
P_i	$\sim i^{-\tau}\exp(-ci\varepsilon^{1/\sigma})$	τ	2.5	2.2
		σ	0.5	0.46
$\langle s^2\rangle_z^{1/2}$	$\sim\varepsilon^{-\nu}$	ν	0.5	0.88
$\langle s_i^2\rangle^{1/2}$	$\sim i^{1/d_f}$	d_f	4	2.5
$DP_{w,sol}$	$\sim\varepsilon^{-\gamma'}$	γ'	1	
W_{gel}	$\sim\varepsilon^{\beta}$	β	1	0.4

Source: Reference 76.

intermediate results ($\gamma < 1.5$)[88] and results clearly higher than the classical prediction (the best data giving $\gamma = 1.6$),[89] and the system styrene–divinylbenzene giving a percolation exponent ($\gamma = 1.8$).[90] On the balance, then, nonclassical behavior seems to be dominant. None of the values given should be considered to be the norm for their respective systems, because differences in the ranges of ε, in the precision with which p_c (for example) is known exist between the different studies. In measurement of critical exponents, moreover, an unspecified range of ε close to 0 always needs to be examined, with two accompanying difficulties: the error in p_c becomes increasingly significant, and characterization techniques become increasingly unreliable.

It should be emphasized that percolation results should apply only in some region near the gel point, which may be brief or extended depending on the particulars of the system (such as dilution), and that percolation theory makes no pretense of predicting important quantities such as p_c.[91] Mean field behavior, however, may still not apply outside the critical regime, particularly when steric limitations affect the reactivity of functional groups or phase separation occurs. Such problems occur for systems such as phenolic resins, amino resins, and sol–gel systems[1] and are particularly bad for crosslinking free-radical copolymerizations. The most easily recognizable feature in this last case is a severe delay in gel point by one or two orders of magnitude,[89,92] due most likely to the combined effects of (1) cyclization,[93–101] (2) a decreased reactivity of the pendant double bond, either from a true chemical substitution effect (as might be the case in divinylbenzene) or a physical shielding,[88,101–103] and (3) heterogeneity resulting from extensive cyclization and microphase separation.[104,105] Therefore it is easy to appreciate that these systems are quite difficult to model.

The limitations of the statistical approach in the proper formulation of the mean field description bear remembering. The simple statistical approaches are not faithful to the true process of structural growth in a number of situations. This has been pointed out most notably in connection with substitution effect.[58,106–108] Such criticism is valid for the typical statistical approaches, which do ignore the history of reaction (which is important in the presence of substitution effects), but this condemnation of the statistical approach is based on statistical models that in hindsight are obviously approximate. Exact statistical models can often be formulated, either by a "superspecies" approach[109] as has been shown for substitution effects in A_3 homopolymerization,[110] or by properly writing the recursive relations as the integral equations which they become in the presence of such effects, as has been shown for substitution effects in free-radical crosslinking polymerizations.[54] The intricacies of correct statistical modeling when these complications are present are beyond the scope of this text, and thus practically the reader needs to be aware first of the many kinds of situation in which the statistical approach will not faithfully represent the structure and second that solution by the kinetic equations is probably recommended in those cases.

5.8 Long-Chain Branching

We now return to a topic introduced in Section 5.1.1: long-chain branching, as it occurs in free-radical polymerizations of monomers such as ethylene, vinyl chloride, and vinyl acetate. We especially focus on vinyl acetate, since other systems involve other complications—for example, poly(vinyl chloride) is insoluble in its monomer, hence usually proceeds by precipitation polymerization, and the transfer mechanism is much more complicated. We recall from Section 5.1.1 that long-chain branching occurs by chain transfer to polymer and by terminal double-bond polymerization.

Transfer to polymer occurs when very reactive radicals abstract labile protons (or halogens), as is the case in the polymerization of vinyl acetate. This transfer mechanism leads to trifunctional branch points. The same high reactivity and lability also yield a high rate of transfer to monomer, which not only leads to short primary chains but leaves a terminal double bond, either on the end of the former living chain (as in the case of vinyl chloride) or at the beginning of the now living chain (as is the case with vinyl acetate). A terminal double bond also remains on one of the chains engaged in a disproportionation reaction. Propagation through these terminal double bonds also leads to a trifunctional branch point.

Although the branch points are trifunctional, there may be many of these on a given molecule, and so one can form a highly branched structure as shown in Figure 5.15a. Such a polymer is characterized (though hardly completely) by its degree of polymerization (n monomer units) and the number of branch points, b ($b < n$).* This molecule may bear more than one radical, as shown in Figure 5.15b. Multiple radicals may come from several incidents of polymer transfer within a radical lifetime, for example. We might reason that such multiradical species will be negligible as a result of the short lifetime of a radical, but if we suspect

---

*In this and in the inequalities to follow, the upper bound may actually be less than that stated (e.g., it may not be possible to have $b = n - 1$). We will not explicitly define what the real upper bound is; we only state that the upper bound is at most n (or b in later inequalities).

Figure 5.15.
Attributes possible for polymers with long-chain branching: (a) branching, (b) polyradicals, and (c) multifunctional chains.

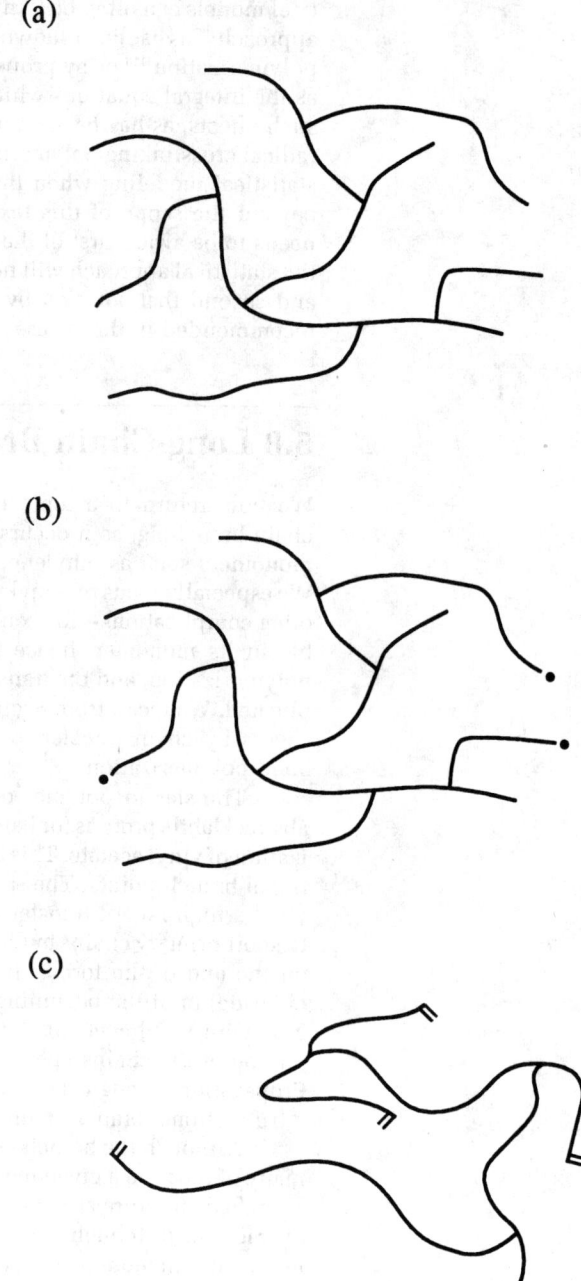

(a)

(b)

(c)

that a molecule of diverging size might arise in our reaction, such an infinite molecule could easily bear many radicals. Thus in principle we allow for r radicals ($r < b$). Finally, a chain may bear more than one terminal double bond, as shown in Figure 5.15c. Chains with two terminal double bonds can be formed, for example, by the coupling of chains begun by transfer to monomer (in the case of vinyl acetate), or by a disproportionation reaction between the same two chains (in which case only one of the dead chains bears two terminal double bonds). Further reactions can lead to chains of higher functionality—for example, a similarly formed chain propagating through one of the double bonds to yield a species with three double bonds. A molecule may thus have a number of double bonds d ($d < b$).

Thus a molecule may be described as $P_{n,b,r,d}$: that is, a molecule comprising n monomer units, having b branch points, and bearing r radicals and d terminal double bonds. Note that even this scheme does not completely define the molecule, for it does not contain the "sequence distribution" of branch points. Nonetheless, this description does contain all the information necessary to properly form the kinetic equations.[111,112] Although free-radical network formation was first studied using combinatorial techniques,[27,113] the analysis of long-chain branching is better approached with kinetic equations, since branching introduces a history to the reaction difficult to capture in statistical approaches. Indeed, since the 1950s most works dealing with the subject have been based on deterministic techniques.[56a,57,111,112, 114–128] (Sophisticated methods based on statistical ideas can properly reproduce the history of formation.[129]) All these attempts vary in the level of complexity of the kinetics (i.e., in the kind of assumption invoked). For instance, some assume constant monomer concentration and a most probable distribution for the "primary" chains (termination by transfer or disproportionation).[114,115] More complete kinetic schemes, including the Trommsdorff effect, have been presented more recently.[122–125] We take as our example a special case neglecting initiation and termination steps,[117] for which analytical results in terms of conversion are obtained. Section 5.8.2 presents a more intricate scheme permitting gelation.

5.8.1 Transfer-Dominated Polymerization

We first examine a case[117] in which only four reactions occur: propagation, transfer to monomer, transfer to polymer, and terminal double-bond polymerization:

$$R_{n,b} + M \xrightarrow{k_p} R_{n+1,b} \tag{5.8.1}$$

$$R_{n,b} + M \xrightarrow{k_{tr,m}} P_{n,b} + R_{1,0} \tag{5.8.2}$$

$$R_{n,b} + P_{m,c} \xrightarrow{k_{tr,p}} P_{n,b} + R_{m,c+1} \tag{5.8.3}$$

$$R_{n,b} + P_{m,c} \xrightarrow{k_p^*} R_{n+m,b+c+1} \tag{5.8.4}$$

Initiation and therefore termination reactions are neglected, and the chain length is controlled totally by transfer (hence the designation "transfer-dominated"). It is easily seen that in this case, which could be experimentally generated by very low amounts of initiator (and only with great patience, since polymerization would be very slow), every polymer molecule would have one and only one terminal double bond. Therefore, the last index can be dropped. The third index for the number of radicals is likewise dropped, so that reactions between radicals are neglected or prohibited.

We first deal with only the size distribution, hence drop the second index, as well. We can write balances for a batch reactor as follows:

$$\frac{dR_1}{dt} = k_{tr,m}M\lambda_0 - k_pMR_1 - k_{tr,m}MR_1 - k_{tr,p}\mu_1R_1$$
$$- k_p^*R_1\lambda_0 + k_{tr,p}P_1\lambda_0 \tag{5.8.5}$$

$$\frac{dR_n}{dt} = k_pMR_{n-1} - k_pMR_n - k_{tr,m}MR_n - k_{tr,p}\mu_1R_n$$
$$- k_p^*R_n\lambda_0 + k_{tr,p}nP_n\lambda_0 + k_p^*\sum_{s=1}^{r-1}P_{r-s}R_s \tag{5.8.6}$$

$$\frac{dP_n}{dt} = k_{tr,m}MR_n + k_{tr,p}\mu_1 - k_{tr,p}nP_n\lambda_0 - k_p^*P_n\lambda_0 \tag{5.8.7}$$

where the moment λ_0 equals the (constant) radical concentration, and the moment μ_1 is the dead polymer mass, proportional to the conversion or mass yield x.

If we reduce all concentrations by the initial monomer concentration (neglecting a change of notation for this) and define the conversion increment as $dx = k_p M_0(1 - x)\lambda_0 dt$, the transfer constants $C_m = k_{tr,m}/k_p$ and $C_p = k_{tr,p}/k_p$, and the relative terminal double bond reactivity $K = k_p^*/k_p$, we can express equations (5.8.5)–(5.8.7) as evolution equations for $H(s)$, the live chain generating function, and $G(s)$, the dead chain generating function:

$$\frac{\partial H(s)}{\partial x} = 0 = C_m s + \frac{H(s)}{\lambda_0}s - \frac{H(s)}{\lambda_0} - C_m\frac{H(s)}{\lambda_0} - C_p\frac{x}{1-x}\frac{H(s)}{\lambda_0}$$
$$- K\frac{\mu_0}{1-x}\frac{H(s)}{\lambda_0} + C_p\frac{1}{1-x}\frac{\partial G(s)}{\partial \ln s} + K\frac{H(s)}{\lambda_0}G(s) \tag{5.8.8}$$

$$\frac{\partial G(s)}{\partial x} = C_m\frac{H(s)}{\lambda_0} + C_p\frac{x}{1-x}\frac{H(s)}{\lambda_0} - C_p\frac{1}{1-x}\frac{\partial G(s)}{\partial \ln s} - K\frac{1}{1-x}G(s) \tag{5.8.9}$$

Thus we have two coupled partial differential equations to solve, if we were to attempt to solve for the generating functions themselves. Rather than attempt that, we instead will satisfy ourselves with the moments of the distribution (specifically the dead chain distribution), found by repeated differentiation of the preceding equations with respect to $\ln(s)$ followed by evaluation at $s = 1$.

Solution of these equations leads to an apparent difficulty in moment closure which is an artifact of the way the problem has been formulated.[126] The equation for μ_i is found to depend on μ_{i+1}. The latter is eliminated by the (quasi–steady state) equation for λ_i and all terms incorporating λ_i/λ_0 are found to cancel, while terms containing λ_{i-1}/λ_0 are found from the equation for λ_{i-1}, now solvable because μ_{i-1} is known. Thus the moment equations can be solved sequentially, despite the apparent problem. In the present case, one obtains:

$$\frac{d\mu_0}{dx} = C_m - K\frac{\mu_0}{1-x} \tag{5.8.10}$$

$$\frac{d\mu_1}{dx} = 1 \tag{5.8.11}$$

$$\frac{d\mu_2}{dx} = \frac{2\left[1 + K\dfrac{x}{1-x}\right]\left[1 + K\dfrac{x}{1-x} + C_p\dfrac{x}{1-x}\mu_2\right]}{C_m + C_p\dfrac{x}{1-x}} \tag{5.8.12}$$

Notice that if there is no transfer to polymer or propagation through terminal double bonds $(C_p = K = 0)$, the kinetic scheme is one of dominant transfer to monomer for which the instantaneous polydispersity is nearly 2.

Equations (5.8.10) and (5.8.11) can be readily solved to yield the DP_n $(K \ne 1)$:

$$DP_n = \frac{1-K}{C_m}\frac{x}{(1-x)^K - (1-x)} \tag{5.8.13}$$

Note that the number-average degree of polymerization is independent of the transfer to polymer, since C_p does not appear. This is expected, because transfer to polymer neither increases nor decreases the number of polymer molecules in the system, unlike the reactions of transfer to monomer and terminal double bond polymerization. On the other hand, DP_w is expected to depend on the amount of transfer to polymer, and does, as is seen in the following equation:

$$DP_w = \frac{2}{C_m}\frac{1}{x}\frac{(1-(1-a)x)^{2(K-a)/(1-a)}}{(1-x)^{2K}}\int_0^x \frac{(1-\overline{x})^{2K+1}}{(1-(1-a)\overline{x})^{(2K+1-3a)/(1-a)}}\left(1 + K\frac{\overline{x}}{1-\overline{x}}\right)^2 d\overline{x} \tag{5.8.14}$$

where $a = C_p/C_m$. Analytical solutions are available for the following cases:

$K = 0$

$C_p = 0$

$K = 1$

$K = 1$ \quad and \quad $a = 1$

In general, though, integration of equation (5.8.14) is numerical. The values of C_m and K, however, can be determined experimentally using equation (5.8.13). An integration of equation (5.8.14) for various values of C_p would yield an optimal C_p, upon comparison with experimental DP_w data.[117,130] For vinyl acetate, this procedure led to $C_m = 1.9 \times 10^{-4}$, $C_p = 1.2 \times 10^{-4}$, and $K = 0.80$.[117] The evolution of DP_n and DP_w according to these parameters is shown in Figure 5.16, which also gives DP_w for two other values of C_p, to show the effect of transfer to polymer on polydispersity.

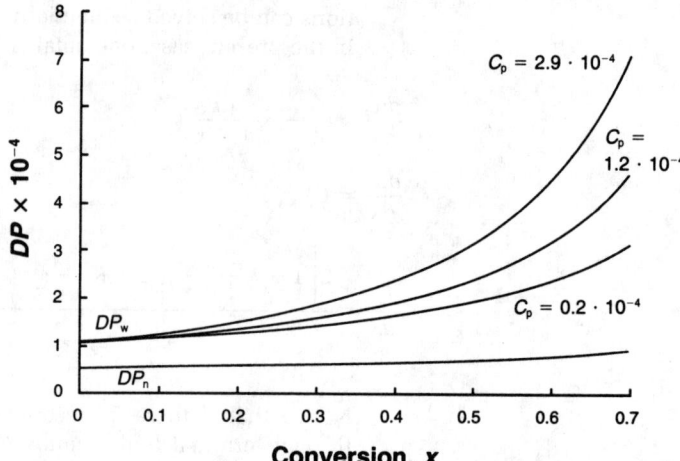

Figure 5.16.
Prediction growth of degree of polymerization for vinyl acetate: $C_m = 1.9 \times 10^{-4}$, $K = 0.8$.

Note the relative constancy of DP_n and the great increase in DP_w over the conversion range of 70%.

The foregoing description of the polymers formed under this kinetic scheme is hardly complete. Obviously, the complete molecular weight distribution is lacking, but more importantly there are many quantities peculiar to the branching nature of this polymerization. Thus even if we were to find the chain length distribution P_n, that would not tell us how many branches a given molecule possesses, or anything else about architecture of the molecule. We could ask for the distribution $P_{n,b}$, the concentration of chains containing n monomers and b branches. It is clear, though, that if the calculation of the chain length distribution is difficult, the calculation of the branching distribution would be worse yet, even for the simple kinetics under study. Thus we look for the average branching densities.[130] The number-average and weight-average branching densities B_n and B_w are given as follows:

$$B_n = \frac{\displaystyle\sum_{n=1}^{\infty} \sum_{b=1}^{n-1} b\, P_{n,b}}{\mu_0} \qquad (5.8.15)$$

$$B_w = \frac{\displaystyle\sum_{n=1}^{\infty} \sum_{b=1}^{n-1} nb\, P_{n,b}}{\mu_1} \qquad (5.8.16)$$

This is a counting process for the *number* of branches. The first equation is simply the total number of branches divided by the total number of molecules. The second equation counts the number of branches associated with chains of a certain weight. It is *not* the weight of the branches, since we do not know their length. It assigns branches in larger molecules a heavier weight. Since properties like intrinsic viscosity are more affected by the longer chains, B_w is a more useful index (but not *the* index) of structure than B_n.

From the foregoing kinetic scheme, one obtains for B_n:

$$\frac{d(\mu_0 B_n)}{dx} = \frac{C_p x + K\mu_0}{1 - x} \tag{5.8.17}$$

This can be integrated analytically to yield:

$$B_n = \frac{C_p(1 - K)}{C_m} \frac{-\ln(1 - x) - x}{(1 - x)^K - (1 - x)} + \frac{1 - (1 - x)^K + Kx}{(1 - x)^K - (1 - x)} \tag{5.8.18}$$

The first term is the contribution due to transfer to polymer, the second due to terminal double-bond polymerization. These are plotted in Figure 5.17, using the estimated parameters just listed. We can find B_w by numerical integration of the following equation:

$$\frac{d(x B_w)}{dx} = \frac{\left(1 - K\dfrac{x}{1 - x}\right) [C_p x(1 + B_w) + K\mu_2(1 + B_n)]}{C_m(1 - x) + C_p x} + C_p \frac{x}{1 - x} DP_w + K\frac{x}{1 - x} \tag{5.8.19}$$

$$+ \frac{K\lambda_0(1 + B_n)\left[1 + C_p \dfrac{x}{1 - x} DP_w + K\dfrac{x}{1 - x}\right]}{C_m(1 - x) + C_p x}$$

What is clear from our kinetic scheme is that despite the branching, network formation does not occur. The system does not gel. One can make arguments for this, noting that transfer to polymer does not "crosslink" chains, but only grafts them. Terminal double-bond polymerization acts in much the same fashion. Furthermore, the form of the differential equation for the second moment differs from that for A_f homopolymerization. The latter, though separable, is a nonlinear first-order ordinary differential equation—in particular, what is called a Ricatti equation. Equation (5.8.12), on the other hand, is a linear equation,

Figure 5.17.
Branching number for vinyl acetate: $C_m = 1.9 \times 10^{-4}$, $K = 0.8$.

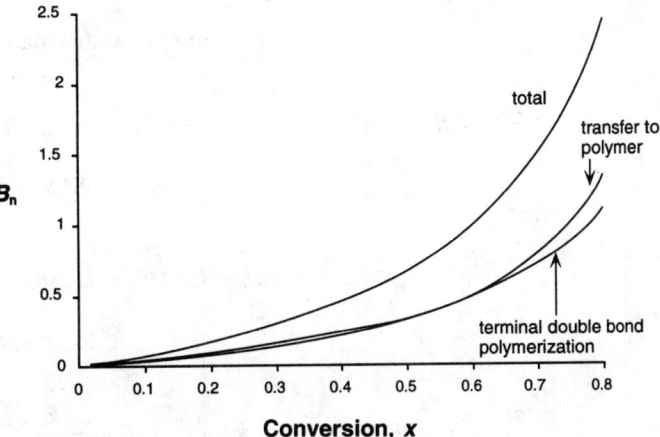

a fact we used to recast it in the form given by equation (5.8.14) for DP_w.

If we return to the point made about the system yielding only grafting, we ask whether a mechanism exists by which poly(vinyl acetate) can indeed crosslink. This turns out to be so if combination is allowed[56a] or if multifunctional polymers can be formed—by disproportionation, for example.[128] Thus we must turn from a transfer-dominated situation to one in which chains can be ended by a termination reaction. We will exclude terminal double-bond polymerization here, hence neglecting the cross-linking afforded by multifunctional chains. In this case, we will see that gelation occurs only if combination occurs.

5.8.2 Polymerization with Terminal Double-Bond Polymerization Prohibited

In the following scheme, terminal double bonds are assumed to be unreactive and radical species do not react between themselves except to terminate either by combination or disproportionation. Initiation is by the usual thermal decomposition mechanism of Chapter 3, and propagation, transfer to monomer, and transfer to polymer occur as in equations (5.8.1), (5.8.2), and (5.8.3), respectively. In addition, transfer to solvent is allowed:

$$R_{n,b} + S \xrightarrow{k_{tr,s}} P_{n,b} + R_{1,0} \tag{5.8.20}$$

as well as termination by both disproportionation and combination:

$$R_{n,b} + R_{m,c} \xrightarrow{k_{td}} P_{n,b} + P_{m,c} \tag{5.8.21}$$

$$R_{n,b} + R_{m,c} \xrightarrow{k_{tc}} P_{n+m,b+c} \tag{5.8.22}$$

We can now write balances for a batch reactor as follows:

$$\frac{dR_1}{dt} = 2fk_dI + k_{tr,m}M\lambda_0 + k_{tr,s}S\lambda_0 - k_pMR_1 - k_{tr,m}MR_1 - k_{tr,s}SR_1 - k_{tr,p}\mu_1R_1 \tag{5.8.23}$$

$$- (k_{tc} + k_{td})R_1\lambda_0 + k_{tr,p}P_1\lambda_0$$

$$\frac{dR_n}{dt} = k_pMR_{n-1} - k_pMR_n - k_{tr,m}MR_n - k_{tr,s}SR_n - k_{tr,p}\mu_1R_n \tag{5.8.24}$$

$$- (k_{tc} + k_{td})R_n\lambda_0 + k_{tr,p}nP_n\lambda_0$$

$$\frac{dP_n}{dt} = k_{tr,m}MR_n + k_{tr,s}SR_n + k_{tr,p}\mu_1 + k_{td}R_n\lambda_0 + \frac{k_{tc}}{2}\sum_{m=1}^{n-1} R_mR_{n-m} - k_{tr,p}nP_n\lambda_0 \tag{5.8.25}$$

If we again reduce all concentrations by the initial monomer concentration and apply the definitions given before (the radical concentration is assumed constant), we can write:

$$\frac{\partial H(s)}{\partial x} = 0 = \frac{2fk_dI}{\lambda_0 k_p(1-x)}s + C_m s + C_s\frac{S}{1-x}s + \frac{H(s)}{\lambda_0}s - \frac{H(s)}{\lambda_0} - C_m\frac{H(s)}{\lambda_0}$$

$$- C_s\frac{S}{1-x}\frac{H(s)}{\lambda_0} - C_p\frac{x}{1-x}\frac{H(s)}{\lambda_0} - \frac{k_{td}+k_{tc}}{k_p}\frac{1}{1-x}H(s) + C_p\frac{1}{1-x}\frac{\partial G(s)}{\partial \ln s} \quad (5.8.26)$$

$$\frac{\partial G(s)}{\partial x} = C_m\frac{H(s)}{\lambda_0} + C_s\frac{S}{1-x}\frac{H(s)}{\lambda_0} + C_p\frac{x}{1-x}\frac{H(s)}{\lambda_0}$$

$$+ \frac{k_{td}}{k_p}\frac{H(s)}{1-x} + \frac{1}{2}\frac{k_{tc}}{k_p}\frac{H^2(s)}{\lambda_0(1-x)} - C_p\frac{1}{1-x}\frac{\partial G(s)}{\partial \ln s} \quad (5.8.27)$$

We can again derive the moment equations, the first two of which follow:

$$\frac{d\mu_0}{dx} = C_m + C_s\frac{S}{1-x} + \frac{k_{td}+\frac{1}{2}k_{tc}}{k_p}\frac{\lambda_0}{1-x} \quad (5.8.28)$$

$$\frac{d\mu_1}{dx} = 1 \quad (5.8.29)$$

From these DP_n may be found, and is again analytic:

$$DP_n = \frac{x}{C_m x + \dfrac{2k_{td}+k_{tc}}{2k_p}\lambda_0\ln\left(\dfrac{1}{1-x}\right) + S_0(1-(1-x)^{C_s})} \quad (5.8.30)$$

where S_0 is the initial concentration of transfer agent. DP_w, on the other hand, must be found by numerical solution of the equation for the μ_2:

$$\frac{d\mu_2}{dx} = 2\left(\frac{1 + C_p\dfrac{1}{1-x}\mu_2}{C_m + C_s\dfrac{S}{1-x} + C_p\dfrac{x}{1-x} + \dfrac{k_{td}+k_{tc}}{k_p}\dfrac{\lambda_0}{1-x}}\right)$$

$$+ \frac{k_{tc}}{k_p}\frac{\lambda_0}{1-x}\left(\frac{1 + C_p\dfrac{1}{1-x}\mu_2}{C_m + C_s\dfrac{S}{1-x} + C_p\dfrac{x}{1-x} + \dfrac{k_{td}+k_{tc}}{k_p}\dfrac{\lambda_0}{1-x}}\right)^2 \quad (5.8.31)$$

from which DP_w can be found, since $DP_w = \mu_2/x$. The form of this equation differs qualitatively when combination is present. The second term supplied by the combination reaction supplies the nonlinearity, which makes it a Ricatti equation. It shares that designation with the equation governing the second moment in A_f homopolymerization, although in that case the nonlinear equation is separable, hence easily solvable. This system can in fact gel, as shown in Figure 5.18.

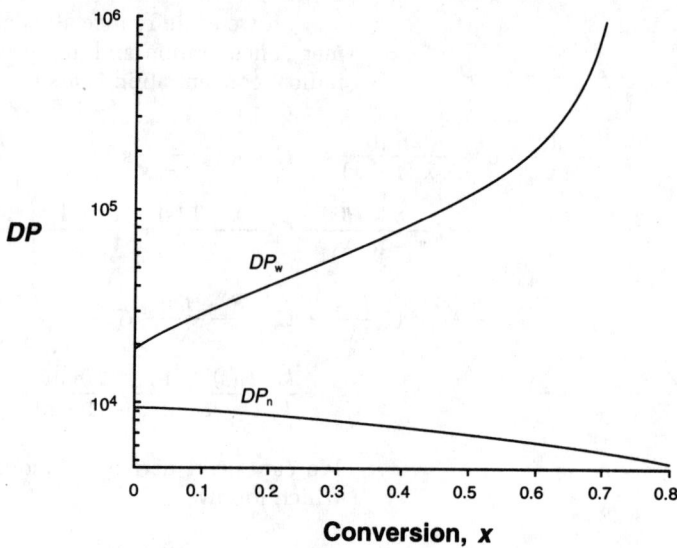

Figure 5.18.
Predicted growth of degree of polymerization for a system undergoing transfer to polymer and termination: $C_p = 10^{-3}$, $k_{tc}\lambda_0/k_p = 10^{-5}$, $k_{td}\lambda_0/k_p = 10^{-4}$. Gelation occurs at approximately $x_c = 0.74$. Recall that radical concentration λ_0 is reduced by monomer concentration. (After Tobita and Hamielec,[56a] used by permission of Hüthig & Wepf Verlag.)

It should be noted that the preceding discussion does not support the argument that gelation relies only on combination occurring, for if terminal double-bond polymerization were allowed, multifunctional chains, which could act as crosslinkers, could arise.[128] The mechanism of termination in vinyl acetate polymerization is somewhat in question, but this is not surprising, since many techniques and analyses for deciding between the two assume that transfer to monomer is negligible, hardly the case with vinyl acetate.[131–133]

Problems

5.1. For the paradigm A_f step homopolymerization, do the following:
(a) Derive equations (5.2.10)–(5.2.12) from (5.2.5), the general kinetic equation.
(b) Show that the size distribution of equation (5.2.9) satisfies the kinetic equation (5.2.5).
(c) Derive the asymptotic size distribution, equation (5.2.20), from the exact size distribution of equation (5.2.9).

5.2. The stepwise system AB_{f-1} is a classic example of a highly branched system that does *not* gel. Show this, by reapplying the steps taken for A_f homopolymerization to this system.
(a) In a j-mer without cycles, how many unreacted A groups are there? How many unreacted B groups?
(b) In A_f homopolymerization, the size distribution was given as P_i, where only one index was needed; that index referred to the number of monomers attached. Is one index sufficient for AB_{f-1} polymerization? Would the same be true for $A_3 + B_2$?
(c) What are the rate equations for the disappearance of A groups and B groups? What is the limiting reagent? What is the relation-

ship between time and conversion p, where p is the conversion of the limiting reagent?

(d) What is the rate of reaction between an i-mer and a j-mer?

(e) Write the rate equation for P_j.

(f) Derive the equations for μ_0, μ_1, and μ_2.

(g) Solve these equations and present equations for DP_n and DP_w in terms of conversion p. (Note that for $f = 2$, all the results should revert to those described in Appendix 2A.)

5.3. Consider the system $A_2 + B_3$ at unbalanced stoichiometry, that is: $r = A_0/B_0 < 1$. At what r will gelation be prevented? Consider the system $A_2 + B_2 + B_3$, at balanced stoichiometry but with a ratio $r' = 3B_3/(3B_3 + 2B_2) < 1$. At what value of r' will gelation be prevented in this system?

5.4. For both of the systems of Problem 5.3, derive the equations for $P(F_A^{out})$, and find the solutions as well.

5.5. Consider the reaction of a poly(propylene oxide) triol with 4,4'-diphenylmethylene diisocyanate at 25°C. Two different poly(propylene oxide) samples were used: one with $M_n = 708$ and the other with $M_n = 2630$. Stoichiometric ratios used were $r = 1.0, 0.95, 0.90, 0.85, 0.80$, and 0.75; r is defined as the ratio of initial alcohol groups to isocyanate groups. Thus 12 different polyurethane samples were made in all. Always assume densities of 1 g/mL.

(a) For the final bulk polyurethane, calculate the modulus G (in MPa) for the final products in bulk (i.e., all 12 cases). Explain the results you find. For which samples is the rubber elasticity equation more valid?

(b) Calculate the swelling $q_m = 1/v_2$ for all 12 cases as well, assuming a solvent–polymer interaction parameter $\chi = 0.55$ and a solvent molar volume of 100 cm³/mol.

5.6. Any hydroxyl group can react with any other in the polymerization of pentaerythritol and 1,4-butanediol via ether linkages to form a network polymer:

$$
\begin{array}{c}
\text{OH} \\
| \\
\text{CH}_2 \\
| \\
\text{HOCH}_2\!-\!\!-\text{C}\!-\!\!-\text{CH}_2\text{OH} \;+\; \text{HO(CH}_2)_4\text{OH} \\
| \\
\text{CH}_2 \\
| \\
\text{OH}
\end{array}
$$

(a) Use the recursive method to derive the relations for M_w and the gel point.

(b) Which assumptions of this derivation are likely to be violated for this reaction and why?

(c) If etherification is limited to 75% conversion, how much diol will be needed to just prevent gelation?

5.7. A reaction not discussed in the text is trimerization of cyanate groups to form triazine rings in the presence of transition metal carboxylates (e.g., zinc octoate) and heat[134,135]:

$$R\text{-}O\text{-}C\equiv N \quad \xrightarrow[\text{ZnOct}]{200°C}$$

where R is aromatic. Assuming that the reaction shown is the only reaction that occurs, derive the relations for M_w, M_n, and p_c for the homopolymerization of:

5.8. Consider a free-radical homopolymerization with chain transfer to a tetrathiol. Let $C_s = 1$ and let the reaction be transfer-dominated throughout (i.e., the majority of the chain-ending steps are transfer, but the chains are still long, as could be achieved by very low initiator concentration and relatively low thiol concentration).
(a) Sketch the structure of the molecules that can be produced. Can this system gel?
(b) What will the polydispersity of the final ($p = 1$) product be? (Keep in mind that k_{tr} is the transfer rate constant with respect to thiol *functional groups*.)

5.9. For long-chain branching, sketch the various sequences of reactions that lead to polyradical species and multifunctional species. Explain from these why the mode of termination and the mode of chain transfer to monomer are important. (In discussing mode of transfer to monomer, indicate whether the terminal double bond survives on the new or old chain.)

References

1. G. Odian, *Principles of Polymerization*, 3rd ed. Wiley, New York (1991).

2. S.-I. Nozakura, Y. Morishima, and S. Murahashi, *J. Polym. Sci., Polym. Chem.*, **10**, 2853 (1972).

3. W. H. Starnes, Jr., H. Chung, and G. M. Benedikt, *Polym. Prepr.*, **34(1)**, 604 (1993).

4. M. H. Litt and K. H. S. Chang, *ACS Symp. Ser.*, **165**, 455 (1981).

5. J. T. Clarke, R. O. Howard, and W. H. Stockmayer, *Makromol. Chem.*, **44–46**, 427 (1961).

6. M. K. Lindeman, The mechanism of vinyl acetate polymerization, in *Vinyl Polymerization*, Vol. 1, Part I, G. E. Ham, Ed. p. 207 Dekker, New York (1967).

7. W. H. Starnes, Jr., F. C. Schilling, K. B. Abbås, R. E. Cais, and F. A. Bovey, *Macromolecules*, **12**, 556 (1979).

8. W. H. Starnes, Jr., F. C. Schilling, I. M. Plitz, R. E. Cais, D. J. Freed, R. L. Hartless, and F. A. Bovey, *Macromolecules*, **16**, 790 (1983).

9. W. H. Starnes, Jr., *Pure Appl. Chem.*, **57**, 1001 (1985).

10. W. H. Starnes, Jr., B. J. Wojciechowski, A. Velazquez, and G. M. Benedikt, *Macromolecules*, **25**, 3638 (1992).

11. W. H. Starnes, Jr., and B. J. Wojciechowski, *Makromol. Chem., Macromol. Symp.*, **70/71**, 1 (1993).

12. W. H. Starnes, Jr., H. Chung, B. J. Wojciechowski, D. E. Skillicorn, and G. M. Benedikt, *Polym. Prepr.*, **34(2)**, 114 (1993).

13. T. Hjertberg and E. M. Sörvik, *Polymer*, **24**, 673, 685 (1983).

14. M.-F. Llauro-Darricades, N. Bensemra, A. Guyot, and R. Petiaud, *Makromol. Chem., Macromol. Symp.*, **29**, 171 (1989).

15. F. Rodriguez, *Principles of Polymer Systems*, 2nd ed. McGraw-Hill, New York (1982).

16. M. E. Adams, D. J. Buckley, R. E. Colborn, W. P. England, and D. N. Schissel, *RAPRA Rev. Rep.*, **6(10)**, report no. 70 (1993).

17. C. E. Scott, Ph.D. thesis, University of Minnesota (1990).

18. F. Ide and A. Hasegawa, *J. Appl. Polym. Sci.*, **18**, 963 (1974).

19. S. Y. Hobbs, R. C. Bopp, and V. H. Watkins, *Polym. Eng. Sci.*, **23**, 380 (1983).

20. (a) S. Cimmino, L. d'Orazio, R. Greco, G. Maglio, C. Malinconico, M. Mancarella, E. Martuscelli, R. Palumbo, and G. Ragosta, *Polym. Eng. Sci.*, **24**, 48 (1984). (b) S. Cimmino, F. Coppola, L. d'Orazio, R. Greco, G. Maglio, M. Malinconico, C. Mancarella, E. Martuscelli, and G. Ragosta, *Polymer*, **27**, 1874 (1986).

21. J. M. J. Fréchet, *Science*, **263**, 1710 (1994).

22. M. R. Krejsa and J. L. Koenig, *Rubber Chem. Technol.*, **66**, 376 (1993).

23. (a) C. J. Brinker, K. D. Keefer, D. W. Schaefer, and C. S. Ashley, *J. Non-Cryst. Solids*, **48**, 47 (1982). (b) C. J. Brinker, K. D. Keefer, D. W. Schaefer, R. A. Assink, B. D. Kay, and C. S. Ashley, *J. Non-Cryst. Solids*, **63**, 45 (1984).

24. J. K. Bailey, C. W. Macosko, and M. L. Mecartney, *J. Non-Cryst. Solids*, **125**, 208 (1990).

25. P. J. Flory, *Principles of Polymer Chemistry*. Cornell University Press, Ithaca, NY (1953).

26. R. M. Ziff and G. Stell, *J. Chem. Phys.*, **73**, 3492 (1980).

27. W. H. Stockmayer, *J. Chem. Phys.*, **11**, 45 (1943).

28. M. Gordon and S. B. Ross-Murphy, *Pure Appl. Chem.*, **43**, 1 (1975).

29.. C. W. Macosko, *Br. Polym. J.*, **17**, 239 (1985).

30. (a) H. H. Winter, *Polym. Eng. Sci.*, **27**, 1698 (1987). (b) H. H. Winter, *Prog. Colloid Polym. Sci.*, **75**, 104 (1987).

31. C. W. Macosko and D. R. Miller, *Macromolecules*, **9**, 199 (1976).

32. T. E. Harris, *The Theory of Branching Processes*. Dover, New York (1989).

33. M. Gordon, *Proc. R. Soc. London*, **A268**, 240 (1962).

34. T. A. Bak and B. Lu, *Chem. Phys.*, **112**, 189 (1987).

35. D. R. Miller, E. M. Valles, and C. W. Macosko, *Polym. Eng. Sci.*, **19**, 272 (1979).

36. D. R. Miller and C. W. Macosko, *Macromolecules*, **11**, 656 (1978).

37. (a) J. L. Spouge, *Macromolecules*, **16**, 121 (1983). (b) J. L. Spouge, *J. Stat. Phys.*, **31**, 363 (1983).

38. W. H. Stockmayer, *J. Polym. Sci.*, **9**, 69 (1952); **11**, 424 (1953).

39. M. Gordon and G. R. Scantlebury, *Trans. Faraday Soc.*, **60**, 604 (1964).

40. D. Durand and C.-M. Bruneau, *Polymer*, **24**, 587 (1983).

41. W. H. Stockmayer, *J. Chem. Phys.*, **12**, 125 (1944).

42. A. Charlesby, *Proc. R. Soc. London*, **A222**, 542 (1954).

43. J. F. Yan, *Macromolecules*, **12**, 260 (1979).

44. K. Nijenhuis, *Makromol. Chem.*, **192**, 603 (1991).

45. D. R. Miller and C. W. Macosko, *J. Polym. Sci., Polym. Phys.*, **25**, 2441 (1987); **26**, 1 (1988).

46. D. R. Miller, *Makromol. Chem., Macromol. Symp.*, **30**, 57 (1989).

47. N. A. Dotson, R. Galván, and C. W. Macosko, *Macromolecules*, **21**, 2560 (1988).

48. K. Dusek and M. Ilavsky, *J. Polym. Sci., Symp.*, **53**, 57, 75 (1975).

49. (a) A. B. Scranton and N. A. Peppas, *J. Polym. Sci., Polym. Chem.*, **28**, 39 (1990). (b) A. B. Scranton, J. Klier, and N. A. Peppas, *Macromolecules*, **24**, 1442 (1991).

50. N.A. Dotson, Ph.D. thesis, University of Minnesota (1991).

51. D. Durand and C.-M. Bruneau, *Eur. Polym. J.*, **21**, 527, 611 (1985).

52. (a) R. J. J. Williams, *Macromolecules*, **21**, 2568 (1988). (b) R. J. J. Williams and C. I. Vallo, *Macromolecules*, **21**, 2571 (1988).

53. R. J. J. Williams, D. P. Fasce, and J. C. Lucas, *Polym. Commun.*, **32**, 226 (1991).

54. N. A. Dotson, *Macromolecules*, **25**, 308, 7080 (1992).

55. (a) H. Tobita, *Macromolecules*, **26**, 836 (1993). (b) H. Tobita, *Makromol. Chem., Theory Simulations*, **2**, 761 (1993).

56. (a) H. Tobita and A. E. Hamielec, *Makromol. Chem., Macromol. Symp.*, **20/21**, 501 (1988). (b) H. Tobita and A. E. Hamielec, *Macromolecules*, **22**, 3098 (1989). (c) T. Xie and A. E. Hamielec, *Makromol. Chem., Theory Simulations*, **2**, 777 (1993).

57. S. Zhu and A. E. Hamielec, *Macromolecules*, **26**, 3131 (1993).

58. K. Dusek, *Polym. Bull.*, **1**, 523 (1979).

59. R. M. Ziff, *J. Stat. Phys.*, **23**, 241 (1980).

60. R. F. T. Stepto, *Polym. Bull.*, **24**, 53 (1990).

61. Đ. R. Miller and C. W. Macosko, *Macromolecules*, **9**, 206 (1976).

62. W. H. Carothers, *Trans. Faraday Soc.*, **32**, 39 (1936).

63. Z. Changren, *J. Appl. Polym. Sci.*, **44**, 383 (1992).

64. M. Falk and R. E. Thomas, *Can. J. Chem.*, **52**, 3285 (1974).

65. (a) H. Tobita and A. E. Hamielec, *Polymer*, **33**, 3647 (1992). (b) S. Zhu and A. E. Hamielec, *Macromolecules*, **25**, 5457 (1992). (c) S. Zhu, A. E. Hamielec, and R. H. Pelton, *Makromol. Chem., Theory Simulations*, **2**, 587 (1993).

66. R. A. Hutchinson, *Polym. React. Eng.*, **1**, 521 (1992/1993).

67. (a) O. Okay, *Polymer*, **35**, 796 (1994). (b) O. Okay, *Macromol. Chem. Theory Simulations*, **3**, 417 (1994).

68. (a) D. S. Argyropoulos, R. M. Berry, and H. I. Bolker, *J. Polym. Sci., Polym. Phys.*, **25**, 1191 (1987). (b) D. S. Argyropoulos, R. M. Berry, and H. I. Bolker, *Makromol. Chem.*, **188**, 1985 (1987). (c) D. S. Argyropoulos, R. M. Berry, and H. I. Bolker, *Macromolecules*, **20**, 357 (1987).

69. A. J. Staverman, *Adv. Polym. Sci.*, **44**, 73 (1982).

70. J. P. Queslel and J. E. Mark, *Adv. Polym. Sci.*, **65**, 135 (1984).

71. P. J. Flory, *Br. Polym. J.*, **17**, 96 (1985).

72. G. Heinrich, E. Straube, and G. Helmis, *Adv. Polym. Sci.*, **85**, 33 (1988).

73. S. Candau, J. Bastide, and M. Delsanti, *Adv. Polym. Sci.*, **44**, 27 (1982).

74. N.A. Neuberger and B. E. Eichinger, *Macromolecules*, **21**, 3060 (1988).

75. P.-G. de Gennes, *Scaling Properties in Polymer Physics*. Cornell University Press, Ithaca, NY (1979).

76. D. Stauffer, A. Coniglio, and M. Adam, *Adv. Polym. Sci.*, **44**, 103 (1982).

77. J. E. Martin and D. Adolf, *Annu. Rev. Phys. Chem.*, **42**, 311 (1991).

78. D. Stauffer, *Phys. Rep.*, **54**, 1 (1979).

79. D. Stauffer, *J. Chem. Soc. Faraday Trans. 2*, **72**, 1354 (1976).

80. E. M. Hendriks, M. H. Ernst, and R. M. Ziff, *J. Stat. Phys.*, **31**, 519 (1983).

81. M. Adam, M. Delsanti, J. P. Munch, and D. Durand, *J. Phys.*, **48**, 1809 (1987).

82. E. V. Patton, J. A. Wesson, M. Rubinstein, J. C. Wilson, and L. E. Oppenheimer, *Macromolecules*, **22**, 1946 (1989).

83. D. Adolf, J. E. Martin, and J. P. Wilcoxon, *Macromolecules*, **23**, 527 (1990).

84. C. Konak, Z. Tuzar, J. Jakes, P. Stepanek, and K. Dusek, *Polym. Bull.*, **18**, 329 (1987).

85. K. Kajiwara, W. Burchard, M. Kowalski, D. Nerger, K. Dusek, L. Matejka, and Z. Tuzar, *Makromol Chem.*, **185**, 2543 (1984).

86. K. Grof, L. Mrkvicková, C. Konák, and K. Dusek, *Polymer*, **34**, 2816 (1993).

87. J. E. Martin, J. Wilcoxon, and D. Adolf, *Phys. Rev. A*, **36**, 1803 (1987).

88. R. S. Whitney and W. Burchard, *Makromol. Chem.*, **181**, 869 (1980).

89. N. A. Dotson, T. Diekmann, C. W. Macosko, and M. Tirrell, *Macromolecules*, **25**, 4490 (1992).

90. J.-P. Munch, M. Ankrim, G. Hild, R. Okasha, and S. Candau, *Macromolecules*, **17**, 110 (1984).

91. D. Stauffer, *Pure Appl. Chem.*, **53**, 1479 (1981).

92. C. Walling, *J. Am. Chem. Soc.*, **67**, 441 (1945).

93. W. Simpson, T. Holt, and R. J. Zetie, *J. Polym. Sci.*, **10**, 489 (1953).

94. M. Gordon, *J. Chem. Phys.*, **22**, 610 (1954).

95. W. Simpson and T. Holt, *J. Polym. Sci.*, **18**, 335 (1955).

96. M. Gordon and R.-J. Roe, *J. Polym. Sci.*, **21**, 27, 39, 57, 75 (1956).

97. H. Wesslau, *Angew. Makromol. Chem.*, **1**, 56 (1967).

98. B. Soper, R. N. Haward, and E. F. T. White, *J. Polym. Sci. A-1*, **10**, 2545 (1972).

99. I. Holdaway, R. N. Haward, and I. W. Parsons, *Makromol. Chem.*, **179**, 1939 (1978).

100. A. C. Shah, I. Holdaway, I. W. Parsons, and R. N. Haward, *Polymer*, **19**, 1067 (1978).

101. D. T. Landin and C. W. Macosko, *Macromolecules*, **21**, 846 (1988).

102. L. Minnema and A. J. Staverman, *J. Polym. Sci.*, **29**, 281 (1958).

103. R. Okasha, G. Hild, and P. Rempp, *Eur. Polym. J.*, **15**, 975 (1979).

104. K. Dusek, H. Galina, and J. Mikes, *Polym. Bull.*, **3**, 19 (1980).

105. K. Dusek, in *Developments in Polymerisation*, Vol. 3, R. N. Haward, Ed., pp. 143–206. Applied Science, London (1982).

106. J. Mikes and K. Dusek, *Macromolecules*, **15**, 93 (1982).

107. S. I. Kuchanov and Ye. S. Povolotskaya, *Polym. Sci. USSR*, **A24**, 2499, 2512 (1982) (*Vysokomol. Soyed. A*, **A24**, 2179, 2190).

108. (a) H. Galina, *Europhys. Lett.*, **3**, 1155 (1987). (b) H. Galina and A. Szustalewicz, *Macromolecules*, **22**, 3124 (1989). (c) H. Galina and A. Szustalewicz, *Macromolecules*, **23**, 3833 (1990). (d) H. Galina, *Makromol. Chem., Macromol. Symp.*, **40**, 45 (1990). (e) H. Galina, K. Kaczmarski, B. Para, and B. Sanecka, *Makromol. Chem., Theory Simulations*, **1**, 37 (1992).

109. D. R. Miller and C. W. Macosko, in *Biological and Synthetic Polymer Networks*, O. Kramer, Ed., p. 219. Elsevier Applied Science, London (1988).

110. C. Sarmoria and D. R. Miller, *Macromolecules*, **24**, 1833 (1991).

111. R. J. Zeman and N. R. Amundson, *Chem. Eng. Sci.,* **20**, 331, 637 (1965).

112. N. G. Taganov, *Polym. Sci. USSR,* **24**, 1767 (1982). (*Vysokomol. Soyed. A,* **A24**, 1552).

113. P. J. Flory, *J. Am. Chem. Soc.,* **69**, 30, 2893 (1947).

114. J. K. Beasley, *J. Am. Chem. Soc.,* **75**, 6123 (1953).

115. C. H. Bamford and H. Tompa, *Trans. Faraday Soc.,* **50**, 1097 (1954).

116. D. J. Stein, *Makromol. Chem.,* **76**, 157, 170 (1964).

117. W. W. Graessley, H. Mittelhauser, and R. Maramba, *Makromol. Chem.,* **86**, 129 (1965).

118. W. W. Graessley, R. D. Hartung, and W. C. Uy, *J. Polym. Sci. A-2,* **7**, 1919 (1969).

119. O. Saito, K. Nagasubramanian, and W. W. Graessley, *J. Polym. Sci. A-2,* **7**, 1937 (1969).

120. K. Nagasubramanian and W. W. Graessley, *Chem. Eng. Sci.,* **25**, 1549, 1559 (1970).

121. J. C. Hyun, W. W. Graessley, and S. G. Bankoff, *Chem. Eng. Sci.,* **31**, 945 (1976).

122. A. E. Hamielec and J. F. McGregor, in *Polymer Reaction Engineering,* K. H. Reichert and W. Geiseler, Eds. Hanser, Munich (1983).

123. J. Villermaux and L. Blavier, *Chem. Eng. Sci.,* **39**, 87 (1984).

124. T. W. Taylor and K. H. Reichert, *J. Appl. Polym. Sci.,* **30**, 227 (1985).

125. A. G. Mikos, C. G. Takoudis, and N. A. Peppas, *Macromolecules,* **19**, 2174 (1986).

126. D. J. Arriola, Ph.D. thesis, University of Wisconsin (1989).

127. T. Y. Xie and A. E. Hamielec, *Makromol. Chem., Theory Simulations,* **2**, 455 (1993).

128. S. Zhu and A. E. Hamielec, *J. Polym. Sci., Polym. Phys.,* **32**, 929 (1994).

129. (a) H. Tobita, *Polym. React. Eng.,* **1**, 357, 379 (1992/1993). (b) H. Tobita, *J. Polym. Sci., Polym. Phys.,* **31**, 1363 (1993). (c) H. Tobita, *J. Polym. Sci., Polym. Phys.,* **32**, 901, 911 (1994).

130. W. W. Graessley and H. M. Mittelhauser, *J. Polym. Sci. A-2,* **5**, 431 (1967).

131. C. H. Bamford and A. D. Jenkins, *Nature,* **176**, 78 (1955).

132. C. H. Bamford, R. W. Dyson, and G. C. Eastmond, *Polymer,* **10**, 885 (1969).

133. B. L. Funt and W. Pasika, *Can. J. Chem.,* **38**, 1865 (1960).

134. A. M. Gupta, *Macromolecules,* **24**, 3459 (1991).

135. A. M. Gupta and C. W. Macosko, *Macromolecules,* **26**, 2455 (1993).

6

REACTOR CONFIGURATION

6.1 Introduction

It can be reasoned from the preceding chapters that the polymerization process can determine to a great extent the properties of the polymeric product obtained. For example, temperature control during a free-radical polymerization can be crucial; an excursion in temperature due to the exothermic reaction, even if within the bounds of safety, might give a very different, and perhaps unacceptable, molecular weight distribution. For polycondensations, obtaining product of high molecular weight implies efficient removal of by-product, especially for polyesters for which the equilibrium is very unfavorable toward polymerization. Thus, the kind of reactor and how well it generates interfacial area exposed to the gas phase will determine the rate of reaction.

This kind of complication comes about because of the manifold characteristics and properties of a polymeric product, which are coupled to physical processes occurring during polymerization, such as diffusion of species of low molecular weight, entanglements of polymer chains, glass or crystalline transitions, and heats of reaction. The distinctiveness of polymerization, as mentioned in Chapter 1, gives the process a much more prominent role in determining the product than in small-molecule reactors.[1] For the reaction engineer, this means that scale-up is not trivial; a process implemented in the lab may give a very different product in a commercial scale reactor.

The influences of process on product are as varied as the characteristics affected and the physical sources. To introduce the subject of process and its effects, we present in this chapter approaches to the modeling of reactor configurations other than batch. We also give some general ideas of the effects of reactor configuration on polymer characteristics, such as chain length distribution and composition, by reexamining the polymerizations discussed in the preceding chapters. Denbigh[2] appears to have been the first to realize the importance of reactor type on polymer characteristics, and the topic is well explored in the subsequent literature. For the moment our example of polycondensation is sufficient to demonstrate the significance of process: a true batch polyesterification will yield only oligomer, whereas the usual reactor configuration, which is semibatch because by-product is being removed,[3] will yield the desired high molecular-weight product.

The text has thus far been concerned mainly with batch polymerizations, to permit the identification of the aspects of polymer reactor engineering peculiar to polymers (e.g., the various distributions that characterize the polymer) and observation of how those are related to the chemical mechanisms alone. In addition, the batch reactor configuration is commonly used (almost exclusively so for specialty polymers),[4] and so the analyses are of immediate practical value. However, continuous processes have an obvious economic appeal, and semibatch processes are important for many other products. Since the groundwork with batch polymerization has been laid, we proceed to describing how reactor configuration affects the characteristics of the polymeric product. Chain length and its distribution will be the focus of the chapter, although a few comments regarding free-radical copolymerizations and to branching polymerizations become necessary as we work our way through the various reactors, beginning with the homogeneous continuous stirred tank reactor (HCSTR).

6.2 Homogeneous Continuous Stirred Tank Reactor (HCSTR)

In continuous stirred tank reactors, which are familiar from general reactor engineering,[5] reactants are continually fed and product continually removed to ensure that the system maintains a steady state. Agitation serves to maintain mixing good enough to result in identical compositions of product and of the material in the reactor. Such perfect mixing results in an exponential residence time distribution, $f(\theta')$[5-9]:

$$f(\theta')d\theta' = \frac{1}{\theta} e^{-\theta'/\theta} \, d\theta' \qquad (6.2.1)$$

The word "homogeneous" further implies not only that all fluid elements are equally likely to be anywhere in the reactor, regardless of their residence time, but the length scale of any heterogeneity is vanishingly small. This means that in an HCSTR, the environment of any given functional group is the same as overall environment in the reactor (a "mean field" is assumed).

We now turn to the idealized polymerization schemes presented in earlier chapters and see how this reactor configuration alters the chain length and its distribution.[10] The approach will be from kinetic equations because the statistical approach in simple form cannot account faithfully for the residence time distribution (although more complicated statistical derivations do[11]). This effectively limits straightforward application of the statistical methods to batch reactors.

6.2.1 Anionic Polymerization

We first consider an anionic polymerization performed in an HCSTR. The polymerization conforms to that of Section 3.2, so that initiation is instantaneous and no transfer or termination reactions occur. As shown in Figure 6.1, the entering stream of volumetric flow rate Q_0 consists only of initiated chains of length *1* at a concentration $I_0 = P_{1,0} = P_0 = P$, and monomer at a concentration M_0. The subscript zero refers not to an initial condition, but to the stream number, which is a useful indexing for a cascade as we will see later. Here we deal only with a single HCSTR and omit the subscript 'one' from the exit stream, which flows at rate Q (distinguished from polydispersity by context) and consists of chains of different lengths P_i and monomer M.

The balance equations for monomer and polymer chains are:

$$\frac{d(VM)}{dt} = Q_0 M_0 - QM - VkPM \tag{6.2.2}$$

$$\frac{d(VP_i)}{dt} = Q_0 P_{i,0} - QP_i + k(P_{i-1} - P_i)M \tag{6.2.3}$$

Under steady state conditions, the time derivatives vanish. Furthermore, if no volume change occurs upon reaction (as might be nearly true for reaction in a solvent), $Q_0 = Q$ and $I_0 = I$. The balance equations for monomer and polymer then simplify to the following:

$$-\frac{M_0 - M}{\theta} = -kMI_0 \tag{6.2.4}$$

$$-\frac{P_{i,0} - P_i}{\theta} = k(P_{i-1} - P_i)M \tag{6.2.5}$$

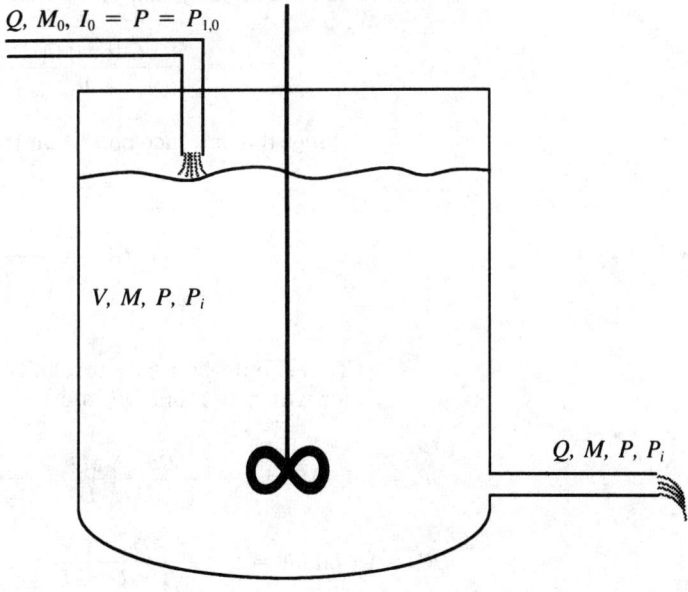

Figure 6.1.
Schematic of anionic polymerization in an HCSTR.

$Q, M_0, I_0 = P = P_{1,0}$

V, M, P, P_i

Q, M, P, P_i

where the average residence time θ is defined as the ratio of the reaction volume V to the volumetric flow rate[5-9]:

$$\theta = \frac{V}{Q} \qquad (6.2.6)$$

Equation (6.2.4) can be solved for the monomer concentration to yield:

$$\frac{M}{M_0} = 1 - p = \frac{1}{1 + kI_0\theta} \qquad (6.2.7)$$

The denominator suggests a useful dimensionless number: the Damköhler number (of the first kind), which is the ratio of the rate of reaction to the rate of exit, or equivalently of the residence time to the time scale of reaction.[8] The Damköhler number is here defined as $Da = kI_0\theta$, and so conversion p can be rewritten from equation (6.2.7) as follows:

$$p = \frac{Da}{1 + Da} \qquad (6.2.8)$$

This result is different from the batch reactor case, for which $p = 1 - e^{-Da}$, where $Da = kI_0t$. It can be shown by series expansion that conversion is less in the HCSTR than in the batch reactor at the same Da. The average rate of reaction exhibited by the batch reactor (p/t) is greater than that for an HCSTR because the batch reactor experiences a history of higher monomer concentrations and thus faster kinetics, whereas the HCSTR, operating at steady state, sees only the low monomer concentration at conversion p.

The decreased reaction rate is of relatively minor importance, however, compared to how the chain length distribution is affected. The evolution equation for the chain length distribution, equation (6.2.5), can be written in transformed space as:

$$-\frac{G_0(s) - G(s)}{\theta} = k(s - 1)G(s)M \qquad (6.2.9)$$

Under the entrance condition that $G_0(s) = sI_0$, the solution is:

$$G(s) = \frac{\left(\dfrac{I_0}{1 + k\theta M}\right)s}{1 - \left(\dfrac{k\theta M}{1 + k\theta M}\right)s} \qquad (6.2.10)$$

This should be easily recognized as a geometric distribution,[12] for which DP_n and DP_w are:

$$DP_n = 1 + kM\theta = 1 + \left(\frac{Da}{1 + Da}\right)\frac{M_0}{I_0} = 1 + p\frac{M_0}{I_0} \qquad (6.2.11)$$

$$DP_w = 1 + 2kM\theta = 1 + 2\left(\frac{Da}{1 + Da}\right)\frac{M_0}{I_0} = 1 + 2p\frac{M_0}{I_0} \qquad (6.2.12)$$

We know the geometric distribution to be broader than the Poisson distribution. Because of the homogeneity of reaction and steady state operation, the increased polydispersity cannot be attributed to any sort of drift dispersion. Thus, the statistical dispersion has been increased owing to the broad residence time distribution in the HCSTR. The distribution of residence times constitutes an additional random element, and so statistical dispersion depends not only on the reaction mechanism but also on the reactor type and configuration.

6.2.2 AB Step Polymerization

If a polymerization that would yield a product of a narrow Poisson distribution in a batch reactor yields a product of geometric distribution in an HCSTR, how will a polymerization characterized by a geometric distribution in a batch reactor behave in an HCSTR? To answer this, we consider the step polymerization of AB-type monomers, as presented in Section 2.2. The entering stream consists only of AB monomer at a concentration $P_{1,0}$. Assuming from the start steady state operation and no volume change upon reaction, the evolution equation for a particular species P_i can be written by reference to equations (2.2.7) and (6.2.5):

$$-\frac{P_{i,0} - P_i}{\theta} = k \sum_{j=1}^{i-1} P_j P_{i-j} - 2kP_i P \tag{6.2.13}$$

where $P = G(1)$, and the first term on the right-hand side does not exist for $i = 1$. In transformed space this is written:

$$-\frac{G_0(s) - G(s)}{k\theta} = G^2(s) - G(s)G(1) \tag{6.2.14}$$

Equation (6.2.14) can be solved with the entrance condition that $G_0(s) = sP_{1,0}$. We first solve for $G(1)$, which is directly related to conversion:

$$G(1) = \frac{-1 + \sqrt{1 + 4kP_{1,0}\theta}}{2k\theta} \tag{6.2.15}$$

Since the relevant characteristic reaction time in this case is $1/kP_{1,0}$, the Damköhler number is given by $Da = kP_{1,0}\theta$, which allows equation (6.2.15) to be rewritten:

$$\frac{G(1)}{P_{1,0}} = 1 - p = \frac{-1 + \sqrt{1 + 4Da}}{2Da} \tag{6.2.16}$$

From our experience with anionic polymerization, we can assume that conversion here will be less than would be expected in a batch polymerization at time $t = \theta$. From equation (2.2.18), we recall that for the batch polymerization $p = Da/(1 + Da)$, where $Da = kP_{1,0}t$. By comparing series expansions, we can confirm our intuition that the conversion is indeed less in the HCSTR than in the batch reactor, for exactly the same reason given for anionic polymerization.

The chain length distribution is of greater interest, though. Solving equation (6.2.14) for $G(s)$ yields:

$$\frac{G(s)}{P_{1,0}} = \frac{\sqrt{1 + 4Da} - \sqrt{1 + 4(1 - s)Da}}{2Da} \qquad (6.2.17)$$

In terms of real variables, this distribution may be written as follows[8,12,13]:

$$P_i = P_{1,0} \frac{(2i - 2)!}{i!(i - 1)!} (\sqrt{1 + 4Da})^{-(2i-1)} (Da)^{i-1} \qquad (6.2.18)$$

$$= P_{1,0} (1 - p) \frac{(2i - 2)!}{i! (i - 1)!} \frac{p^{i-1}}{(1 + p)^{2i-1}} \qquad (6.2.19)$$

This distribution is characterized by the following degrees of polymerization and polydispersity Q:

$$DP_n = \frac{1}{1 - p} \qquad (6.2.20)$$

$$DP_w = \frac{1 + p^2}{(1 - p)^2} \qquad (6.2.21)$$

$$Q = \frac{1 + p^2}{1 - p} \qquad (6.2.22)$$

This distribution has the remarkable characteristic of unbounded polydispersity as $p \to 1$, and a greater polydispersity than would exist in a batch reactor at any given conversion. Again, the residence time distribution has contributed to the statistical dispersion. Reversibility, however, will tend to narrow the distribution toward a most probable distribution[14] for reasons to be explained in Section 6.2.3.

The extent to which the statistical dispersion has been increased can be judged also from the limiting form of the distribution given in equation (6.2.19):

$$\lim_{i \to \infty} P_i \propto i^{-3/2} \left(\frac{4p}{(1 + p)^2}\right)^i \qquad (6.2.23)$$

The form of the distribution is similar to that of equation (5.2.20) for A_f homopolymerization: a power law modified by an exponential part which vanishes at $p = 1$ (rather than at a gel point). All moments of the distribution are bounded until $p = 1$, at which point all moments higher than $\mu_{1/2}$ diverge. Thus the residence time distribution plays a role similar to branching, but here all the polymers are necessarily linear, and gelation does not occur. Instead both DP_n and DP_w diverge at complete conversion, $p = 1$, as in the case of the geometric distribution arising in a batch reactor, although since DP_w diverges faster the polydispersity is unbounded, rather than 2.

6.2.3 Free-Radical Homopolymerization

By comparison with the two preceding examples and with our earlier work on free-radical homopolymerization, we can immediately write the evolution equation for $H(s)$, the growing chain generating function. Neglecting transfer to monomer and transfer agents, we write:

$$-\frac{H_0(s) - H(s)}{\theta} = 2fk_d Is - k_p M(1 - s)H(s) - k_t H(s)H(1) \qquad (6.2.24)$$

[cf. equation (3.5.4)]. Similarly, we can write the balance equation for monomer:

$$-\frac{M_0 - M}{\theta} = -2fk_d I - k_p MH(1) \qquad (6.2.25)$$

We take as an entrance condition $H_0(s) = 0$, so that initiation occurs only within the reactor. Solving equation (6.2.24) for the radical concentration, $H(1)$:

$$H(1) = \frac{\sqrt{1 + 8fk_d Ik_t \theta^2} - 1}{2k_t \theta} \qquad (6.2.26)$$

for which initiator depletion has been neglected. The solution to equation (6.2.25) reads:

$$\frac{M}{M_0} = 1 - p = \frac{1 - 2fk_d I\theta/M_0}{1 + k_p H(1)\theta} \qquad (6.2.27)$$

Under common operating conditions, equation (6.2.26) can be simplified by rewriting it first in the following way:

$$H(1) = H(1)_{\theta=\infty} \left\{ \left(1 + \frac{1}{8fk_d Ik_t \theta^2}\right)^{1/2} - \left(\frac{1}{8fk_d Ik_t \theta^2}\right)^{1/2} \right\} \qquad (6.2.28)$$

where the dominant term $H(1)_{\theta=\infty}$ is defined as in a batch reactor ($\theta = \infty$):

$$H(1)_{\theta=\infty} = \left(\frac{2fk_d I}{k_t}\right)^{1/2} \qquad (6.2.29)$$

The difference between square roots in equation (6.2.28), which corrects for finite θ, will not be significant if $k_t H(1)\theta \gg 1$. This inequality ensures that the time scale for exit is much longer than the lifetime of a radical. If this were not so, no significant amount of polymer could be formed, since high conversions require $k_p H(1)\theta > 1$, and generally $k_t \gg k_p$ experimentally. Thus we will assume that the radical concentration conforms to that of a batch reactor at the same conversion. Thus $H(1)$ depends on residence time θ implicitly through conversion, which affects the rate constants, k_t in particular.

The transformed distribution $H(s)$ is given by:

$$H(s) = \frac{\left(\dfrac{2fk_dI}{k_pM + k_tH(1) + 1/\theta}\right)s}{1 - \left(\dfrac{k_pM}{k_pM + k_tH(1) + 1/\theta}\right)s} = H(1)\frac{(1-q)s}{1-qs} \tag{6.2.30}$$

which is a geometric distribution with a parameter q given by:

$$q = \frac{k_pM}{k_pM + k_tH(1) + 1/\theta} \tag{6.2.31}$$

This is the same result as for batch polymerizations [cf. equation (3.5.17)] except that an additional "termination" mechanism—namely, exit from the reactor at a rate $1/\theta$—has been added. However, since for any significant polymerization to occur it must be that $k_tH(1) \gg 1/\theta$, in practical situations the residence time distribution has little effect on the growing chain distribution. The life of a chain is so short that the residence time distribution is essentially "invisible." No increase in statistical dispersion occurs as in the two preceding cases, for which the lifetime of the chain was identified with the residence time.[2] The same occurs when reversibility and interchange reactions occur in a step polymerization in an HCSTR; the "life" of a chain is, in essence, decreased and the distribution tends toward a most probable distribution, as would be obtained in a batch reactor.[14]

Even more important, though, steady state operation results in an absence of drift dispersion in the dead chain (product) distribution:

$$\frac{G(s)}{\theta} = k_{td}H(s)H(1) + \frac{k_{tc}}{2}H^2(s) \tag{6.2.32}$$

In an HCSTR operating at steady state, accumulation of polymer produced under different conditions cannot occur because there are no differing conditions. The same logic would apply in other cases; for example, for a free-radical copolymerization, a polymer of uniform composition may be obtained even at high conversion by performing the reaction in an HCSTR.[15–19]

It is difficult to generalize about the conversion in an HCSTR compared with that for a batch reactor. From equation (6.2.27), conversion is:

$$p = (1+d)\frac{Da}{1+Da} \tag{6.2.33}$$

where $Da = k_pH(1)_{\theta=\infty}\theta$ and $d = k_tH(1)/k_pM_0$. (Within the long-chain approximation, which ignores the term related to consumption through the initiation step, we may neglect the factor of d.) If depletion of monomer is the sole source of drift, then clearly the conversion in an HCSTR will be lower than that in a batch reactor at equal Da, since the latter has experienced a history of higher rates. On the other hand, in the presence of a severe Trommsdorff effect, the conversion at equal Da may be higher in the HCSTR, and the degrees of polymerization likewise higher, since the HCSTR has not traveled through a history of lower reaction rates [due to lower $H(1)$] and lower degree of polymerization (due to higher k_t).

Another difficulty arises here: *isothermal multiplicity of steady states*. In general, for even simple reaction schemes, an HCSTR can exhibit several steady states at different temperatures, since the heat removed term is fairly linear with temperature, while the generation term exhibits a sigmoidal shape.[5] In a free-radical polymerization, however, multiple steady states may be obtained even under isothermal conditions.[1] To see this, consider the monomer balance under the long-chain hypothesis:

$$-\frac{M_0 - M}{\theta} = -k_p M R = -k_p M_0 (1 - p) H(1) \qquad (6.2.34)$$

or

$$\frac{1}{\theta} p = k_p H(1)(1 - p) = \frac{R_p}{M_0} \qquad (6.2.35)$$

This equation has only one solution if k_p and $H(1)$ are independent of p, but we know that this is rarely the case, especially because of the Trommsdorff effect. This problem can in a general way be approached graphically, as in Figure 6.2. It is clear that under the appropriate residence time θ there are three solutions, and the middle one is inherently unstable. The physics behind this multiplicity is uniquely polymeric: a rapid drop in a diffusion-controlled rate constant caused by chain entanglement, which increases the rate of reaction.

6.2.4 A Note on Nonlinear Polymerizations

Since for the case of irreversible step polymerizations we observed an extreme broadening in the chain length distribution

Figure 6.2.
Isothermal multiplicity for a free-radical homopolymerization occurring in an HCSTR. The points of intersection between the straight line [left-hand side of equation (6.2.35)] and the curve [right-hand side of equation (6.2.35)] indicate solutions (steady states) to equation (6.2.35). The middle solution is of course unstable.

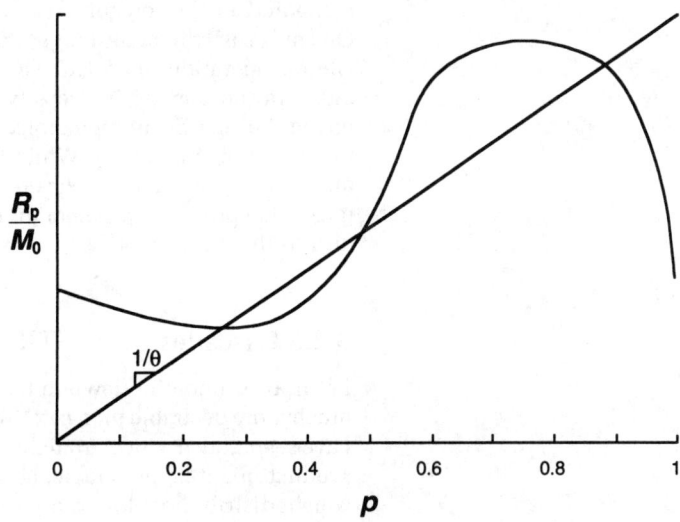

and a change in the form of the distribution to one resembling that for gelling systems, we might suspect even worse for a true gelling system, such as an A_f step homopolymerization. Similar concerns arise for branching polymerizations, such as poly(vinyl acetate), which are far more likely than nonlinear polymerizations to be performed in an HCSTR. Nonetheless, to see the complexities that can arise, we here briefly discuss the simplest case, A_f homopolymerization in an HCSTR.[20,21] More complicated treatments of this system, with the complications appearing either in reaction scheme[22] or in reactor configuration,[23,24] have been examined. The problems of crosslinking free-radical copolymerization[25] and long-chain branching[26-32] in an HCSTR have been examined as well.

For nonlinear polymerizations at steady state in an HCSTR, it is found that there is a critical Damköhler number Da_c beyond which DP_w becomes complex (i.e., has an imaginary component). This nonsensical condition suggests that steady state operation is impossible at residence times greater than that corresponding to the critical Damköhler number. Since, however, steady state cannot be reached because gelation occurs during the start-up if $Da > Da_c$, the apparent impossibility is resolved.

Strangely, though, at steady state DP_w is quite low, even at Da_c; the polydispersity Q is bounded. The rather mild polydispersity is misleading, because in fact there is no residence time sufficiently low that there is not some high moment of the distribution that is divergent. At Da_c, the third moment μ_3 is divergent, and thus DP_z; such divergence, however, is not indicative of gelation. This odd behavior arises because the form of the distribution is different from that of the Stockmayer distribution, which one ideally would obtain in a batch reactor; the power law form survives, but without the exponential cutoff.[21] The power law exponent, moreover, is a function of Da, allowing for progressively lower moments to be divergent as Da increases. This behavior points to the risk of characterizing a distribution by a finite number of moments.

Experimentally, one gets a polymer which is very difficult to characterize.[33] This is no doubt due in part both to the physics embodied in the foregoing equations and to mixing problems. One may easily imagine that in this case good mixing, a prerequisite for operation of an HCSTR, would be difficult to maintain either in the start-up if viscosity diverges, or under steady state (given the significant high molecular-weight tail, which leads to the divergent moments). While the inability to achieve perfect mixing is expected to be a severe problem for branching systems, it can also present a problem for other polymerizations. We will turn to this topic shortly.

6.2.5 Cascades of HCSTRs

From an economic viewpoint, continuous production is often much more desirable than batchwise production. However, as we have seen, the products from the two are different, and batchwise product might be preferable, because of the narrower molecular weight distribution, for example. However, HCSTRs can be operated to gain the advantages of continuous operation while approaching batch behavior. It is well known, for example, that N identical CSTRs in series give progressively narrower residence time distributions as N increases.[8,9] If the total residence time is θ, so that the residence time in each particular reactor is θ/N, the overall residence time distribution is given by:

$$f(\theta')d\theta' = \frac{N}{\theta} \frac{1}{(N-1)!} \left(\frac{\theta'}{\theta/N}\right)^{N-1} e^{-\theta'/(\theta/N)} \, d\theta' \qquad (6.2.36)$$

which becomes narrower as N increases. (More complicated expressions can be derived when the reactors are not identical, i.e., do not share the same residence time.[9]) This suggests the possibility that the increase in statistical dispersion can be diminished by a cascade of HCSTRs.

Extending the previous analysis of anionic polymerization for a cascade of HCSTRs (see Figure 6.3) is straightforward. The monomer and polymer balance equations (6.2.7) and (6.2.10) can be rewritten as follows:

$$M_1 = \frac{1}{1 + kI_0(\theta/N)} M_0 \qquad (6.2.37)$$

$$G_1(s) = \frac{1}{1 + k(1-s)M_1(\theta/N)} G_0(s) \qquad (6.2.38)$$

These equations can be applied interactively as one progresses down the cascade; at the mth reactor:

$$M_m = \frac{1}{1 + kI_0(\theta/N)} M_{m-1} \qquad (6.2.39)$$

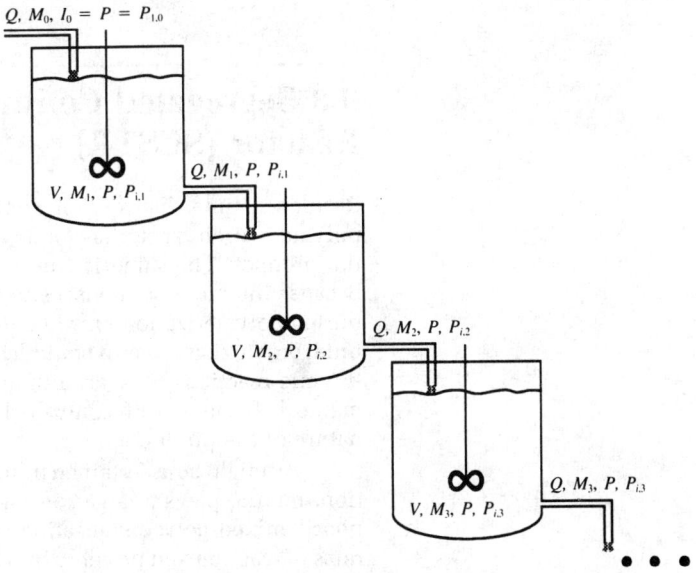

Figure 6.3.
Anionic polymerization occurring in a cascade of identical HCSTRs.

$$G_m(s) = \frac{1}{1 + k(1 - s)M_m(\theta/N)} G_{m-1}(s) \qquad (6.2.40)$$

These equations can be recast with reference to the feed conditions as follows:

$$M_N = M_0 \left\{ \frac{1}{1 + kI_0(\theta/N)} \right\}^N \qquad (6.2.41)$$

$$G_N(s) = sI_0 \prod_{m=1}^{N} \left\{ 1 + k(1 - s)M_0 \left\{ 1 + kI_0 \frac{\theta}{N} \right\}^{-m} \frac{\theta}{N} \right\}^{-1} \qquad (6.2.42)$$

(We have written these equations for the stream exiting the final reactor, but similar relations can be written for intermediate streams.) For $N = 2$, one gets a convolution of two geometric distributions, which will be narrower than the geometric distribution emerging from a single reactor of the same total residence time. As $N \to \infty$, the distribution approaches a Poisson distribution and the conversion approaches (more rapidly than the distribution) the conversion that would be attained in a batch reactor at time θ.[13]

Thus a cascade of CSTRs can act to restrict the increases in statistical dispersion while still allowing continuous operation. This restriction relies on the long lifetime of the chain; the same would be true for the irreversible stepwise polymerization of Section 6.2.2. For free-radical polymerization, on the other hand, each CSTR would produce a population of chains having a distribution the same as would be obtained instantaneously in a batch reactor at the conversion at which that CSTR operates. A discrete accumulation of dead chains would result, giving a broader (perhaps multimodal) distribution. No such accumulation would attend a reversible stepwise polymerization, which would tend toward a geometric distribution regardless of the number of reactors.

6.3 Segregated Continuous Stirred Tank Reactor (SCSTR)

Residence time distribution affects not only the "yield" of a polymerization reactor (as for any reactor) but also the nature of the product. The same is true for mixing, which is important because the increase in viscosity of several orders of magnitude during polymerization may not allow for good mixing. That this unfavorable relationship could lead to control problems for exothermic reactions is clear, although we will not delve into the matter.[34] Even if well controlled, how does mixing affect the nature of the product?

As might be ascertained from the example of polycondensation, mixing plays a large role in polymerization processes.[8] A poorly mixed polycondensation reactor will be marked by poor rates of reaction and possibly by a broad molecular weight distribution, if different elements of the polymer are not in equilibrium. This is a kind of drift dispersion, since polymers made under different conditions are being combined, just as occurs during the accumulation of dead chains in a free-radical polymerization. This drift dispersion arising from poor mixing may well superimpose on the "natural" drift dispersion of a free-radical polymerization, if, for example, hot spots exist.

Understanding of the role of mixing in polymerization processes is, at this point, quite limited.[35] Imperfect mixing presents a difficulty in modeling so vast that a single, straightforward treatment is not possible; many different models may be proposed.[7,36] For example, the fluid elements of a feed stream (to a semibatch reactor or CSTR) cannot be dispersed immediately, and it has been proposed to model this with "segregated feed zones," small, distinct parts of the reactor that exchange mass with the rest of the reactor by some mass transfer mechanism.[37,38] Numerous models of this sort appear in the chemical engineering literature; applications to polymerization reactors have been more sparse, even though the effects on the various distributions characterizing the product may be much more severe than for small-molecule reactors characterized only by yield. Modeling in this vein should not be dismissed, but we do not deal with it here because a more traditional view better illustrates with much less effort and greater clarity the possible effects of poor mixing.

The traditional paradigm for mixing defines two extremes.[39,40] The first, which we have dealt with in the preceding section, is that in which mixing is perfect: incoming material is dispersed throughout the reactor, *on a molecular level,* on time scales much shorter than the residence time. The second is that in which the material is dispersed throughout the reactor quickly but *not on a molecular level.* Fluid elements maintain their identity on time scales much larger than that of the residence time, and no mass transfer occurs between the hermetic fluid elements. Such a limit gives a collection of batch reactors with exponential residence time distribution. Nothing explicit need be stated about the length scale of the heterogeneity, except that it is large enough to allow all fluid elements to be treated to a good approximation as batch reactors of infinitely large size. It is this limit of the segregated continuous stirred tank reactor (SCSTR) that we describe in this section.

We briefly review the effects of segregation on idealized reactions in a CSTR; reviews and treatments of this type are common.[41] Although the HCSTR and the SCSTR do present two limits, not all examples of poor mixing will be bracketed thereby, as is often implied. Channeling and bypassing fall outside these limits.[7] Segregation of feed zones, for example, may be so extreme that no reaction occurs (initiator may be introduced into a dead zone in which it decomposes and "terminates" before ever encountering monomer). We would expect the behaviors of the SCSTR and the HCSTR to bracket the actual systems in which mixing is not perfect but for which fluid elements are not hermetic. There is one case for which SCSTR treatment should be quite satisfactory: that of an ideal free-radical suspension polymerization occurring in a CSTR, for which the monomer and other reactants have negligible water solubility (see Chapter 7).

Analysis of the SCSTR is attractive because it is straightforward. Because the product is an accumulation of products of (imaginary) batch reactors with a given residence time distribution, the result for an SCSTR is merely an integral of the batch reactor result over the residence time distribution. Integration or averaging, however, is only proper with quantities that have units of inverse volume (or such quantities scaled by a constant value), since we consider that *volume* elements have an exponential residence time distribution. Thus it is appropriate to average quantities such as P_i, W_i, $G(s)$, and μ_k (all quantities of units of inverse volume), and conversion (which is a quantity of inverse volume scaled by an initial concentration). Averaging of the mole fraction distribution is generally inappropriate unless the number of moles is constant (as for anionic polymerization); the same is true for ratios of moments (degrees of polymerization or molecular weights) unless the denominator moment is constant (thus DP_w may be averaged for a step polymerization, but not DP_n; and DP_n may be averaged for an anionic polymerization, but not DP_w). We can apply these simple principles to our idealized reactions schemes: anionic, stepwise, and free radical.[41]

6.3.1 Anionic Polymerization

For an ideal anionic polymerization occurring in an SCSTR, the chain length distribution can immediately be written as follows:

$$P_i = I_0 \int_0^\infty e^{-\tau'} \frac{(\tau')^{i-1}}{(i-1)!} \frac{1}{\theta} e^{-\theta'/\theta} \, d\theta' \qquad (6.3.1)$$

where

$$\tau' = \frac{M_0}{I_0} (1 - e^{-k_p^i I_0 \theta'}) \qquad (6.3.2)$$

Recall that τ is simply conversion multiplied by the initial ratio of monomer to initiator. One then defines an average τ:

$$\tau = I_0 \int_0^\infty \tau' \frac{1}{\theta} e^{-\theta'/\theta} \, d\theta' \qquad (6.3.3)$$

and this is allowed, since τ has units of inverse volume, scaled by a constant. Integration of equation (6.3.3) according to (6.3.4) gives:

$$\tau = \left(\frac{Da}{1 + Da} \right) \frac{M_0}{I_0} \qquad (6.3.4)$$

that is, the same answer obtained in an HCSTR [cf. equations (6.2.8) and (6.2.11)]. Conversion is not affected by segregation because of the pseudo-first-order character of the reaction (the active anion is present in constant concentration). Both the SCSTR and the HCSTR, then, exhibit lower conversion or DP_n

Figure 6.4.
DP_n versus Damköhler number for anionic polymerization in different reactors, with $\tau_{max} = 1000$.

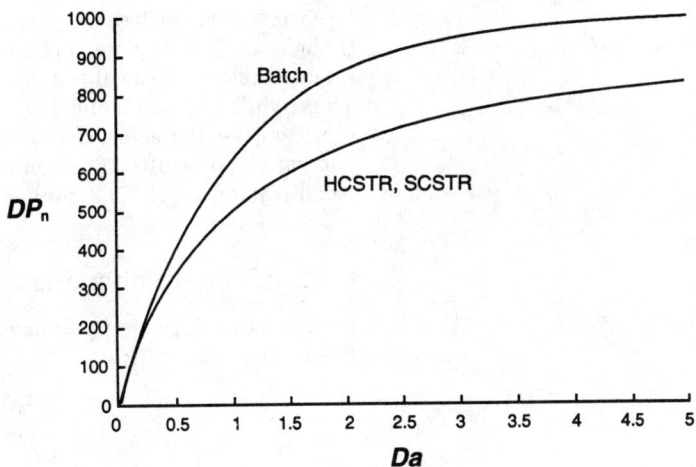

than the batch reactor at equal Damköhler number, as shown in Figure 6.4.

The degrees of polymerization themselves are given as follows:

$$DP_n = 1 + \tau \tag{6.3.5}$$

$$DP_w = \frac{1}{1 + \tau}\left(1 + 3\tau + \frac{2\tau^2}{1 + \tau(I_0/M_0)}\right) \tag{6.3.6}$$

Notice that in equation (6.3.6) that as the maximum τ is reached (τ_{max} or M_0/I_0), the batch reactor result for DP_w is obtained. Similarly, in Figure 6.5, we see that the resulting DP_w (at equal conversion τ or DP_n) moves from HCSTR to batch result as the mean residence time θ increases. This happens because the average residence time increases, a greater fraction of "batch reactors" have a large residence time τ'. Moreover, at long residence times the degree of polymerization approaches a constant value of approximately τ_{max} so that the effect of the residence time distribution becomes invisible. A long residence time θ (large Da) may

Figure 6.5.
DP_w versus τ for anionic polymerization in different reactors, with $\tau_{max} = 1000$.

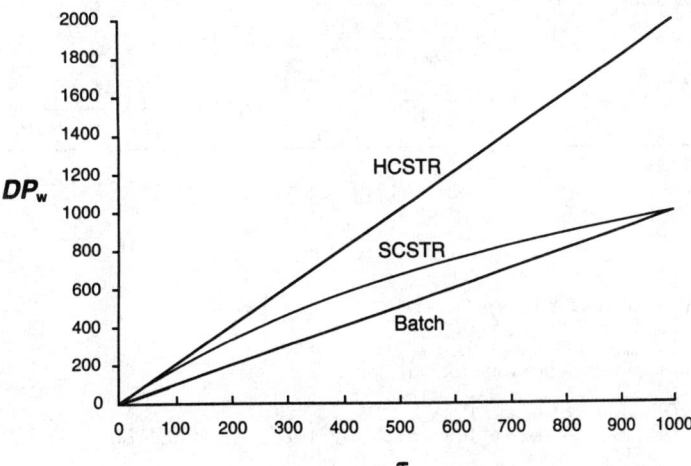

be necessary to reach that equivalence, however. More important, though, an SCSTR gives a chain length distribution of breadth intermediate between those of a batch reactor and an HCSTR. Thus, while having no effect on overall conversion, it has a great effect on the characteristics of the polymer product. Table 6.1 compares the results for anionic polymerization occurring in a batch reactor, an HCSTR, and an SCSTR.

6.3.2 AB Step Polymerization

For the case of AB step polymerization in an SCSTR, the chain length distribution is:

$$P_i = P_{1,0} \int_0^\infty (1 - p')^2 (p')^{i-1} \frac{1}{\theta} e^{-\theta'/\theta} \, d\theta' \qquad (6.3.7)$$

where the conversion p' in a given fluid element is related to the particular residence time θ' of the individual "batch" reactor by:

$$p' = \frac{kP_{1,0}\theta'}{1 + kP_{1,0}\theta'} \qquad (6.3.8)$$

TABLE 6.1 / Comparison of Batch, HCSTR, and SCSTR Results for Ideal Anionic Polymerization

	Batch	*HCSTR*	*SCSTR*
Da	$kI_0 t$	$kI_0\theta$	$kI_0\theta$
$p(Da)$	$1 - e^{-Da}$	$\dfrac{Da}{1 + Da}$	$\dfrac{Da}{1 + Da}$
$DP_n(p)$	$1 + \tau$	$1 + \tau$	$1 + \tau$
$DP_w(p)$	$1 + \tau + \dfrac{\tau}{1 + \tau}$	$1 + 2\tau$	$\dfrac{1}{1 + \tau}\left(1 + 3\tau + \dfrac{2\tau^2}{\tau(I_0/M_0) + 1}\right)$
$DP_n(Da)$	$1 + \tau_{max}(1 - e^{-Da})$	$1 + \tau_{max}\dfrac{Da}{1 + Da}$	$1 + \tau_{max}\dfrac{Da}{1 + Da}$
$DP_w(Da)$	$1 + \tau_{max}(1 - e^{-Da})\dfrac{2 + \tau_{max}(1 - e^{-Da})}{1 + \tau_{max}(1 - e^{-Da})}$	$1 + 2\,\tau_{max}\dfrac{Da}{1 + Da}$	$\dfrac{1 + Da + 3\tau_{max}\,Da + \dfrac{2(\tau_{max}Da)^2}{1 + 2Da}}{1 + Da + \tau_{max}Da}$
P_i	$I_0\,e^{-\tau}\dfrac{\tau^{i-1}}{(i-1)!}$	$I_0\left(\dfrac{\tau_{max}p}{1 + \tau_{max}p}\right)^{i-1}\dfrac{1}{1 + \tau_{max}p}$	$I_0\displaystyle\int_0^\infty e^{-\tau'}\dfrac{(\tau')^{i-1}}{(i-1)!} e^{-\theta'/\theta}\dfrac{d\theta'}{\theta}$

Notes:

$$\tau = \frac{M_0}{I_0}p$$

$$\tau_{max} = \frac{M_0}{I_0}$$

At equal conversion (p)
$DP_{n,batch} = DP_{n,HCSTR} = DP_{n,SCSTR}$
$DP_{w,batch} < DP_{w,HCSTR} < DP_{w,SCSTR}$
At equal residence time (or Da)
$DP_{n,batch} > DP_{n,HCSTR} = DP_{n,SCSTR}$
$DP_{w,batch} < DP_{w,HCSTR};\ DP_{w,SCSTR} < DP_{w,HCSTR}$
$DP_{w,batch} < DP_{w,SCSTR}$ at low Da; $DP_{w,batch} > DP_{w,SCSTR}$ at high Da

The average degrees of polymerization coming from such a polymerization are:

$$DP_n = \frac{Da}{e^{1/Da}E_1(1/Da)} \tag{6.3.9}$$

$$DP_w = 1 + 2Da \tag{6.3.10}$$

where $Da = kP_{1,0}\theta$. Furthermore, $E_1(x)$ signifies the exponential integral defined as follows[42]:

$$E_1(x) \equiv \int_x^\infty \frac{e^{-t}}{t}\,dt \tag{6.3.11}$$

The average conversion, p, found simply from DP_n by the relation $DP_n = 1/(1 - p)$, is clearly not the same as in an HCSTR [cf. equation (6.2.16)]; this is because of the second-order nature of the reaction. In fact, it lies between that of the batch reactor and the HCSTR, as shown in Figure 6.6. The elements in the SCSTR, by virtue of the segregation, experience the faster kinetics, characteristic of lower conversions, that is denied the HCSTR, so the conversion in the former is greater than that in the latter. Because of the residence time distribution, though, the conversion still falls short of that from a batch reactor.

In this case, as in the case of perfect mixing in an HCSTR, the polydispersity is unbounded as $Da \to \infty$, although the increase in Q here is weaker, being logarithmic [$Q \approx 2\ln(Da)$ for the SCSTR at large Da, whereas $Q \approx 2Da^{1/2}$ for the HCSTR]. As for ideal anionic polymerization, the polydispersity of the product from the SCSTR always lies between that of a batch reactor and that of an HCSTR, as seen in Figure 6.7. Figure 6.7 is made on the basis of conversion (as the ordinate DP_n is a simple expression of conversion), but the same trend would be seen on the basis of Damköhler number as well. Table 6.2 presents the various

Figure 6.6.
DP_n versus Damköhler number for AB step polymerization in different reactors: LFTR (Newtonian), the laminar flow tubular reactor with a Newtonian reaction fluid (see Section 6.4.2). Note that differences in conversion at a given Da are exaggerated by plotting DP_n rather than p.

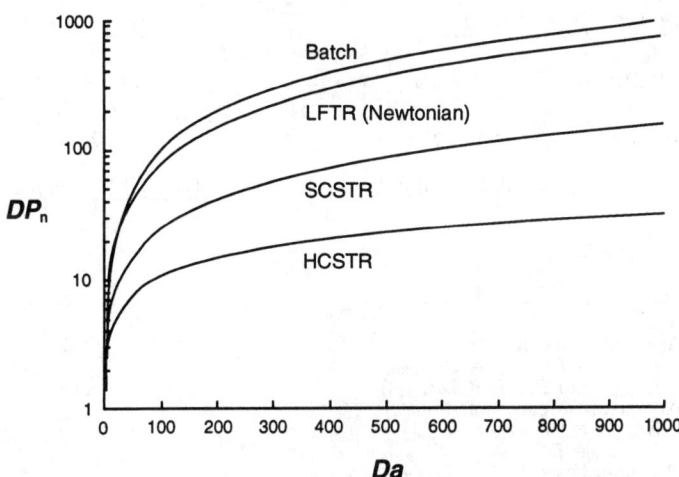

Figure 6.7.
Polydispersity Q versus DP_n for AB step polymerization in different reactors: LFTR (Newtonian), laminar flow tubular reactor with a Newtonian reaction fluid (see Section 6.4.2).

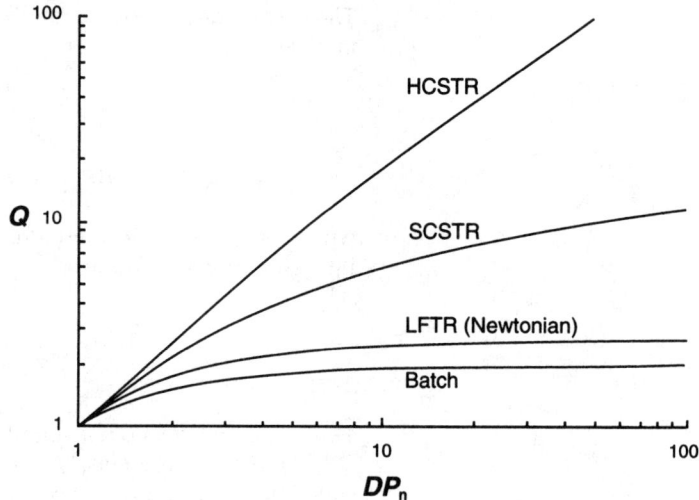

results for stepwise AB polymerization for these three different kinds of reactor and compares the relations for DP and Q.

Batch reactor results are not obtained, however, as average residence time increases, as was the case for anionic polymerization. The difference lies in how the degree of polymerization behaves near complete conversion. As complete conversion is approached in the case of an anionic polymerization, a definite limit to the degrees of polymerization, set by the initial stoichiometry of M_0 to I_0, is reached. Thus, as the mean residence time θ

TABLE 6.2 / Comparison of Batch, HCSTR, and SCSTR Results for AB Step Polymerization

	Batch	*HCSTR*	*SCSTR*
Da	$kP_{1,0}t$	$kP_{1,0}\theta$	$kP_{1,0}\theta$
$p(Da)$	$\dfrac{Da}{1 + Da}$	$1 - \dfrac{\sqrt{1 + 4\,Da} - 1}{2Da}$	$\dfrac{Da - e^{1/Da}E_1(1/Da)}{Da}$
$DP_n(p)$	$\dfrac{1}{1 - p}$	$\dfrac{1}{1 - p}$	$\dfrac{1}{1 - p}$
$DP_w(p)$	$\dfrac{1 + p}{1 - p}$	$\dfrac{1 + p^2}{(1 - p)^2}$	
$DP_n(Da)$	$1 + Da$	$\dfrac{2Da}{\sqrt{1 + 4Da} - 1}$	$\dfrac{Da}{e^{1/Da}E_1(1/Da)}$
$DP_w(Da)$	$1 + 2Da$	$1 + 2Da$	$1 + 2Da$
P_i	$P_{1,0}(1 - p)^2 p^{i-1}$	$P_{1,0}\dfrac{(2i - 2)!}{i!\,(i - 1)!}\dfrac{(1 - p)\,p^{i-1}}{(1 + p)^{2i-1}}$	$P_{1,0}\displaystyle\int_0^\infty (1 - p')^2(p')^{i-1}$
			$e^{-\theta'/\theta}\dfrac{d\theta'}{\theta}$

Notes:
At equal conversion (p)
$DP_{n,\text{batch}} = DP_{n,\text{HCSTR}} = DP_{n,\text{SCSTR}}$
$DP_{w,\text{batch}} < DP_{w,\text{SCSTR}} < DP_{w,\text{HCSTR}}$
At equal residence time (or Da)
$DP_{n,\text{batch}} > DP_{n,\text{SCSTR}} > DP_{n,\text{HCSTR}}$
$DP_{w,\text{batch}} = DP_{w,\text{HCSTR}} = DP_{w,\text{SCSTR}}$
On either basis: $Q_{\text{batch}} < Q_{\text{SCSTR}} < Q_{\text{HCSTR}}$.

becomes larger and an increasingly larger share of the (conceptual) batch reactors exits at nearly complete conversion, the discrepancies in residence time "disappear" and a batch reactor result is obtained. For the irreversible step polymerization of AB monomers, though, the polymer knows no upper bound in degree of polymerization, and thus the longer the residence time the higher the molecular weight ($DP_n \sim \theta'$); no such disappearance of effects is thus possible.

6.3.3 Free-Radical Polymerization

For the case of a free-radical polymerization occurring in an SCSTR, we may again write the expression for the resulting chain length distribution as an integral over the residence time distribution. Because of all the peculiarities of free-radical polymerizations, though, we can write this only in the most general of forms unless we consider very restrictive cases, such as the one treated in Section 6.2, where only monomer depletion leads to instantaneous chain length dependent on conversion and no transfer mechanisms are operative. Here, from equation (3.5.58) we find:

$$
P_i = M_0 d \int_0^\infty \left\{ \left[\frac{(\xi/2)(1 + di) + (1 - \xi)}{i} \right] \left[\frac{1}{1 + d} \right]^i \right.
$$
$$
\left. - \left[\frac{(\xi/2)(1 + di/(1 - p')) + (1 - \xi)}{i} \right] \left[\frac{1 - p'}{1 + d - p'} \right]^i \right\} \frac{1}{\theta} = e^{-\theta'/\theta} \, d\theta' \qquad (6.3.12)
$$

where the relationship between the conversion p' in a given fluid element and time is as in a batch reactor:

$$
p = (1 + d)(1 - e^{-Da}) \qquad (6.3.13)
$$

The average conversion p is related to time as in the HCSTR [as shown in equation (6.2.33)]:

$$
p = (1 + d)\frac{Da}{1 + Da} \qquad (6.3.14)
$$

Both initiator depletion and the Trommsdorff effect have been neglected; therefore, conversion is unaffected by segregation because of the constant concentration of radicals. Coincidentally, DP_n is also the same for both the SCSTR and HCSTR for this case; see Figure 6.8. The coincidence of the HCSTR and SCSTR results for DP_n is purely a result of the same average number of chains produced and the same amount of monomer converted. Table 6.3 gives the results from the different reactors and compares the applicable equations.

Contrary to the two preceding cases, the SCSTR results do not fall between those of the batch reactor and HCSTR with regard to the breadth of the distribution; this is most easily seen in a plot of polydispersity as a function of conversion (Figure

Figure 6.8.
DP_n versus conversion for free-radical polymerization with monomer depletion only, in different reactors: $d = 0.0001$, $\varepsilon = 0$.

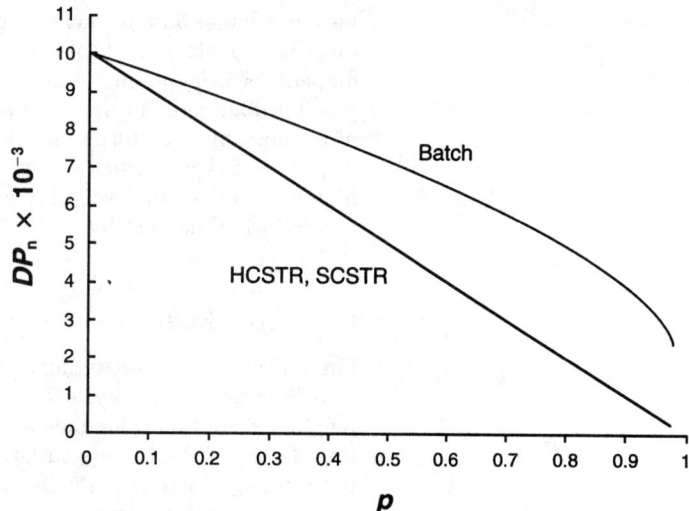

TABLE 6.3 / Comparison of Batch, HCSTR, and SCSTR Results for Ideal Free-Radical Polymerization: No Transfer, Only Monomer Depletion

	Batch	HCSTR	SCSTR
Da	$k_pH(1)t$	$k_pH(1)\theta$	$k_pH(1)\theta$
$p(Da)$	$(1+d)(1-e^{-Da})$	$(1+d)\dfrac{Da}{1+Da}$	$(1+d)\dfrac{Da}{1+Da}$
$DP_n(p)$	$\dfrac{p}{(1-\varepsilon/2)d\ln\left[\dfrac{1+d}{1+d-p}\right]}$	$\dfrac{1+d-p}{d(1-\varepsilon/2)}$	$\dfrac{1+d-p}{d(1-\varepsilon/2)}$
$DP_w(p)$	$1+\varepsilon+\dfrac{1+\varepsilon/2}{d}(2-p)$	$1+\varepsilon+\dfrac{1}{d}(2+\varepsilon)(1-p)$	$1+\varepsilon+\dfrac{2+\varepsilon}{d}\dfrac{1+d(1-p)}{1+d+p}$
$DP_n(Da)$	$\dfrac{(1+d)(1-e^{-Da})}{(1-\varepsilon/2)dDa}$	$\dfrac{1+d}{d(1-\varepsilon/2)}\dfrac{1}{1+Da}$	$\dfrac{1+d}{d(1-\varepsilon/2)}\dfrac{1}{1+Da}$
$DP_w(Da)$	$1+\varepsilon+\dfrac{1+\varepsilon/2}{d}(1-d+(1+d)e^{-Da})$	$1+\varepsilon+\dfrac{2+\varepsilon}{d}\dfrac{1-dDa}{1+Da}$	$1+\varepsilon+\dfrac{2+\varepsilon}{d}\dfrac{1+(1-d)Da}{1+2Da}$
P_i	See (3.5.58)	See (3.5.43)	See (6.3.12)

Notes:

$$d = \frac{k_tH(1)}{k_pM_0}$$

ε = fraction of growing chains ended by combination

At equal conversion (p)

$DP_{n,batch} > DP_{n,HCSTR} = DP_{n,SCSTR}$

$DP_{w,batch} > DP_{w,SCSTR} > DP_{w,HCSTR}$

At equal residence time (or Da)

$DP_{n,batch} > DP_{n,HCSTR} = DP_{n,SCSTR}$

$DP_{w,batch} > DP_{w,HCSTR}$; $DP_{w,SCSTR} > DP_{w,HCSTR}$

$DP_{w,batch} > DP_{w,SCSTR}$ at low Da; $DP_{w,batch} < DP_{w,SCSTR}$ at high Da

On either basis:

$Q_{HCSTR} < Q_{batch} < Q_{SCSTR}$.

Figure 6.9.
Polydispersity Q versus conversion for free-radical polymerization with monomer depletion only, in different reactors: $d =$ 0.0001, $\varepsilon = 0$.

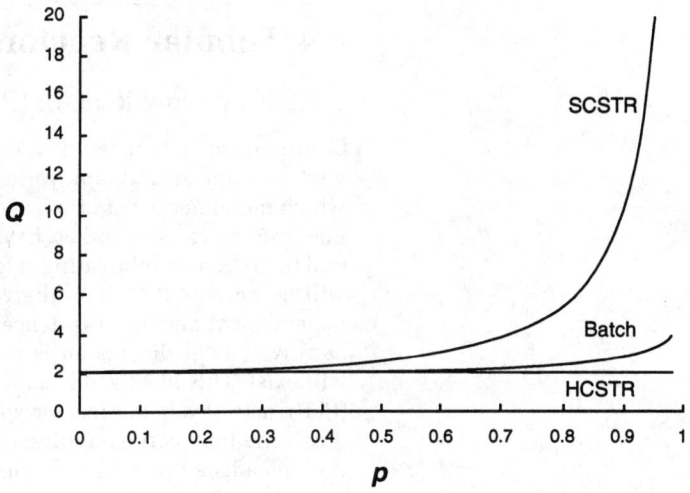

6.9). This effect is a result of the general rule* that for chains with long lifetimes, segregation narrows the distribution, while for those with short lifetimes, segregation broadens the distribution.[41] Thus for the case of reversible step polymerization, where both the batch and HCSTR should give geometric distributions, the SCSTR will give a broader distribution because the hermetic elements cannot equilibrate with one another. Figure 6.9 serves as an immediate warning with respect to the dangers of attempting to counteract the effects of drift dispersion by operation in a continuous stirred tank reactor. Mixing must be very good, because segregation may make the chain length distribution much broader than would have been the case for a batch reactor. Because such different results are obtained as a result of mixing, it has been suggested that the breadth of the distribution in a CSTR be used as a measure of the degree of segregation.[43] There are experimental difficulties, though,[44] given the need, for example, to eliminate any other causes of distributional broadening (e.g., mixing zones, side reactions). Similar results are found for the composition distribution in free-radical copolymerizations; segregation leads to broader composition distribution than either a batch or an HCSTR would give.[15–17,19,45] Thus, copolymer composition distribution has also been suggested as a measure of segregation, although the same experimental difficulties are encountered here, in addition to the difficulty in measuring the copolymer composition distribution.[17]

Generalizations such as this and the preceding one comparing batch and HCSTR results apply only for linear polymerizations; nonlinear and branching polymerizations often provide exceptions.

6.4 Tubular Reactors

6.4.1 Plug Flow Reactor (PFR) and PFR with Recycle

Continuous reaction can be accomplished in tubular reactors as well as in the stirred tank variety. Consider a tubular reactor in which monomer is fed into one end and (partially) reacted polymer product exits the other. Under conditions of turbulent flow, or if there exists a lubricating layer at the wall, the velocity profile will be nearly "flat" (i.e., the velocity will exhibit little radial dependence) and the residence time distribution will be very narrow if axial dispersion is negligible. Plug or "piston" flow will exist. This idealized reactor, known as the plug flow reactor (PFR), acts as a batch reactor with the axial distance x traveled down the tube corresponding to the time t in the batch reactor through a factor of the fluid velocity v ($x = vt$). There is no need to analyze this kind of reactor separately.

A modification of the PFR worth mentioning is one in which a given fraction of the exit stream is recycled, as shown in Figure 6.10 (for an ideal anionic polymerization). The recycle ratio r is defined as the ratio of the volumetric flow rate of the recycle stream to that of either the feed or outlet stream (in our case equivalent):

$$r = \frac{Q_r}{Q} \qquad (6.4.1)$$

The recycle ratio thus ranges from 0 to ∞. The volumetric flow rate into the PFR is the sum of Q_r and Q, or $(1 + r)Q$. The residence time distribution is discrete, at multiples of the time taken to traverse the reactor once, t, given by:

$$t = \frac{\theta}{1 + r} \qquad (6.4.2)$$

where θ is V/Q. It is easy to reason that the residence time distribution is geometric with parameter $r/(1 + r)$ and that the average residence time is θ, independent of r. In the limit $r \rightarrow \infty$, t vanishes, so that the contents of the reactor are homogeneous and the residence time distribution is exponential.[46] These conditions merely describe an HCSTR. Thus, the PFR with recycle at finite r provides a model intermediate between the batch reactor ($r = 0$) and the HCSTR ($r \rightarrow \infty$).[6,8,15]

To make this clearer, we return to the example of ideal anionic polymerization, for which the streams in Figure 6.10

Figure 6.10.
Schematic of anionic polymerization in a PFR with recycle.

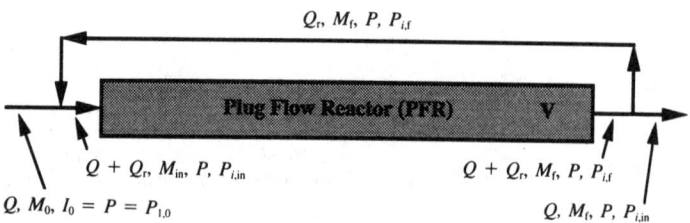

have been labeled. It has been assumed that no volume change occurs upon reaction, that the original stream comprises only monomer and initiated monomer, and that the volume of the recycle piping (or the residence time therein) is negligible compared to that of the reactor itself. The monomer concentration M_{in} and the generating function of the product $G_{\text{in}}(s)$ flowing into the reactor are given by:

$$M_{\text{in}} = \frac{M_0 + rM_f}{1 + r} \tag{6.4.3}$$

$$G_{\text{in}}(s) = \frac{sI_0 + rG_f(s)}{1 + r} \tag{6.4.4}$$

Given these, the final monomer concentration M_f and product generating function $G_f(s)$ are given by generalizations of equations (3.2.6) and (3.2.29):

$$M_f = M_{\text{in}}\, e^{-kI_0 t} \tag{6.4.5}$$

$$G_f(s) = G_{\text{in}}(s)e^{-(1-s)\tau} \tag{6.4.6}$$

Solving equations (6.4.3) through (6.4.6) for M_f and $G_f(s)$ yields:

$$M_f = \frac{M_0\, e^{-kI_0 t}}{(1 + r) - re^{-kI_0 t}} \tag{6.4.7}$$

$$G_f(s) = \frac{sI_0\, e^{-(1-s)\tau}}{(1 + r) - re^{-(1-s)\tau}} \tag{6.4.8}$$

From equation (6.4.8), the moments of the distribution can be found:

$$\mu_0 = I_0 \tag{6.4.9}$$

$$\mu_1 = I_0(1 + (1 + r)\tau) \tag{6.4.10}$$

$$\mu_2 = I_0\{1 + (1 + r)(3\tau + (1 + 2r)\tau^2)\} \tag{6.4.11}$$

[cf. equations (3.2.5), (3.2.30), and (3.2.31)]. The degrees of polymerization and the polydispersity Q are thus given by:

$$DP_n = 1 + (1 + r)\tau \tag{6.4.12}$$

$$DP_w = \frac{1 + (1 + r)(3\tau + (1 + 2r)\tau^2)}{1 + (1 + r)\tau} \tag{6.4.13}$$

$$Q = \frac{1 + (1 + r)(3\tau + (1 + 2r)\tau^2)}{(1 + (1 + r)\tau)^2} \tag{6.4.14}$$

[cf. equations (3.2.32), (3.2.33), and (3.2.34)].

For $r = 0$, the generating function of equation (6.4.8) and the moments and degrees of polymerization that follow clearly define a Poisson distribution, as must be. For $r > 0$, we must know how the eigenzeit transform variable, τ (as defined in equation (3.2.7), depends on r. Here, the monomer concentration M, at a point through the reactor corresponding to a time $t' \leq t$, is given by:

$$M = \frac{M_0\, e^{-kI_0 t'}}{(1 + r) - re^{-kI_0 t}} \tag{6.4.15}$$

Thus:

$$\tau = \int_0^t kM(t')dt' = \frac{kM_0}{(1 + r) - re^{-kI_0 t}} \int_0^t e^{-kI_0 t'} dt' = \frac{(M_0/I_0)(1 - e^{-kI_0 t})}{(1 + r) - re^{-kI_0 t}} \tag{6.4.16}$$

From this equation for t, it can be shown that as $r \to \infty$, the distribution becomes the geometric distribution described in equation (6.2.10). The rapidity with which the HCSTR result is obtained can be judged from Figure 6.11.

6.4.2 Laminar Flow Tubular Reactor (LFTR)

The high viscosities encountered in polymerization may prevent the attaining of a Reynolds number (Re) sufficiently high for turbulent flow. The flow will be laminar, so that the velocity profile in the tube will not be flat and the residence time distribution will be broadened somewhat. The form of the velocity profile is determined by the rheological constitutive equation which the polymeric fluid obeys. In the simplest of these, of course, the viscosity is independent of the shear rate: a Newtonian liquid. For Newtonian fluids, if Re ($= 2Rv_{ave}/\nu$, where R is the radius of the tube, v_{ave} the average fluid velocity, and the ν the kinematic viscosity) is less than 2100, the flow will be laminar.[47] In this case, the velocity profile assumes the well-known parabolic shape:

$$v(r) = 2v_{ave} \left[1 - \left(\frac{r}{R}\right)^2 \right] \tag{6.4.17}$$

From this velocity profile, one can derive the residence time distribution $f(\theta')d\theta'$ if diffusion is neglected.[8,48,49] The frac-

Figure 6.11.
Polydispersity Q as a function of recycle ratio r for an anionic polymerization in a PFR with recycle. M_0/I_0 = 10,000, $Da = kI_0\theta = 1$. For such large M_0/I_0, the polydispersity is only a weak function of Da; DP_n and DP_w separately, however, depend strongly on Da.

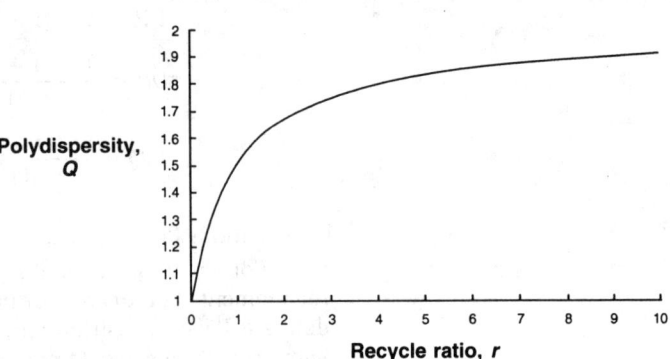

tion of material with a residence time θ' corresponds to the fraction of fluid at the radius r corresponding to that velocity at which the length of the tube is traversed in time θ':

$$f(\theta')d\theta' = \frac{v(r)\,2\pi r\,dr}{v_{ave}\,\pi R^2} = \frac{v(r)2r\,dr}{v_{ave}\,R^2} \qquad (6.4.18)$$

It is clear that the ratio of the velocity at the radius r to the average velocity is the inverse of the ratio of the specific residence time θ' to the average residence time θ:

$$\frac{v(r)}{v_{ave}} = \frac{\theta}{\theta'} \qquad (6.4.19)$$

The highest velocity is obtained at the centerline, where $v(0)/v_{ave} = 2$. Thus the minimum value of θ' is $\theta/2$. The ratio $2r\,dr/R^2$ can also be expressed in terms of the residence times by rearranging the equation (6.4.17):

$$\left(\frac{r}{R}\right)^2 = 1 - \frac{v(r)}{2v_{ave}} = 1 - \frac{\theta}{2\theta'} \qquad (6.4.20)$$

From which it is easily found that:

$$\frac{2r\,dr}{R^2} = \frac{\theta\,d\theta'}{2\theta'^2} \qquad (6.4.21)$$

The residence time distribution is thus given as follows:

$$f(\theta')d\theta' = 0 \qquad \theta' < \frac{\theta}{2} \qquad (6.4.22)$$

$$= \frac{\theta^2}{2\theta'^3}\,d\theta' \quad \theta' \geq \frac{\theta}{2}$$

This distribution can be contrasted with the exponential residence time distribution of the CSTR in two ways. First, fluid elements do not travel through the reactor infinitely fast, hence the cutoff at low residence times. Second, the distribution has a power-law, rather than exponential, character; therefore the distribution has a long tail, which can be important.[7]

Polymeric fluids, however, are generally non-Newtonian and exhibit a shear-dependent viscosity η.[50] Often this is satisfactorily described by the commonly used power-law model[51]:

$$\eta = m\dot{\gamma}^{n-1} \qquad (6.4.23)$$

where $\dot{\gamma}$ is the shear rate, and n is the power law index, which, if equal to unity, reverts to a Newtonian fluid. Most polymeric fluids are shear thinning ($n < 1$), although some may be shear thickening or dilatant ($n > 1$). For such fluids also, the regime in which the flow will be laminar has been explored.[52] Here, the velocity profile arising from the tubular Poiseuille flow is given as follows:

$$v(r) = \frac{3n + 1}{n + 1} v_{\text{ave}} \left[1 - \left(\frac{r}{R} \right)^{(n+1)/n} \right] \qquad (6.4.24)$$

Following the same procedure used earlier, we find that the residence time distribution is:

$$f(\theta')d\theta' = 0 \qquad\qquad\qquad \theta' < \theta \frac{(n + 1)}{3n + 1} \qquad (6.4.25)$$

$$= \frac{2n}{3n + 1} \left\{ 1 - \frac{n + 1}{3n + 1} \frac{\theta}{\theta'} \right\}^{(n-1)/(n+1)} \frac{\theta^2}{(\theta')^3} d\theta' \qquad \theta' \geq \theta \frac{(n + 1)}{3n + 1}$$

In graphic representation it is traditional (and easier) to plot the cumulative distribution function[53] $F(\theta')d\theta'$[5,7,49]:

$$F(\theta') = \int_0^{\theta'} f(\theta'')d\theta'' \qquad (6.4.26)$$

This function thus describes the fraction of fluid elements with a residence time less than or equal to the time θ'. For the batch reactor or the PFR, we thus have:

$$F(\theta') = 0 \qquad \theta' < \theta \qquad (6.4.27)$$

$$= 1 \qquad \theta' \geq \theta$$

For a CSTR:

$$F(\theta') = 1 - e^{-\theta'/\theta} \qquad (6.4.28)$$

For the laminar flow tubular reactor with a Newtonian fluid and with a power law fluid, the cumulative distribution functions are:

$$F(\theta') = 0 \qquad\qquad\qquad \theta' < \frac{\theta}{2} \qquad (6.4.29)$$

$$= 1 - \frac{1}{4} \left(\frac{\theta}{\theta'} \right)^2 \qquad \theta' \geq \frac{\theta}{2}$$

$$F(\theta') = 0 \qquad\qquad\qquad \theta' < \theta \frac{(n + 1)}{3n + 1} \qquad (6.4.30)$$

$$= \left\{ 1 - \frac{n + 1}{3n + 1} \frac{\theta}{\theta'} \right\}^{2n/(n+1)} \left\{ 1 + \frac{2n}{3n + 1} \frac{\theta}{\theta'} \right\} \quad \theta' \geq \theta \frac{(n + 1)}{3n + 1}$$

In Figure 6.12 we show the plots for the batch reactor, the continuous stirred tank reactor (homogeneous or segregated), and the laminar flow tubular reactor (LFTR) with a Newtonian fluid and with a power law fluid, with indices $n = 0.5$ and $n = 0.1$. Note that the laminar flow tubular reactors have a breadth of distribution between the broadest (CSTR) and the narrowest (batch or PFR) and that the breath of the distribution decreases with decreasing

Figure 6.12.
Cumulative residence time distribution $F(\theta')$ for different reactors. (After Middleman,[49] used by permission of McGraw-Hill.)

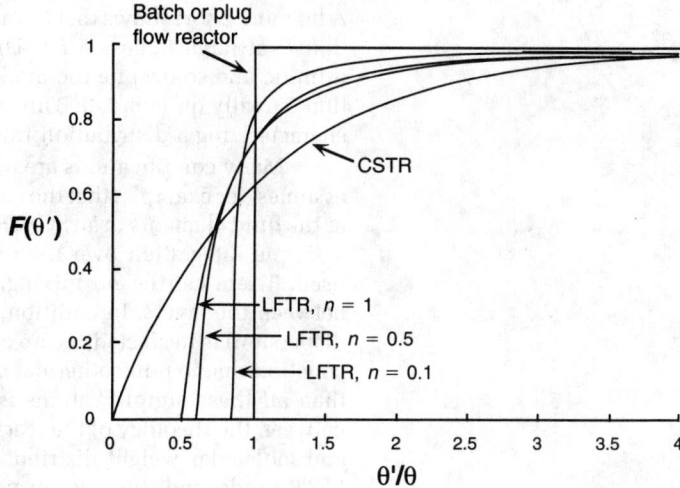

power law index. Nonetheless, a long tail persists in the distribution for these reactors.

We mention only one case here: AB step polymerization occurring in a Newtonian fluid.* Averaging the expressions for the first three moments over the residence time distribution of equation (6.4.22) gives the following expressions for the average degrees of polymerization:

$$DP_n = \frac{1}{1 - Da + (Da^2/2)\ln(1 + 2/Da)} \tag{6.4.31}$$

$$DP_w = 1 + 2Da \tag{6.4.32}$$

The conversion p is then easily found from the expression for DP_n, since $DP_n = 1/(1 - p)$.

From Figures 6.6 and 6.7, which showed DP_n versus Da and Q versus DP_n for the LFTR as well as the other three reactors examined, it is clear that the behavior of the LFTR is intermediate between a batch reactor and the HCSTR and SCSTR in terms of both conversion and polydispersity. In some respects, the behavior of the LFTR in this case is similar to that of the batch reactor; for example, in contrast to the results for both the HSCTR and the SCSTR, the polydispersity is bounded, although by a value of 8/3 at high Da rather than 2. This similarity is misleading, however; the long residence times near the wall of the tube, which mathematically gives the power-law character of the residence time distribution, result in a product with a correspondingly long tail in the molecular weight distribution. It is straightforward to show that the third moment of the distribution, μ_3, is always divergent regardless of Da, as are all higher moments.

---

*A more complete cataloguing of reaction schemes is given elsewhere.[48] Caution is indicated, however, because in several places in this older work it appears that quantities not appropriate to integrate are integrated (see the introduction to Section 6.3). A more complete list of residence time distributions in rheologically complicated fluids, and in many different reactor types, is also given elsewhere.[50]

A bit more effort shows that the asymptotic chain length distribution P_i at large i goes as i^{-4}. Thus, DP_z, DP_{z+1}, ..., are always infinite, and so despite the modest polydispersity, the distribution is really quite broad. This is another example of the risk of characterizing a distribution from a finite number of moments.

Many complications are overlooked in the LFTR model. It assumes, for example, that the concentric laminae are as hermetic as the fluid elements in an SCSTR; this assumption is what allows a simple integration over the residence time distribution to be used. There can be no mixing, no diffusion, and no reaction between the layers. In addition, diffusion in general (e.g., axial dispersion) is neglected, as are complications in the constitutive equation due to nonisothermal conditions. Perhaps more severe than all these simplifications is the omission of the clear link between the rheology of the reacting mixture and the conversion and molecular weight distribution and so forth. Moreover, the LFTR model indicates, for example, that there will be a radial gradient in conversion, which will give higher viscosities at the wall. Thus, what appears as a mere long tail in the residence time distribution may in practice lead to fouling. The problem of fouling has recently been examined for free-radical polymerizations in tubular reactors.[54] Fouling is seen to become worse as the molecular weight of the dead polymer increases and as the conversion increases; the orientation of the tubes is crucial (a vertical flow downward being best with respect to fouling), and the no-slip boundary condition [assumed in equation (6.4.17)] is not applicable. These problems are sufficient to show the simplicity of the LFTR model.

Tubular reactors in practice are used most commonly in production of low density polyethylene, which is a high pressure (~2000 atm), free-radical process.[4] Copolymers of ethylene with vinyl acetate, for example, can also be made by such a process.[4] Fairly detailed models of such polymerizations have been devised.[55,56] In practice, a pulsed flow is used to reduce fouling and to give a more nearly plug flow, a practice that can also be applied for emulsion polymerization (a particular kind of heterogeneous free-radical polymerization described in Chapter 7).[57] Homogeneous free-radical polymerizations besides those involving ethylene have been modeled in tubular reactors,[54] as have Ziegler–Natta polymerizations (see Chapter 7).[58]

6.5 Reactive Extrusion

The high viscosities of polymeric products can make not only mixing, but also mere movement, difficult. Tubular reactors may become fouled, but even in the absence of fouling, reliance on pressure-driven Poiseuille flow to convey a reacting mixture down a tube may demand a prohibitively high pressure gradient. This problem is commonly avoided by using the drag flow (perhaps with a superimposed pressure gradient) afforded by fitting the tube (or barrel) with a rotating screw, so that the polymer is pushed down the length of the barrel. Such an operation, called *extrusion,* is a common processing step, used in more than 60% of plastics.[59]

Many polymer products are extruded more than once during the processing history.[60] For example, the production of the polymer and the manufacture of an article from that polymer often are carried out by separate companies. Thus, a polymer may be extruded into pellets (a common form for shipment) after synthesis, and later melted and extruded into final shape (e.g., rods or pipes) in a second extruder by the processor. Needless to say, these two operations may be quite different, since the feed to the first is a melt (*melt extrusion*), while that to a second is a solid, which requires melting (*plasticating extrusion*). In both cases a shaping is taking place; in the first, to pellets for ease of storage and transport, and in the second to the desired object. As such, extrusion is purely a processing operation and thus is not a concern in this book; other texts handle the rheological and engineering aspects in great detail.[60-62]

Extruders are also used to perform more than mere shaping; devolatilization, compounding, adding plasticizers, and so forth can also be done in an extruder. One can also do *reactive extrusion.*[63] The use of the extruder itself as the reactor, thus eliminating a piece of equipment, is restricted by an inability to handle fluid of low viscosity (since the extruder is designed for highly viscous materials) and by poor heat transfer.[64] Thus the ideal application for reactive extrusion is the modification of polymers, since the medium is highly viscous at all times, and since little heat of reaction is generated. Examples of modification performed in extruders are halogenation of isoprene units in isobutylene–isoprene copolymers (butyl rubber),[65] hydrolysis of acetate groups in copolymers of ethylene and vinyl acetate to give vinyl alcohol units,[66] functionalization or internal plasticization of poly(vinyl chloride) with ester groups,[67] and modifications of polystyrene with starch.[68] Grafting and reactive blending, as considered in Section 5.1.2, can be carried out in extruders.[69] Examples include grafting of maleic anhydride onto polyethylene,[70] the reactive compatibilization of polyolefins and nylon through maleic anhydride functionalities,[71-74] blending of nylons and polyesters through amide–ester interchange,[75] and blending of other incompatible polymers such as polystyrene and polyethylene,[76] fully aromatic polyesters (polyarylates) and polycarbonate,[77] and poly(phenylene oxide) and nylon.[78] Extruders may also be employed for controlled degradation (by either mechanical or chemical means) or for reequilibration of condensation polymers to give a more uniform molecular weight distribution.*

Performing a complete polymerization in an extruder is, as mentioned before, hampered by poor heat transfer and poor transport of the initial fluids of low viscosity, as well as by upper limits on residence time (about 20 min).[64] The problem of poor

---

Random scission, as should occur for chemical degradation, will result in a product tending toward a geometric chain length distribution regardless of the initial distribution.[79] If the scission is not random, but is dependent on chain length or position in the chain, as might be the case for mechanical degradation, this simple result no longer applies.[80]

Figure 6.13.
Schematic of a single-screw
extruder as might be used
for plasticating extrusion.
(From reference 61, used by
permission of Van
Nostrand Reinhold.)

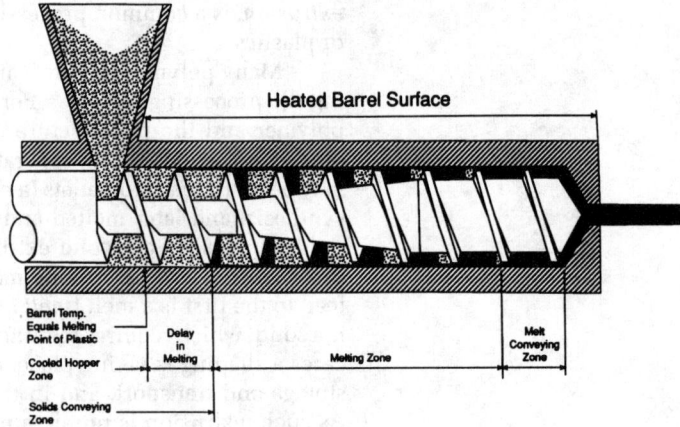

heat transfer is avoided in polymerizations with low heat of
polymerization, such as ring-opening polymerizations; polymerizations to form nylon 6 and polyoxymethylene have been done
in extruders.[81-83] Extruders have also been tried, though, for anionic polymerizations,[84] production of polyurethanes,[69,85,86] and
free-radical polymerizations.[69,87-89] Reaction in a solvent avoids
in part the problem of exothermicity, at the cost of exacerbating
the problem of low viscosity. The problems of both exothermicity
and low viscosity in these polymerizations can of course be
avoided by placing the extruder after a CSTR or tubular reactor
in which much of the polymerization occurs.

The simplest kind of extruder is the single-screw extruder
(Figure 6.13); but even for this we find a multitude of screw
designs (not shown). The polymer travels down the helical path
described by the channel between the screw flights. Unsurpris-
ingly, the flow within this channel is quite complicated compared to the flow in the laminar flow tubular reactor discussed.[61]
The simplest approximation is to unwind the helix, presenting
for consideration a straight channel over which a wall is dragged
at an angle θ equal to the helical pitch of the screws (the angle
between the screw flight and a radial line in the tube),[90] as indicated in Figure 6.14. The simplified flow, which is found to be

Figure 6.14.
Schematic of the simplified
view of the channel in a
single-screw extruder.
(From reference 91, used by
permission of the Society of
Plastics Engineers.)

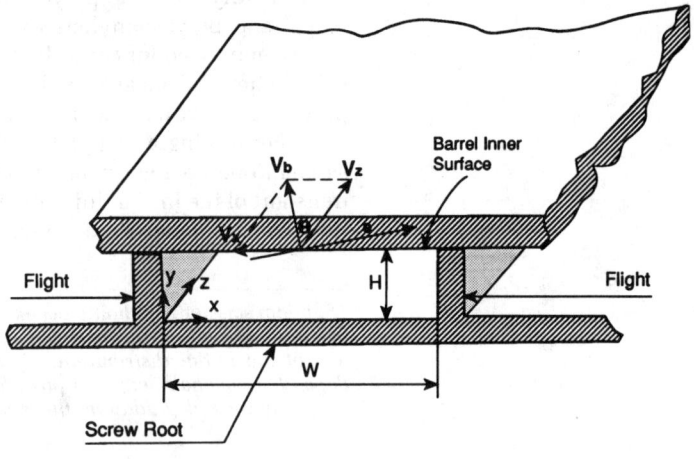

helical within the channel,[91] may of course be complicated by leakage flows occurring through the slight clearance between the screw and the barrel.

The residence time distribution for a Newtonian fluid of constant viscosity in a single-screw extruder of constant pitch and channel depth has been shown to be[91]:

$$f(\theta')d\theta' = \frac{9}{\theta}\left[\frac{a^3(a - 1 - \sqrt{(1 + 3a)(1 - a)})^3}{(6a^2 - 4a - 1)\sqrt{(1 + 3a)(1 - a)} + 3a - 1}\right]d\theta' \qquad (6.5.1)$$

or

$$F(\theta') = \frac{1}{2}[3a^2 - 1 + (a - 1)\sqrt{(1 + 3a)(1 - a)}] \qquad (6.5.2)$$

where the residence time θ' is implicitly related to the dimensionless distance $a\ (= y/H)$ as follows:

$$\theta' = \frac{\theta}{6}\frac{3a - 1 + 3\sqrt{(1 + 3a)(1 - a)}}{a[1 - a + \sqrt{1 + 3a)(1 - a)}]} \qquad (6.5.3)$$

where a ranges from 2/3 to unity. The average residence time, θ, is given by:

$$\theta = \frac{2L}{v_b\sin\theta\cos\theta\,(1 + \Phi)} \qquad (6.5.4)$$

where v_b is the tangential velocity of the tube over a screw flight, L is the length, and Φ is the dimensionless ratio between the pressure and drag flows:

$$\Phi = -\frac{H^2}{6\mu v_b\cos\theta}\frac{\partial P}{\partial z} \qquad (6.5.5)$$

This distribution is shown in Figure 6.15, and is seen to be narrower than that of the LFTR with a Newtonian fluid, instead

Figure 6.15.
Cumulative residence time distribution $F(\theta')$ for idealized single-screw extruder as compared with other reactors. (After reference 91, used by permission of the Society of Plastics Engineers.)

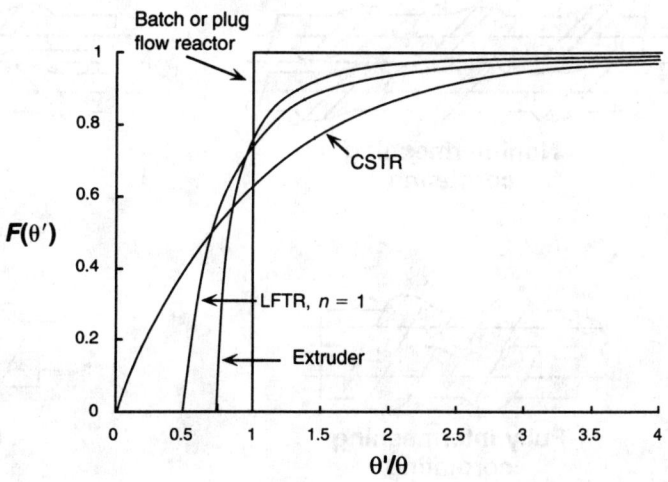

more closely resembling that of a power-law fluid. However, the residence time distribution decays more rapidly than that in an LFTR: as $(\theta')^{-4}$ rather than $(\theta')^{-3}$ (this latter holding for Newtonian or power-law fluids). The power-law decay, although stronger, is not exponential, and so we still expect the long tail of the distribution to give divergent moments in every case; DP_{z+1} and higher averages will be divergent. The foregoing analysis has been extended to power-law fluids, and it is found that contrary to the LFTR case (where shear-thinning behavior leads to a narrowing of the residence time distribution), a broadening of the residence time distribution occurs with decreasing index n.[92] Thus, the analysis is rather optimistic.

However well suited the single-screw extruder is for polymer modification reactions in which the viscosity does not change appreciably, it is inappropriate for polymerizations in which orders-of-magnitude changes in viscosity are expected. First, because it is designed to convey highly viscous material, the low-viscosity monomer is not conveyed down the channel effectively. Conversely, the high-viscosity polymer may tend to build up on the screw because it is not participating in the main flow.[88] One might expect the leakage flows to be more severe as well. Thus, although a single-screw extruder may be acceptable if the feed is partially reacted,[93] this design is unacceptable for polymerizations from monomer.

Although many of the problems just outlined can be addressed by using barrel sections differing in channel depth, pitch, and so forth, twin-screw extrusion offers advantages.[94] Two factors not found in single-screw extrusion are introduced: the proximity of the screws and the rotation of the screws relative to one another (see Figure 6.16). Polymer buildup is avoided only by the intermeshing type, the flights of the one screw fitting into the channel of the other. The material is conveyed down the figure-eight-shaped cavity of the barrel in the twisted C-shaped sections, which either alternate between the two screws (corotation) or are more independent (counterrotation). The counterro-

Figure 6.16.
Various types of twin-screw extruder. (From reference 88, used by permission of the author.)

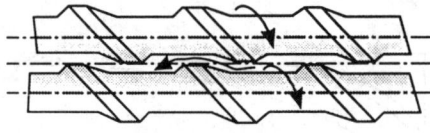

Nonintermeshing corotating

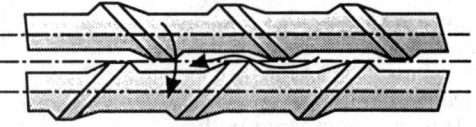

Nonintermeshing counterrotating

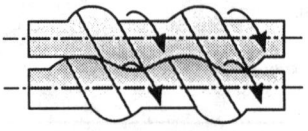

Fully intermeshing corotating

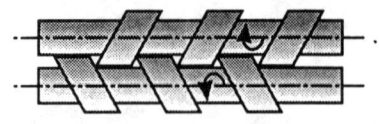

Fully intermeshing counterrotating

tating twin-screw extruder can be made with much smaller mechanical clearances, but there are still various leakage flows.[94] It has also been more often studied[85-88,95] and is generally analyzed as a series of CSTRs (not at steady state) connected by the leakage flows.[96]

6.6 Semibatch or Multistep Reactors

Many polymerization processes are neither batch nor continuous, but rather intermediate between the two. These processes may be termed semibatch or semicontinuous. Polymerization reactors in which some transport of mass occurs during the polymerization (either reactants or inert agents are added or product or by-product removed) are good examples. Two trivial examples are familiar from Chapter 2: removal of condensate by-product to effect molecular weight buildup, and removal of nylon 6/10 from the interface in the well-known "nylon rope trick." Alternatively, polymerizations may occur in several stages.

A reaction may be done in a semibatch or semicontinuous mode, or in stages for countless reasons. Common examples include:

to promote more desirable reaction path;

to achieve a desired morphology;

to adjust or maintain stoichiometry;

to obtain a desired chain length, composition, or sequence length distribution;

to modify the polymer product in some way; and

for control purposes.

The reader may certainly find other desired ends in the literature or in practice. The number of differing goals motivating the use of a semibatch or multistep scheme immediately suggests a similar number of different particular implementations. Thus, we cannot generalize the semibatch reactor. Even if we were to limit ourselves to a semicontinuous feed, we would still have an infinitely variable $Q(t)$ and feed composition. Instead, we examine, anecdotally, particular examples of semibatch or multistep operation to understand how particular goals are achieved.

6.6.1 Poly(ethylene terephthalate)

As discussed in Section 2.3.2, the most important polyester is poly(ethylene terephthalate), which is used in fibers but increasingly in a number of plastics applications, perhaps the most visible being for bottles for beverages. Synthesis does not proceed directly from terephthalic acid and ethylene glycol, but rather from terephthalic acid or dimethyl terephthalate reacted with ethylene glycol to produce water or methanol as by-product. The product bis(2-hydroxyethyl)terephthalate (often termed the "monomer") is then reacted and ethylene glycol is released. This route not only ensures stoichiometric balance (by giving an A_2-type polymerization, although with many impurities) but also avoids low rates of polymerization and (if dimethyl terephthalate is used) solubility problems.[4,97] The polymerization is thus staged to promote a more desirable pathway, and, if it were run batchwise (which is rare), it would be a semibatch process.

6.6.2 Salting of Monomers for Production of Polyamides

As for polyesters, imbalanced stoichiometry also adversely affects production of polyamides. Here, though, the rates and equilibrium constants are such that the reaction proceeds very rapidly without transamidation, and even without acid catalyst. Because the two-step process used for poly(ethylene terephthalate) does not ensure stoichiometric balance, another method must be used. In the production of nylons such as nylon 6/6, stoichiometric balance is obtained by a first stage in which a 1:1 ammonium salt is formed in a preliminary reactor.[97] Stoichiometric imbalance, if later desired to control the molecular weight, can be intentionally engineered with good precision.

6.6.3 Mass Polymerization of Vinyl Chloride

As we noted in Table 1.1, poly(vinyl chloride) is one of the most important commodity plastics, trailing only the various kinds of polyethylene. Most of the PVC produced in the United States is made by suspension polymerization, with a lesser amount made from emulsion polymerization; in Europe, the production is more evenly split between the two.[98,99] These routes are heterogeneous or multiphase polymerizations, which are discussed in detail in Chapter 7. They are often preferred over homogeneous polymerizations because they afford good control of heat evolution and because they produce a desired morphology of the product, which for poly(vinyl chloride) is more important than the details of molecular weight distribution and so forth, on which we have placed so much emphasis. The disadvantages of these processes lie in the presence of contaminants and the production of wastes (such as the contaminated wastewater from the common suspension polymerization).

Control of heat evolution is of course expected to be a problem in free-radical polymerizations in general, as we noted in Chapter 3, but the problems are especially severe for poly(vinyl chloride) because the polymer is insoluble in its monomer. Thus, except for the rare solution polymerization performed in a good solvent, polymer precipitates from the solution from the beginning of the reaction. Polymerization then continues in the polymer particles, which are swollen by monomer, but with a decrease in the rate of termination afforded by the occlusion of the radicals. Thus the polymerization is apparently autocatalytic from the very beginning, and as a rule a bulk polymerization is neither recommended nor used. Nonetheless, one bulk (or *mass*) polymerization process has been described and used industrially for over 30 years in one form or another.[100] The exception is a process that allows for an absence of contaminants, which imparts a greater clarity than possible with a suspension-produced polymer.

The polymerization is a two-stage, semibatch polymerization accommodating the physical state of the polymer and producing a desired morphology.[101] The two steps conform to the two stages of the polymerization itself: that in which the polymer particles first appear ($< \approx 20\%$ conversion), and that in which only the polymer particles swollen with the monomer remain (the rest of the reaction). These two stages require different agitation and thus are best carried out in different reactors.[99–103] In the first reactor, a prepolymerizer, vinyl chloride, is dead-end-polymerized to a conversion of approximately 8–10% (the depletion of initiator allows for storage of the prepolymer if necessary). The prepolymer is then transferred to the second reactor, where additional monomer and initiator are added. The reaction here is carried to 75–80% conversion, then is stopped by releasing the pressure, which causes the volatile monomer to be removed. It is in fact the volatility of the monomer that is used to control the heat of reaction in both reactors; the latent heat of vaporization of the monomer is allowed to eliminate the hot spots that may appear in the reactor. The monomer is then condensed and returned to the reactor. Limiting the conversion of the monomer is necessary for heat control to be performed in this way.

6.6.4 ABS Resins

ABS resins (mentioned in Section 5.1.2) are like PVC in that morphology, and thus final properties, can be quite varied, permitting the materials to be used in a number of different markets.[104] The desired properties of impact strength and toughness arise from a particular heterogeneous morphology, which corresponds to distinct butadiene and styrene–acrylonitrile microphases.[4] Such morphology can be built up by several different processes, all of which are semibatch.[104] In the *mass polymerization* route, prepolymerization of styrene and acrylonitrile is carried out to 20–40% conversion in the presence of dissolved, uncrosslinked or lightly crosslinked polybutadiene to obtain phase inversion [at early conversions there is a continuous polybutadiene/monomer phase and a discrete poly(styrene-*co*-acrylonitrile)/monomer phase]. The polymerization can then be completed in the bulk, or the remainder of the polymerization can be done in suspension (see Section 7.2) in the *mass/suspension* polymerization route. The polybutadiene may be made on site by emulsion polymerization, and the polymer then extracted into the styrene/acrylonitrile phase. Emulsion polymerization can also be used for the entire process, with a core–shell type of process utilizing polybutadiene seed particles (see Section 7.3.7).

6.6.5 Polycarbonate

As mentioned in Section 2.3.2, the production of polycarbonate from bisphenol A and phosgene is an interfacial process, and it is moreover semibatch. The interface present in these reactions is that between the continuous phase of methylene chloride and the dispersed aqueous phase in which the bisphenol A is dissolved along with sodium hydroxide. Phosgene is bubbled through the methylene chloride phase in which it is soluble, and reacts at the interface. The polymer rapidly becomes insoluble in the aqueous phase with increasing length, and so travels to the continuous methylene chloride phase. The by-product HCl reacts with the sodium hydroxide in the dispersed phase to give salt. Analyses of such a process have been presented.[105]

The advantage to doing the polymerization in such a way is that polymer of high molecular weight can be made. Not only is the by-product HCl consumed in the production of salt and in a phase different from the polymer, but semibatch loading of the phosgene (A_2) circumvents the problem of stoichiometry by working intentionally off-stoichiometry, increasing r gradually until it approaches unity. If the reaction is fast compared to the time scale for the addition of phosgene, then at all times the degrees of polymerization will be given by equations (2.4.12) and (2.4.13), and high molecular weights will be approached as r approaches unity.

6.6.6 Seeded Anionic Polymerization

As discussed in Section 3.3.4.1, lack of instantaneous initiation in an ionic polymerization leads to the Gold distribution, which is broader than the Poisson distribution. One suggested response to this disadvantage is to "seed" the initiator solution with a small amount of monomer.[106] In this way supposedly the slow initiation step is completed, and upon semibatch addition of the remainder of the monomer, the polymer chains grow simultaneously and a narrow Poisson distribution can be achieved. It has been shown, however, both by theory and experiment, that seeding does not work for irreversible polymerizations because the initiator is not consumed quantitatively by the monomer.[107] It may be of use for reversible anionic polymerizations, though, and staged feeds of different monomers may be used to produce block copolymers.

6.6.7 Control Strategies for Free-Radical (Co)Polymerization

One of the problems noted with batch free-radical polymerizations was the tendency of various sources of drift dispersion to lead to large polydispersities, which may result in poorer mechanical properties. One solution to this problem is to operate in an HCSTR; but avoiding segregation at high viscosity is difficult. Another solution is a programmed temperature history.[108,109] However, semibatch operation provides even more flexibility.

Because of the possibly large number of reactants in the system, numerous different strategies for control can be devised. One may have feed of monomer,[110,111] initiator,[110,112,113] or solvent.[111,114] One may add monomer, for example, to counteract depletion and to delay the Trommsdorff effect. Initiator may be added to counteract effects working to increase the chain length (although at the risk of contributing to a runaway reaction). Solvent may be added to counteract the Trommsdorff effect, although at the cost of a later separation step. For a copolymerization, all these operational policies have effects on composition and sequence length distributions.[115] To counteract drift dispersion, the more rapidly depleted monomer may be replaced in semibatch fashion.[116]

A common suggestion, and one easily arrived at, is to devise a scheme such that the various reactants are present in the same ratios throughout the polymerization.[117] When such an approach is used, however, one "reactant" in the system, the product polymer itself, is necessarily increasing in concentration, and its presence affects both branching and diffusion-limited reactions such as termination. Thus, more realistic kinetic schemes are necessary for any practical evaluation of a semibatch scheme. Also associated with semibatch operation of free-radical polymerizations is the difficulty that plagues attempts to attain HCSTR behavior. That is, the high viscosities encountered may make mixing of the added component difficult and may increase the polydispersity, making the situation worse. Of course, as in stepwise polymerizations,[118] semibatch operation may also be used to intentionally increase the polydispersity of the polymer.

6.6.8 Modifications of Poly(vinyl acetate)

Modification of polymers is necessarily a staged process; grafting and reactive compatibilization are examples that have been mentioned. A useful intermediate to other polymers is poly(vinyl acetate), which is of course a useful polymer in its own right, particularly as latices for paints and glues.[4] The most obvious reaction of poly(vinyl acetate) is to hydrolyze the ester to form poly(vinyl alcohol) (see Figure 6.17a). This reaction is significant because it is the only route to poly(vinyl alcohol), the monomer vinyl alcohol being tautomeric with the more stable acetaldehyde. Poly(vinyl alcohol) is soluble in water, and so one can imagine many uses as thickeners and so forth. It is also used as a suspending agent in suspension polymerizations, and in Japan as the basis for fibers.

Poly(vinyl alcohol) is also itself an intermediate to two other useful polymers. The first of these (Figure 6.17b), is the formation of poly(vinyl butyral), which is useful as a laminating agent for safety glass.[4] The second (Figure 6.17c), is to perform an esterification to form a highly unsaturated polymer, poly(vinyl cinnamate). An attempt to make this linear polymer directly would entail polymerizing a doubly unsaturated monomer and, just as in the cases of divinylbenzene or ethylene glycol dimethacrylate, branched polymers would be formed and gelation would occur. By this three-step route is formed a linear polymer [or at least one branched only to the same degree as the original poly(vinyl acetate)] with pendant unsaturations at nearly every unit. This highly reactive polymer can be used as a photoresist.[4]

Figure 6.17.
Modifications of poly(vinyl acetate). (a) Hydrolysis of poly(vinyl acetate). (b) Formation of poly(vinyl butyral). (c) Esterification to form poly(vinyl cinnamate).

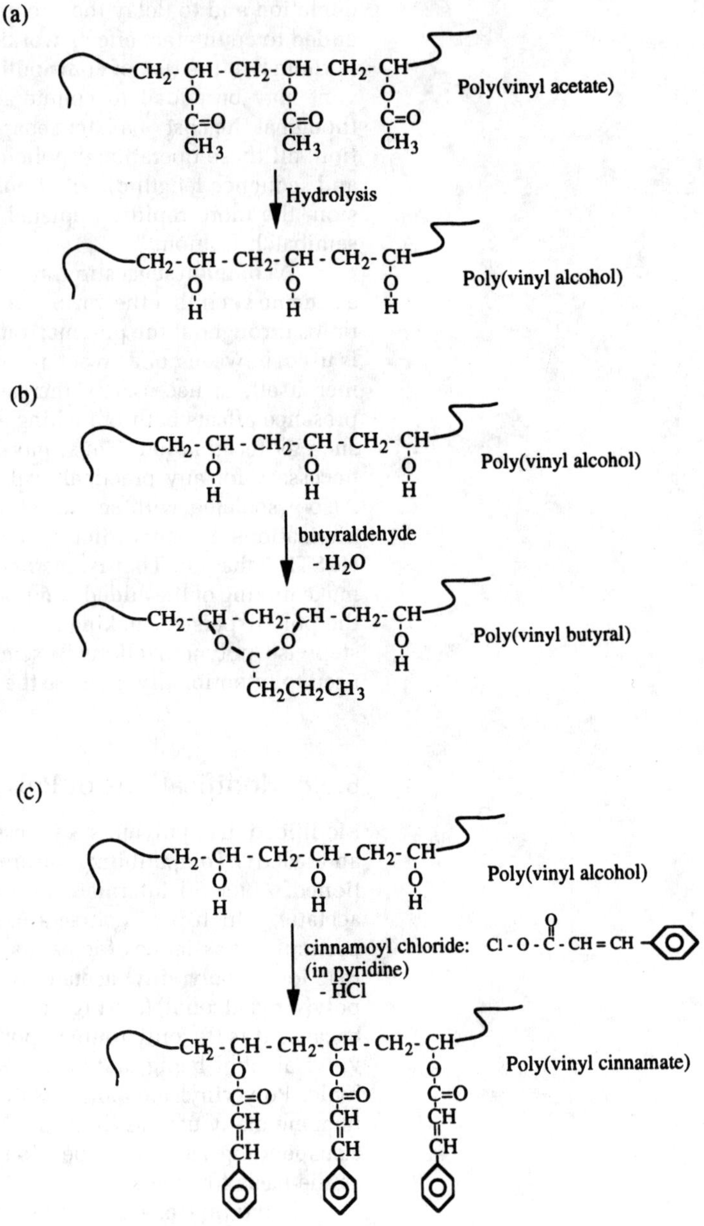

(a)

CH_2- CH - CH_2- CH - CH_2- CH Poly(vinyl acetate)

Hydrolysis

CH_2- CH - CH_2- CH - CH_2- CH Poly(vinyl alcohol)

(b)

CH_2- CH - CH_2- CH - CH_2- CH Poly(vinyl alcohol)

butyraldehyde
- H_2O

CH_2- CH - CH_2- CH - CH_2- CH Poly(vinyl butyral)

(c)

CH_2- CH - CH_2- CH - CH_2- CH Poly(vinyl alcohol)

cinnamoyl chloride: Cl - O - C - CH = CH
(in pyridine)
- HCl

CH_2 - CH - CH_2- CH - CH_2- CH Poly(vinyl cinnamate)

Problems

6.1. As shown in Section 3.3.4.1, noninstantaneous initiation in anionic polymerization can lead to chain length distributions broader than the Poisson. What sort of a distribution is obtained in an HCSTR when initiation is not instantaneous? Is it broader than a geometric distribution?

6.2. As shown in Section 6.2, ideal anionic polymerization in a batch reactor yields polymer with a Poisson distribution, while that in an HCSTR yields a product with a most probable distribution.

(a) Recall that in a batch reactor, in the presence of a transfer agent that could not reinitiate polymerization (leading to the formation of dead chains), the living chains were still Poisson-distributed, although their concentration decreased with time. In an HCSTR, what will the distribution of living chains be?

(b) Recall also that in the batch reactor, the dead chain distribution was broader than the living chain distribution (i.e., broader than a Poisson distribution). In an HCSTR, what will the distribution of dead chains be? Is it broader than the distribution of living chains?

6.3. Consider a free-radical copolymerization occurring in an HCSTR. Does the Mayo–Lewis equation still hold, relating the exit f_1 to the exit F_1? Consider the copolymerization of styrene (1) and acrylonitrile (2) at 60°C. Can this polymerization exhibit a multiplicity of steady states, even if the following rate constants [L/(mol · s)] are conversion independent?

$k_{11} = 176$

$k_{12} = 440$

$k_{21} = 49,000$

$k_{22} = 1960$

$k_{t11} = 7.2 \times 10^7$

$k_{t12} = 7.82 \times 10^8$

$k_{t22} = 2.37 \times 10^8$

The reaction is initiated by benzoyl peroxide with an entering concentration I_0 of 0.01 mol/L; at this temperature benozyl peroxide has a dissociation rate constant, k_d, of 8.4×10^{-6} s^{-1} and an efficiency of 0.6.

6.4. Derive equations (6.3.5) and (6.3.6), the relations for the DP_n and DP_w from an anionic polymerization occurring in an SCSTR. Likewise prove equations (6.3.9) and (6.3.10) for step polymerization in an SCSTR, and equations (6.4.31) and (6.4.32) for step polymerization in a LFTR with a Newtonian fluid.

6.5. Qualitatively, what sort of product should be expected from a cascade of SCSTRs? (Note that the type of polymerization is unspecified.)

6.6. Consider a step growth polymerization with either imbalanced stoichiometry or a monofunctional agent (chain stopper) being performed in an SCSTR. For long average residence times, what will the chain length distribution be, qualitatively?

6.7. Consider an anionic polymerization occurring in an LFTR; consider the reacting mixture to be Newtonian. Derive equations for DP_n and DP_w. As the length of the tube increases, the resulting product is the same as the product from what other reactor? Explain this result.

6.8. As noted in other problems, there exists the possibility of eliminating or reducing dispersion drift in free-radical polymerizations with the use of a chain transfer agent with $C_s = 1$. How can the same goal be achieved with a transfer agent with $C_s > 1$?

6.9. For an $A_2 + B_2$ stepwise polymerization, we know that balanced stoichiometry is essential to obtaining high molecular weights. However, if the A_2 reagent is present with an unknown amount of inert diluent, it will not be known how to balance the stoichiometry. Devise a semibatch operation to overcome this problem. What are possible pitfalls?

6.10. Consider the start-up of an anionic polymerization in an HCSTR. The reactor is initially charged with monomer of concentration M_0 and initiated monomer of concentration I_0. At time $t = 0$, inflow and outflow (at volumetric rates Q) begins, the inflow having a monomer concentration M_0 and initiator concentration I_0 (i.e., the initial and entering conditions are the same).
(a) What is the monomer concentration in the reactor as a function time? What is the time scale for approach to the steady state value?
(b) How do DP_n and DP_w evolve toward their steady state values? Do these degrees of polymerization approach steady state values faster or slower than the overall conversion?

6.11. Consider the crosslinking of a commercial silicone sealant, the crosslinking reaction involving water diffusing into the sealant from the surrounding air. For an infinite slab of thickness L, derive the equation for the crosslinking density as a function of position x and time t.

Assume that each reaction yields a crosslink and that the crosslinking reaction is pseudo-first-order, so that it occurs proportionally to the local water concentration with a rate constant k. Furthermore, assume that the water diffuses into the slab with a constant diffusion coefficient D. Assume that initially the water concentration c in the sealant is zero for all x, except at the surface $(x = L)$, which is in equilibrium with the atmosphere and so is maintained at all times at $c(L) = c_s$. The substrate for the sealant is impermeable, and so $dc/dx = 0$ at $x = 0$ (the interface between the substrate and the sealant).

References

1. F. J. Schork, P. B. Deshpande, and K. W. Leffew, *Control of Polymerization Reactors.* Dekker, New York (1993).

2. K. G. Denbigh, *Trans. Faraday Soc.,* **43**, 648 (1947).

3. A. E. Hamielec and H. Tobita, in *Ullmann's Encyclopedia of Industrial Chemistry,* Vol. A21, pp. 305–428. VCH Publishers, New York (1992).

4. F. Rodriguez, *Principles of Polymer Systems,* 2nd ed. McGraw-Hill, New York (1982).

5. C. G. Hill, Jr., *An Introduction to Chemical Engineering Kinetics and Reactor Design.* Wiley, New York (1977).

6. P. V. Danckwerts, *Chem. Eng. Sci.,* **2**, 1 (1953).

7. E. B. Nauman, *J. Macromol. Sci., Rev. Macromol. Chem.,* **C10**, 75 (1974).

8. J. A. Biesenberger and D. H. Sebastian, *Principles of Polymerization Engineering*. Wiley, New York (1983).

9. L. Chen and G.-H. Hu, *AIChE J.*, **39**, 1558 (1993).

10. J. A. Biesenberger and Z. Tadmor, *Polym. Eng. Sci.*, **6**, 299 (1966).

11. N. A. Dotson, *Macromolecules*, **22**, 3690 (1989).

12. J.A. Biesenberger and Z. Tadmor, *J. Appl. Polym. Sci.*, **9**, 3409 (1965).

13. J.A. Biesenberger, *AIChE J.*, **11**, 369 (1965).

14. A. Kumar, R. K. Agarwal, and S. K. Gupta, *J. Appl. Polym. Sci.*, **27**, 1759 (1982).

15. T. T. Szabo and E. B. Nauman, *AIChE J.*, **15**, 575 (1969).

16. K. F. O'Driscoll and R. Knorr, *Macromolecules*, **2**, 507 (1969).

17. J. C. Mecklenburgh, *Can. J. Chem. Eng.*, **48**, 279 (1970).

18. F. Rodriguez, *J. Appl. Polym. Sci.*, **29**, 3999 (1984).

19. M. Atiqullah and E. B. Nauman, *Chem. Eng. Sci.*, **45**, 1267 (1990).

20. C. Cozewith, W. W. Graessley, and G. ver Strate, *Chem. Eng. Sci.*, **34**, 245 (1979).

21. E. M. Hendricks and R. M. Ziff, *J. Colloid Interface Sci.*, **105**, 247 (1985).

22. S. K. Gupta, *Polym. Eng. Sci.*, **26**, 1314 (1986).

23. S. K. Gupta, S. S. Bafna, and A. Kumar, *Polym. Eng. Sci.*, **25**, 332 (1985).

24. S. K. Gupta, S. Nath, and A. Kumar, *J. Appl. Polym. Sci.*, **30**, 557 (1985).

25. G. ver Strate, C. Cozewith, and W. W. Graessley, *J. Appl. Polym. Sci.*, **25**, 59 (1980).

26. K. Nagasubramanian and W. W. Graessley, *Chem. Eng. Sci.*, **25**, 1549, 1559 (1970).

27. R. A. Jackson, P. A. Small, and K. S. Whiteley, *J. Polym. Sci., Polym. Chem.*, **11**, 1781 (1973).

28. J. C. Hyun, W. W. Graessley, and S. G. Bankoff, *Chem. Eng. Sci.*, **31**, 945 (1976).

29. A. Chatterjee, W. S. Park, and W. W. Graessley, *Chem. Eng. Sci.*, **32**, 167 (1977).

30. W. Baade, H. U. Moritz, and K. H. Reichert, *J. Appl. Polym. Sci.*, **27**, 2249 (1982).

31. T. W. Taylor and K. H. Reichert, *J. Appl. Polym. Sci.*, **30**, 227 (1985).

32. (a) H. Tobita, *Polym. React. Eng.*, **1**, 357, 379 (1992/1993). (b) H. Tobita, *J. Polym. Sci., Polym. Phys.*, **31**, 1363 (1993). (c) H. Tobita, *J. Polym. Sci., Polym. Phys.*, **32**, 911 (1994).

33. G. ver Strate, C. Cozewith, and W. W. Graessley, *Polym. Prepr.*, **20(2)**, 149 (1979).

34. For a discussion of distributions in thermal history in a nonpolymeric context, see E. B. Nauman, *Chem. Eng. Sci.*, **32**, 359 (1977).

35. J. Villermaux, in *Fourth International Workshop on Polymer Reaction Engineering, DECHEMA Monographs,* Vol. 127, p. 3. VCH Publishers, New York (1992).

36. D. P. Rao and L. L. Edwards, *Chem. Eng. Sci.,* **28**, 1179 (1973).

37. J. Villermaux, A Simple Model for Partial Segregation in a Semi-Batch Reactor. Paper 114a, American Institute of Chemical Engineers, meeting, San Francisco (1989).

38. G. Tosun, *AIChE J.,* **38**, 425 (1992).

39. P. V. Danckwerts, *Chem. Eng. Sci.,* **8**, 93 (1958).

40. T. N. Zweitering, *Chem. Eng. Sci.,* **11**, 1 (1959).

41. Z. Tadmor and J. A. Biesenberger, *Ind. Eng. Chem. Fundam.,* **5**, 336 (1966).

42. M. Abramowitz and I. A. Stegun, Eds. *Handbook of Mathematical Functions.* Dover, New York (1972).

43. R. W. Dunn and C. C. Hsu. Paper presented at First International Symposium on Chemical Reaction Engineering, Washington, DC (1970); referred to in reference 7.

44. J. H. Duerksen, A. E. Hamielec, and J. W. Hodgins, *AIChE J.,* **13**, 1081 (1967).

45. R. Thiele and J. Breme, *Int. Polym. Process.,* **3**, 48 (1988).

46. B. Fu, H. Weinstein, B. Bernstein, and A. B. Schaffer, *Ind. Eng. Chem., Process Design Dev.,* **10**, 501 (1971).

47. R. B. Bird, W. E. Stewart, and E. N. Lightfoot, *Transport Phenomena.* Wiley, New York (1960).

48. R. Cintron-Cordero, R. A. Mostello, and J. A. Biesenberger, *Can. J. Chem. Eng.,* **46**, 434 (1968).

49. S. Middleman, *Fundamentals of Polymer Processing.* McGraw-Hill, New York (1977).

50. H. Z. Li and L. Choplin, in *Fourth International Workshop on Polymer Reaction Engineering, DECHEMA Monographs,* Vol. 127, p. 21. VCH Publishers, New York (1992).

51. R. B. Bird, R. C. Armstrong, and O. Hassager, *Dynamics of Polymeric Liquids:* Vol. 1: *Fluid Mechanics,* 2nd ed. Wiley, New York (1987).

52. D. W. Dodge and A. B. Metzner, *AIChE J.,* **5**, 189 (1959).

53. S. M. Ross, *Introduction to Probability Models,* 3rd ed. Academic Press, Orlando, FL (1985).

54. (a) M. F. Cunningham, Ph.D. thesis, University of Waterloo (1990). (b) M. F. Cunningham, K. F. O'Driscoll, and H. K. Mahabadi, *Can. J. Chem. Eng.,* **69**, 630 (1991). (c) M. F. Cunningham, K. F. O'Driscoll, and H. K. Mahabadi, *Polym. React. Eng.,* **1**, 229, 245 (1992/1993).

55. C. H. Chen, J. G. Vermeychuk, J. A. Howell, and P. Ehrlich, *AIChE J.,* **22**, 463 (1976).

56. G. Verros, M. Papadakis, and C. Kiparissides, *Polym. React. Eng.,* **1**, 427 (1992/1993).

57. D. A. Paquet, Jr., and W. H. Ray, *AIChE J.,* **40**, 73, 88 (1994).

58. J. J. Zacca and W. H. Ray, *Chem. Eng. Sci.,* **48**, 3743 (1993).

59. N. G. McCrum, C. P. Buckley, and C. B. Bucknall, *Principles of Polymer Engineering.* Oxford University Press, Oxford (1988).

60. Z. Tadmor and C. G. Gogos, *Principles of Polymer Processing.* Wiley, New York (1979).

61. Z. Tadmor and I. Klein, *Engineering Principles of Plasticating Extrusion.* Van Nostrand Reinhold, New York (1970).

62. C. Rauwendaal, *Polymer Extrusion,* 3rd ed. Hanser, New York (1994).

63. M. Xanthos, Ed. *Reactive Extrusion.* Hanser, Munich (1992).

64. D. B. Todd, *Chem. Eng. Prog.,* **88(8)**, 72 (1992).

65. R. C. Kowalski, *Chem. Eng. Prog.,* **85(5)**, 67 (1989).

66. (a) A. Bouilloux, J. Druz, and M. Lambla, *Polym. Process Eng.,* **4**, 235 (1986). (b) M. Lambla, J. Druz, and A. Bouilloux, *Polym. Eng. Sci.,* **27**, 1221 (1987).

67. P. Cassagnau, M. Bert, and A. Michel, *J. Vinyl Technol.,* **13**, 114 (1991).

68. R. Chinnaswamy and M. A. Hanna, *Starch,* **43**, 396 (1991).

69. K. J. Ganzeveld, Ph.D. thesis, Rijksuniversitet, Groningen, the Netherlands (1992).

70. K. J. Ganzeveld and L. P. B. M. Janssen, *Polym. Eng. Sci.,* **32**, 467 (1992).

71. C. E. Scott, Ph.D. thesis, University of Minnesota (1990).

72. F. Ide and A. Hasegawa, *J. Appl. Polym. Sci.,* **18**, 963 (1974).

73. S. Y. Hobbs, R. C. Bopp, and V. H. Watkins, *Polym. Eng. Sci.,* **23**, 380 (1983).

74. (a) S. Cimmino, L. d'Orazio, R. Greco, G. Maglio, M. Malinconico, C. Mancarella, E. Martuscelli, R. Palumbo, and G. Ragosta, *Polym. Eng. Sci.,* **24**, 48 (1984). (b) S. Cimmino, F. Coppola, L. d'Orazio, R. Greco, G. Maglio, M. Malinconico, C. Mancarella, E. Martuscelli, and G. Ragosta, *Polymer,* **27**, 1874 (1986).

75. L. Z. Pillon and L. A. Utracki, *Polym. Eng. Sci.,* **24**, 1300 (1984).

76. P. van Ballegooie and A. Rudin, *Polym. Eng. Sci.,* **28**, 1434 (1988).

77. A. Golovoy, M. F. Cheung, and H. van Oene, *Polym. Eng. Sci.,* **27**, 1642 (1987).

78. J. R. Campbell, S. Y. Hobbs, T. J. Shea, and V. H. Watkins, *Polym. Eng. Sci.,* **30**, 1056 (1990).

79. E. W. Montroll and R. Simha, *J. Chem. Phys.,* **8**, 721 (1940).

80. R. M. Ziff and E. D. McGrady, *Macromolecules,* **19**, 2513 (1986).

81. G. Menges and T. Bartilla, *Polym. Eng. Sci.,* **27**, 1216 (1987).

82. W. Michaeli, U. Berghaus, and G. Speuser, *Int. Polym. Process.,* **6**, 163 (1991).

83. D. B. Todd, *Polym.-Plast. Technol. Eng.,* **28**, 123 (1989).

84. (a) W. Michaeli, A. Grefenstein, and W. Frings, *Adv. Polym. Technol.,* **12**, 25 (1993). (b) W. Michaeli, H. Höcker, U. Berghaus, and W. Frings, *J. Appl. Polym. Sci.,* **48**, 871 (1993).

85. A. Bouilloux, C. W. Macosko, and T. Kotnour, *Ind. Eng. Chem. Res.,* **30**, 2431 (1991).

86. K. J. Ganzeveld and L. P. B. M. Janssen, *Polym. Eng. Sci.,* **32**, 457 (1992).

87. N. P. Stuber and M. Tirrell, *Polym. Process Eng.,* **3**, 71 (1985).

88. N. P. Stuber, Ph.D. thesis, University of Minnesota (1986).

89. K. J. Ganzeveld and L. P. B. M. Janssen, *Can. J. Chem. Eng.,* **71**, 411 (1993).

90. J. F. Carley, R. S. Mallouk, and J. M. McKelvey, *Ind. Eng. Chem.,* **45**, 974 (1953).

91. G. Pinto and Z. Tadmor, *Polym. Eng. Sci.,* **10**, 279 (1970).

92. D. Bigg and S. Middleman, *Ind. Eng. Chem. Fundam.,* **13**, 66 (1974).

93. B. Siadat, M. Malone, and S. Middleman, *Polym. Eng. Sci.,* **19**, 787 (1979).

94. L. P. B. M. Janssen, *Twin Screw Extrusion.* Elsevier, New York (1978).

95. L. Chen, Z. Pan, and G.-H. Hu, *AIChE J.,* **39**, 1455 (1993).

96. L. P. B. M. Janssen, R. W. Hollander, M. W. Spoor, and J. M. Smith, *AIChE J.,* **25**, 345 (1979).

97. G. Odian, *Principles of Polymerization,* 3rd ed. Wiley-Interscience, New York (1991).

98. J. V. Koleske and L. H. Wartman, *Poly(Vinyl Chloride).* Gordon and Breach, New York (1969).

99. L. F. Albright, *Processes for Major Addition-Type Plastics and Their Monomers,* 2nd ed. Krieger, Malabar, FL (1985).

100. A. Krause, *Chem. Eng.,* **72(26)**, 72 (1965).

101. J.-C. Thomas, *Hydrocarbon Process.,* **47(11)**, 192 (1968).

102. N. Fischer and L. Goiran, *Hydrocarbon Process.,* **60(5)**, 143 (1981).

103. M. W. Allsopp, Bulk processes for the manufacture of PVC, in *Manufacture and Processing of PVC,* R. H. Burgess, Ed., pp. 39–61. Applied Science, London (1982).

104. M. E. Adams, D. J. Buckley, R. E. Colborn, W. P. England, and D. N. Schissel, *RAPRA Rev. Rep.,* **6(10)**, report no. 70 (1993).

105. P. L. Mills, *Chem. Eng. Sci.,* **41**, 1045, 2939 (1986).

106. M. Morton, *Anionic Polymerization: Principles and Practice.* Academic Press, New York (1983).

107. M. Szwarc, M. van Beylen, and D. van Hoyweghen, *Macromolecules,* **20**, 445 (1987).

108. W. H. Ray and C. E. Gall, *Macromolecules,* **2**, 425 (1969).

109. M. Tirrell and K. Gromley, *Chem. Eng. Sci.,* **36**, 367 (1981).

110. R. F. Hoffman, S. Schreiber, and G. Rosen, *Ind. Eng. Chem.,* **56(5)**, 51 (1964).

111. B. M. Louie and D. S. Soong, *J. Appl. Polym. Sci.,* **30**, 3707, 3825 (1985).

112. S.-A. Chen and N.-W. Huang, *Chem. Eng. Sci.,* **36**, 1295 (1981).

113. H. Zamani, M. Daroux, J. L. Greffe, and J. Bordet, *Chem. Eng. Commun.,* **17**, 297 (1982).

114. U. Budde and K.-H. Reichert, in *Polymer Reaction Engineering,* K.-H. Reichert and W. Geisler, Eds., p. 140. VCH Publishers, New York (1989).

115. U. Englemann and G. Schmidt-Naake, *Macromol. Theory Simulations,* **3**, 219 (1994).

116. K. Y. Choi, *J. Appl. Polym. Sci.,* **37**, 1429 (1989).

117. H. Nishimura and F. Yokoyama, *Kagaku Kogaku,* **32**, 601 (1968).

118. H. Tobita and Y. Ohtani, *Polymer,* **33**, 801 (1992).

7

HETEROGENEOUS POLYMERIZATION

7.1 Introduction

In the text thus far, we have considered polymerizations only in systems that are spatially homogeneous (with the exception of the effects of segregation due to poor mixing). However, many polymerization processes are intrinsically heterogeneous. For example, when bishydroxyethyl terephthalate is made from ethylene glycol and terephthalic acid, the original mixture is a slurry because of the limited solubility of the acid in the glycol; only as bishydroxyethyl terephthalate is formed can the acid completely dissolve and the mixture become homogeneous. The opposite case, an initially homogeneous system becoming heterogeneous, is more common. Many networks, such as those formed by free-radical mechanisms, are heterogeneous partly as a result of local deswelling during synthesis, frozen in by further crosslinking. Segmented polyurethanes, as discussed in Chapter 4, are specifically designed to microphase-separate to give a physical network. Composition drift during the copolymerization of styrene and acrylonitrile can lead to chains of different composition which are immiscible.

Perhaps the best example of intrinsic heterogeneity is *precipitation polymerization,* a kind of free-radical polymerization in which the polymer is insoluble in its monomer.[1] Several important polymers fall into this category: poly(vinyl chloride), poly(vinylidene chloride), polytetrafluoroethylene, and poly-acrylonitrile. The same is true for low-density polyethylene made in the high-pressure process, for which the ethylene is a good solvent for the polymer only if the pressure and temperature are sufficiently high.[2] Precipitation obviously affects morphology, but also it also leaves its mark on kinetics. These polymerizations appear autocatalytic from the beginning, because of the occlusion of active radicals in the polymer-rich phase. The dependence of the rate of polymerization on initiator concentration shifts from $I^{1/2}$ toward I^1. The effect is especially severe for acrylonitrile, since little monomer swells the polymer; radicals build up and can be detected by electron spin resonance spectroscopy.[3] The effects in polymerization of vinyl chloride are less severe both because of the swelling of the polymer by the monomer and because of the greater mobility of the radical due to the high rates of chain transfer to monomer and to polymer (which is always extensive because of precipitation). All these complications come about because of the polymeric nature of the product.

Heterogeneity may either be a desirable trait or one to be avoided if possible. Immiscibility in styrene–acrylonitrile copolymers is generally avoided. Microphase separation in segmented polyurethanes, on the other hand, is desired and intentional. Likewise, closely related to precipitation polymerization is *dispersion polymerization*,[4,5] in which a nonsolvent for the polymer is added to a monomer which alone would be a solvent for its polymer. As free-radical polymerization proceeds, the mixed solvent becomes progressively worse until the polymer is insoluble in the mixture. The reaction then appears to be autocatalytic because of the occlusion of radicals as in precipitation polymerization. The particles thus formed are often stabilized sterically by block copolymers. Heterogeneity here is both engineered and intentional.

In this chapter we discuss three heterogeneous polymerizations that offer various advantages in reaction engineering or in the form or characteristics of the final product. These common polymerization processes are suspension polymerization, emulsion polymerization, and polymerization by coordination catalysts. They are discussed roughly in the order in which they were developed, as well as in order of increasing complexity.

7.2 Suspension Polymerization

Bulk free-radical polymerization presents two severe problems to the reaction engineer: the large increase in viscosity and the sizable reaction exotherm. These problems are coupled; polymer–monomer systems have low thermal conductivity, and high viscosity exacerbates the problem by suppressing forced convection, impeding mechanical agitation, and increasing viscous heating. How can efficient heat transfer be effected in these situations? *Solution polymerization* provides one response. The inert diluent acts not only as a heat sink but also to decrease the amount of heat released, the rate of reaction, and the severity of the viscosity rise (which allows better mixing and delays or eliminates the Trommsdorff effect). Heat transfer may be even better if the solvent is volatile and the latent heat of vaporization is used as a mechanism for heat removal. The product of this process is a polymer solution; this may be desirable, but often will not be, so that the advantages in the engineering of the polymerization are paid for in expensive and difficult recovery steps (e.g., by precipitation or devolatilization).

The advantages of solution polymerization can be enjoyed without the separation problem by performing a *suspension polymerization*,[5–9] in which the polymerization occurs within monomer droplets dispersed in a solvent in which the monomer is not soluble, often water. The viscosity within the dispersed monomer droplets is similar to that in a bulk reactor, but in the absence of coagulation the viscosity of the dispersion is much lower. The heat generated by reaction within the droplets is removed efficiently, because of the high surface-to-volume ratio, to a vast heat sink. The product is obtained as spheres easily separated from the aqueous phase, which obviates the need to pelletize the polymer for storage and shipping and may require less energy as well.[8]

Suspension polymerization commonly consists of 25–50% by volume monomer(s) in water. All reagents reside in the organic or "oil" phase: monomer(s), initiators, transfer agents, and so forth. The remaining necessary component, present in the aqueous phase, is the stabilizer, which is generally either an inorganic powder, a water-soluble polymer, or a combination of the two. Commercial suspension polymerizations are performed batchwise in a stirred tank reactor,[5,9] although in the laboratory and even on the pilot plant scale continuous processes have been attempted.[9] Agitation, necessary to maintain the dispersion, becomes more important in preventing aggregation of the monomer droplets, which through the course of polymerization may become rather sticky. This risk of aggregation implies a risk of fouling, which in turn prevents commercial use of a continuous process, for it is difficult to operate continuously in a true steady state while polymer is accumulating on walls and impeller.

7.2.1 Correspondence Between Suspension and Bulk Polymerizations

Ideally, each suspended monomer droplet can be considered to be a separate batch reactor, although good heat transfer may render such drops more nearly isothermal than the corresponding bulk batch polymerization. To the extent that this is so, the polymer characteristics on length scales less than that of the particle size (tacticity, sequence and chain length distributions, branching, and morphology) should correspond fairly well to a bulk polymerization with the same thermal history. The correspondence between a suspension polymerization and a homogeneous bulk polymerization under the same conditions has been shown for styrene[10] and vinyl chloride.[11]

This equivalence is qualified, however, by the effects of heat and mass transfer between the monomer droplets and the aqueous phase. For example, not all beads necessarily experience the same temperature history, due to thermal gradients in the reactor or to differences in the heat transfer from the beads resulting from different bead size. If such differences occur, then the overall polymer will correspond not to the product of one batch reactor but to the accumulated product of several batch reactors of differing thermal history (another source of drift dispersion).

The effects of mass transfer, though, are more widely recognized. Complete insolubility of each phase in the other is an oversimplification; the monomers and other organic reagents may have a slight solubility in the water; likewise, the water may be soluble in the organic monomeric phase (which may lead to loss of clarity). In the former case polymerization can take place in the continuous phase, the resulting polymer being of much lower molecular weight, so that the molecular weight distribution is bimodal.[10] A difference in water solubility of comonomers in a copolymerization may cause the apparent reactivity ratios to differ from the bulk values[9] as has been observed for the methyl acrylate–acrylonitrile,[12] methyl methacrylate–methacrylic acid,[13] and styrene—p-acetoxystyrene.[14] In all these cases there is an apparent decrease in the reactivity of each radical toward the more hydrophilic monomer due to the relative absence of that monomer in the polymerizing phase. Transport of the water-soluble monomer between phases during the polymerization may cause the droplets to act not as batch reactors but as semibatch reactors; the common system vinyl chloride–vinyl acetate is affected in such a way.[15] Similarly, it is speculated that the slight presence of the initiator in the continuous aqueous phase is responsible for the higher extent of conversion attained in suspension polymerization than in the corresponding bulk process.[16] This discrepancy occurs because the initiator present in the continuous phase of low viscosity is not subject to the diffusion limitations that diminish the initiator efficiency f so drastically (see Section 3.6.3), so that those initiators may be a good external source of radicals for the polymer particles. An increase in molecular weight and polydispersity with increasing bead size in suspension copolymerizations of styrene and acrylonitrile has been noted.[17] This obviously cannot be the effect of higher temperatures in the larger beads, which would lead instead to lower molecular weights. The exact source of these observations is not clear, though it may be related to the large difference in the water solubilities of the two.

These exceptions should be kept in mind, but as a first and often good approximation, suspension polymerizations can be modeled as bulk polymerizations. There is thus no need for any further discussion here of molecular weights and so forth. Rather, it is more profitable to concentrate on the unique factor, the particles, and the stabilization and characteristics of them.

7.2.2 Stabilization

One of the risks of a suspension polymerization is that the monomer droplets will coagulate during the process, forming a network of aggregated spheres that cannot be redispersed, from which heat transfer will be poor, and because of which the reactor will have to be cleaned out. Coagulation is a risk because at significant conversion levels short of the point at which the system becomes glassy, the monomer droplet will be sticky.[5–9,18] In this regime of intermediate conversions, there exists the danger of coagulation, which is avoided through mechanical agitation and the use of stabilizers. Mechanical agitation is of course necessary simply to maintain the dispersion and to prevent the two phases from separating.

Stabilizers adsorb on the monomer droplets as a thin layer that acts to prevent coalescence and possibly to decrease the interfacial tension between the two phases. These stabilizers fall into two categories: water-soluble organic polymers (sometimes referred to as protective colloids) and inorganic powders. Hydrophilic polymers may be either natural or synthetic. Of the latter, the most well known is poly(vinyl alcohol); other examples are poly (acrylic acid), its copolymers and salts, and poly(vinylpyrrolidinone). Natural polymers used include gelatin, cellulose derivatives, and starch. Inorganic powders used as stabilizers include salts of magnesium, calcium, aluminum, and so forth, more specifically magnesium hydroxide, calcium phosphate, aluminum hydroxide, talc, and hydroxyapatite; longer lists can be found elsewhere.[7,9] The inorganic powders may be preferred because of their lower cost, decreased pollution, and improved surface characteristics, as discussed below.

7.2.3 Particle Size

Particle size is one of two characteristics unique to suspension polymerization which cannot be derived from the analogous bulk polymerization. Typical average particle sizes in a suspension polymerization are on the order of 10 μm to 1 mm. There is a particle size distribution about the average just as there is a chain length distribution, and often it is desirable for it to be as narrow as possible. Particle size is an obvious design parameter for products used in particulate form, such as ion-exchange resins. For production of expandable polystyrene (used in making polystyrene foams) a "blowing agent" of low molecular weight is added during the polymerization. Here the particle size will determine the amount of agent taken up, and the particle size distribution will determine how uniformly that agent is distributed. There can thus be a variety of motivations for attaining a certain particle size and distribution. For polymerizations of vinyl chloride, there is a link between particle size and porosity because both properties are determined in part by the relative degree of coalescence. The porosity is important because it determines the ease with which unreacted monomer can be removed (vinyl chloride is a suspected carcinogen) and plasticizers added.

Particle size is determined by the processes of drop breakup and coalescence, and as such it is a problem in reaction engineering, not chemistry.[5,9] Breakup occurs only in the region very near the impeller, while coalescence occurs in the rest of the reactor, which serves to recirculate material back to the impeller. Coalescence increases with increasing volume fraction of the dispersed phase, but decreases with the amount of stabilizer. Coalescence (to the extent it occurs) and drop breakup depend on factors both fluid dynamic and interfacial: impeller diameter and speed, circulation frequency (which depends on geometry and overall dimension of the reactor compared to that of the impeller), coalescence frequency, volume fraction of the dispersed phase, the densities and viscosities of the dispersed and continuous phases, and the interfacial tension. Increasing the impeller speed is expected to decrease the particle size, as would increasing the amount of stabilizer, since this lowers the interfacial tension, and smaller droplets generally imply a more uniform distribution. Nonetheless, often at sufficiently high impeller speed or concentration of stabilizer the trend is reversed and an increase in the particle size is noted.

The processes of drop breakup and coalescence occur concurrently with the polymerization process, in which a number of the factors just mentioned change dramatically, notably the viscosity of the dispersed phase and its adhesive properties, which affect coalescence. Thus the course of the polymerization affects the droplet size and its distribution in a complicated fashion. One aspect is clear, though: at a certain point, high viscosity or the reaching of the glass transition temperature of the dispersed phase will prevent further breakup or coalescence. Once that *particle identity point* has been passed, the particle size and its distribution cannot be altered except by separation.

7.2.4 Surface Characteristics

The characteristics of the surface of the particles constitute the second distinctive feature of suspension polymerization compared to bulk.[9] Indeed, one of the main distinctions between the two types of stabilizer is made by observing whether they tend to leave a skin on the polymer beads. The water-soluble polymers will in general leave a skin on the surface, while the inorganic powders that do remain can be easily removed with a dilute acid wash. Ease of cleaning is obviously important for the removal of unreacted monomer or the introduction of plasticizer, or in cases in which the interior of the particle must be accessible in its final use (as is the case with ion-exchange resins).This advantage of the inorganic powders as stabilizers is joined by the advantages of lower impurity, less expense, and decreased pollution. Nevertheless, the water-soluble polymers still find great use for polymers not intended for use as a particulate.

7.2.5 Industrial Applications

Because of its many advantages, suspension polymerization is widely used for the polymerization of some of the more important monomers we have mentioned. Some representative applications, dealt with in more detail in other sources,[9,19,20] are briefly reviewed below.

7.2.5.1 Poly(vinyl chloride): Powder Suspension Polymerization

Poly(vinyl chloride) is one of the most important commodity plastics, behind only the various polyethylenes in terms of annual production. The vast majority of the polymer is made by suspension polymerization, with the remainder by emulsion and by the bulk process described in Chapter 6. (In Europe the emulsion process is used more frequently, but the suspension process is still dominant.[19,21]) Suspension polymerization of vinyl chloride, however, is a special case because it is a precipitation polymerization and so, when performed in suspension, is heterogeneous on two different scales. As in the bulk process, polymer precipitates as a powder within the monomer droplet from the beginning of the reaction; hence we refer to this process as a *powder suspension polymerization.* The morphological characteristics of the polymer are as important, if not more so, than the finer details of molecular structure, inasmuch as the morphology determines the ability to eliminate volatiles and take up plasticizers, as well as affecting the bulk mechanical properties. Other distinctions from the idealized suspension polymerization process stem from the moderate solubility of vinyl chloride in water and from the need to run these processes under pressure because of the low boiling point of vinyl chloride. Nonetheless, the progress of a vinyl chloride polymerization does correspond well to a bulk process,[11] although one should realize that this "corresponding process" is not the two-stage mass polymerization used industrially (see Section 6.6.3). A recent series of papers presented a fairly comprehensive modeling effort accounting for many of the peculiarities of suspension (and bulk) polymerization of vinyl chloride.[22]

7.2.5.2 Poly(methyl methacrylate) and Polystyrene: Pearl Polymerization

Perhaps more straightforward are the polymerizations of methyl methacrylate and styrene through the suspension process. These reactions are referred to as *pearl* or *bead polymerizations* because transparent polymeric spheres are obtained as product rather than the opaque product poly(vinyl chloride); the difference, of course, arises from the solubility of the polymers in the parent monomer solvent. Poly(methyl methacrylate) is also made by bulk process, as is polystyrene. In the latter case, though, there are particular styrenic products made only by suspension polymerization which, like poly(vinyl chloride), yield particles which are themselves heterogeneous. The first of these is expandable polystyrene (EPS), which is used to make polystyrene foams. The added ingredient in this case is the blowing agent (e.g., butane, pentane, hexane, cyclohexane), usually present as occluded droplets within the larger dispersed monomer.[7] The second of these are beads for ion-exchange resins or chromatographic packing, produced through a copolymerization of styrene and divinylbenzene. Such nonlinear free-radical polymerizations result in heterogeneous networks,* because of local deswelling. Heterogeneity is generally desirable here because it allows for the porosity necessary for the end use of these resins. Crosslinked beads based on methyl methacrylate are also made.

7.2.5.3 High Impact Polystyrene and ABS Resins: Mass Suspension Polymerization

Two other products based in part on styrene deserve mention, although separately because of the peculiarities of the process, *mass suspension polymerization,* used. These products are high impact polystyrene (HIPS) and acrylonitrile–butadiene–styrene

---

*See reference 18 for a representative micrograph of the resulting morphology.

(ABS) resins. The process is a two-stage process that involves a polybutadiene latex (made by emulsion polymerization).[9] The rubber is dissolved in the monomer mixture: styrene in the case of HIPS production, or a comonomer mixture of styrene and acrylonitrile in the case of ABS resins. This initial stage of the reaction is a bulk process and is ended at 25–30% conversion, by which time phase inversion has occurred. The viscous solution is transferred to a suspension reactor in which the remainder of the polymerization occurs. Throughout both stages of the process, styrene and acrylonitrile (if present) are grafted onto the polybutadiene latex, stabilizing the interfaces between the two phases.

7.2.5.4 Polyacrylamide: Inverse Suspension Polymerization

An obvious modification of suspension polymerization consists of reversing the phases, so that an aqueous phase is dispersed within an organic medium. Here one wishes to carry out the polymerization of a water-soluble monomer. The best example is the copolymerization of acrylamide with N,N'-methylene bisacrylamide.[20,23] One then of course has all other reactants, such as the initiator, also present in the dispersed aqueous phase.

7.2.5.5 Other Polymers

A number of other polymers are made at least in part by suspension polymerization. While the amount of vinyl acetate to be used for paint latices is polymerized by emulsion polymerization, some is polymerized by suspension polymerization, such as that to be modified as described in Section 6.6.8. Poly(styrene-co-acrylonitrile) (SAN plastic) is polymerized by suspension polymerization as well as by the bulk method.

7.3 Emulsion Polymerization

Emulsion polymerization bears some similarity to suspension polymerization: both involve the polymerization of monomers not soluble in water in an aqueous dispersion. For emulsion polymerizations, however, a surface active agent (surfactant) is present rather than a suspending agent, and the initiator is soluble not in the organic phase, but in the aqueous phase. These two main distinctions have a drastic effect on the course of the reaction, and indeed emulsion polymerization differs greatly from suspension polymerization on two main points. First, polymerization does not occur in the initial monomer droplets; this is obvious from the final polymeric particles, which are much smaller and more numerous than the initial droplets. Second, the kinetics observed bears no resemblance to kinetics in a bulk polymerization.

In Section 7.2 we reasoned that suspension polymerization would obey homogeneous batch reactor kinetics if the effects of mass transfer between the phases were negligible.* The source of the unique kinetics must therefore lie in mass transport between the phases, which also follows obviously from two conditions: the initiator and the monomer are in different phases, and the original monomer droplets do not correspond to the final polymer particles. Thus, whereas in suspension polymerization the continuous and discrete phases ideally do not communicate as far as mass transfer is concerned, in emulsion polymerization the phases must communicate. At the very least, monomer and initiator must find a common place to meet.

The advantages of such an intricate process for polymerization are not as obvious as those for suspension polymerization. Before answering the questions of what those advantages are and how they arise, it is worth inquiring how such a nonintuitive process could have been developed in the first place. A partial answer to this question lies in the reason for attempting the early heterogeneous polymerization—namely, the desire to mimic the form in which natural rubber occurs, a latex. Natural rubber comprises roughly a 33% emulsion of high molecular-weight cis-1,4-polyisoprene in water, stabilized by protective proteins and containing lipids, carbohydrates, and so forth.[24,25] This was well known in the 1930s[24] and had been known at least in part since the work of Faraday in 1826.[26] Given both the natural state of rubber and the state of knowledge before World War II, mimicry was the natural course.† Thus Hohenstein and Mark[10] state that "the more important incentive . . . [was] that all native rubbers occur in the form of latexes. . . ." This was the incentive for the particular route of polymerization; the incentive for producing synthetic rubber in the first place was to relieve dependence on natural sources, which were subject to large fluctuations in prices and wartime shortages (both in World War I and World War II). The early patents, dating from around 1910 at a time of high natural rubber prices, used egg albumin, starch, gelatin, milk, or blood serum in the attempt to reproduce conditions in the rubber-producing plants. Since these polymerizations took weeks and were actually suspension polymerizations initiated by oxygen,[10,26] it must be said that suspension polymerization was born of imitation as well. True emulsion polymerizations were realized in patents dating from the late 1920s and early 1930s, which used soap or other similar substances. It seems that realization of the advantages inherent in the placement of the initiator in the phase opposite that of the monomer probably resulted from chance, since both water- and monomer-soluble peroxides were used as initiators in the early patents.[10]

The rough understanding of how emulsion polymerizations proceed had to wait until the late 1940s, for the initial qualitative description of Harkins,[28] which was fleshed out in

---

*Some conditions were placed on heat transfer as well, but we would expect any inefficient heat transfer to affect emulsion and suspension polymerizations in much the same way.

†Ironically, the mimicry failed technically, since the actual biosynthetic route is not chainwise but, rather, a condensation polymerization of isopentenyl pyrophosphate.[27]

more quantitative terms by Smith and Ewart.[29] Although the theory of emulsion polymerization has become much more intricate in the years since, the theory of Harkins, Smith, and Ewart remains the best point of introduction and of departure. This is both a logical consequence of the adequacy of the classic theory to explain the broad features of the kinetics and a historical consequence of its prominence in the literature. The theory is traditionally, and most easily, given for a batch polymerization, but we will touch on continuous and especially semibatch polymerizations later. Descriptions in standard texts are also recommended,[23,30] as are the more extended reviews that have appeared.[31-37]

7.3.1 Qualitative Description of Emulsion Polymerization

7.3.1.1 Components and Their Locations

The locations of the different components in our system near the beginning of the polymerization are shown in Figure 7.1. Water forms the continuous phase, and monomer the dispersed phase, in an amount comprising around a third of the total volume. Some monomer, however, is present in the aqueous phase according to its small but nonzero solubility there. As specified before, the initiator is present in the continuous aqueous phase. The most common initiators are peroxides such as hydrogen peroxide, persulfates such as potassium persulfate and ammonium persulfate, and azo compounds such as 2,2'-azobis(2-amidinopropane)-dihydrochloride. Redox systems, as mentioned in Section 3.4.1, are also commonly used because of the high initiation rates at relatively low temperatures (e.g., < 50°C).[23]

Many crucial roles are played by the remaining component, the surfactant. Surfactants are amphiphilic molecules, one part preferring one type of environment (e.g., an aqueous environment) and the other a different type (e.g., organic). As such, surfactants in the final product will reside at the surfaces of the polymer particles and will thus stabilize the particles, which is their most important role. Stabilization will be by *electrostatic* or *steric* means, or by some combination of the two. Because a stable emulsion is desired, rather than simply a mechanism for preventing aggregation during polymerization, stronger agents than those used in suspension polymerization are needed. Electrostatic stabilization by ionic surfactants has often been preferred because of the strong repulsion between like charges.[38,39] Although cationic surfactants can be used, anionic surfactants are more popular; these include fatty acid soaps such as sodium laurate (sodium dodecanoate), sodium palmitate (sodium hexadecanoate), sodium stearate (sodium octadecanoate) (or the equivalent forms with potassium), and sulfonates and sulfates (particularly sodium dodecyl sulfate).[23]

Ionic surfactants are not without disadvantages, however. The stability imparted is only kinetic, not thermodynamic, and it is very sensitive to ionic strength (added salt will screen the electrostatic repulsion and destabilize the latex). Steric stabilization can answer both these problems.[40] It is best achieved by

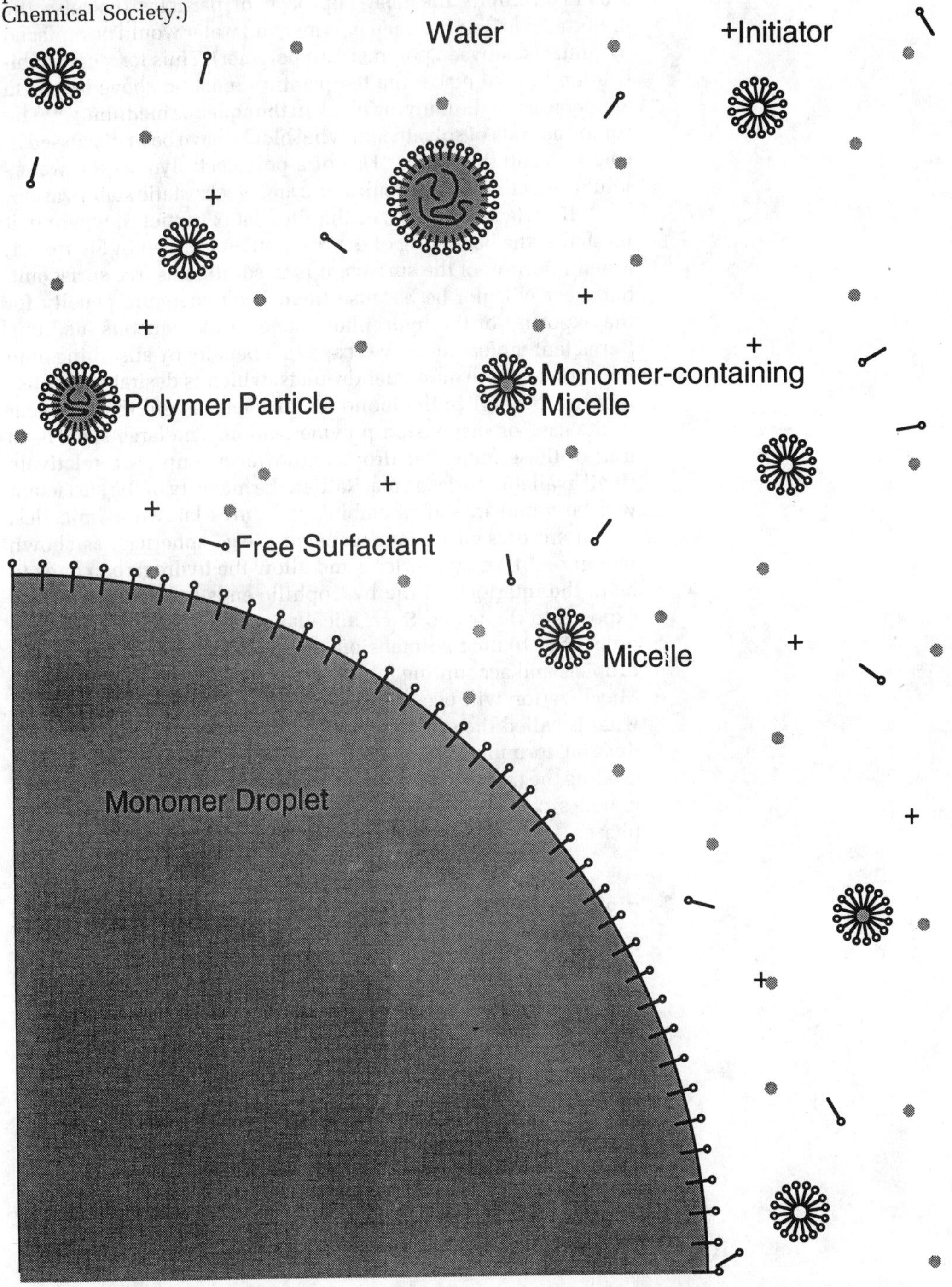

Figure 7.1.
Schematic of emulsion polymerization during interval I. Drawing not to scale in length scale or in volume fraction. (After Harkins,[28] used by permission of the American Chemical Society.)

Water

+Initiator

Polymer Particle

Monomer-containing Micelle

Free Surfactant

Micelle

Monomer Droplet

a block (or graft) copolymer, one block strongly absorbing, or anchoring, to the polymer surface, and the other extending, or buoying, into the aqueous phase. An example of an appropriate block copolymer is poly(styrene-*block*-ethylene oxide), and an appropriate graft copolymer is poly(vinyl acetate-*graft*-vinyl alcohol). The strong anchoring of the first block prevents surface migration to expose bare spots, and the strong solvation of the second prohibits the near approach of particles (because the preferred contacts between polymer and water would be replaced by contacts between polymer and polymer). Thus for steric stabilization to take place, the temperature must be above the theta temperature of the buoying block in the aqueous medium.[41–43] The conformations of solvated buoying blocks have been discussed at length in the literature.[44] Use of a polyelectrolyte as the water-soluble block combines both steric and electrostatic stabilization.

If surfactant stabilizes the final latex particles, where is it located at the beginning of the reaction? As shown in Figure 7.1, a small portion of the surfactant is in solution as free surfactant, but most will not be because there is an energetic penalty for the exposure of the hydrophobic end to the aqueous medium. Surfactant molecules may escape this penalty by adsorbing onto the surface of the monomer droplets, which is desirable because it helps to stabilize the monomer droplets, as did the stabilizer in the case of suspension polymerization. The large size (1–10 μm) of these monomer droplets, however, implies a relatively small available surface area. Rather, the majority of the surfactant will be found in self-assembled structures known as micelles, which are present in some geometry (e.g., spherical, as shown in Figure 7.1, or cylindrical) and allow the hydrocarbon ends to be in the interior and the hydrophilic ends to be at the surface exposed to the water. Such micellar structures are often on the order of 2–10 nm in dimension, much smaller than the monomer droplets and accounting for most of the surfactant in the system. Micellization will occur if the surfactant concentration is above what is called the critical micelle concentration (cmc); most industrial formulations use surfactant at concentrations far exceeding the cmc, which is often quite low (~ 0.001 mol/L). These micelles can also solubilize monomer present in the aqueous phase.

7.3.1.2 Locus of Initiation and Mechanisms of Particle Formation

Two important questions arise from the brief description thus far.

Where does initiation occur?

How do the polymer particles, more numerous and smaller than the monomer droplets, come into being?

The answers to the two questions are related, since the formation of particles (particle nucleation) relies on the initiation of polymer chains, even though initiation continues long after nucleation is complete. The formation of free radicals (e.g., sulfate radicals) almost certainly occurs in the continuous aqueous phase but does not correspond to the initiation of a polymer chain because the radical must encounter monomers with which to react. The locus of the formation of free radicals has been thought to be, in turn, mysterious (prior to the work of Harkins and that of Smith and Ewart), simple, and (finally) rather complicated.[37,45]

Given the location of the sulfate radicals, initiation would appear to be possible either in the bulk of the aqueous phase or at an interface, most likely of a micelle. The traditional view excluded the former possibility because of the low concentration of monomer in the aqueous phase (see Table 7.1) and the high concentration in the swollen micelles (initiation in the monomer droplets was ignored because of the relatively low surface area). Since, however, entry of a charged radical into a micelle bearing like charge is not favorable,[46] it is far more likely that initiation occurs in the aqueous phase, despite the low concentration of monomer there, until a large enough number of monomers is added to impart to the oligomer the attributes of a surfactant (for a hydrophobic monomer like styrene, it is necessary to add on the order of two to three monomers).[37,47] The exact locus of initiation thus depends on the nature of the initiator fragment, the solubility of the monomer, and the structure of the interphase.

The traditional view placing initiation at the monomer-swollen micelles is the original picture of *micellar nucleation*.[28,29] The formation of oligomeric "surfactant" prior to entry modifies the picture somewhat, in particular the rate of nucleation, but still qualifies as micellar nucleation. The oligomer may, on the other hand, simply precipitate from solution to be later stabilized by surfactant. This process for the formation of polymer particles is called *homogeneous nucleation,* and it should become progressively more important with increasing monomer solubility in water and decreasing amounts of surfactant.[46,48-50] Coagulation may also contribute to nucleation. All these processes may contribute in part to the nucleation of particles, which in turn determines the final number (or concentration) of particles and thus their (volume-) average size. Nucleation is thus crucial to the properties of the final product, yet may be sensitive to changes in process (e.g., agitation).

TABLE 7.1 / Water Solubilities of Monomers at Room Temperature[34]

Monomer	Water Solubility (g/L)
Styrene	0.07
Butadiene	0.8
Chloroprene	1.5
Vinyl chloride	7
Vinylidene chloride	7
Ethylene	15
Ethyl acrylate	15
Methyl methacrylate	16
Vinyl acetate	25

7.3.1.3 Progress of the Polymerization by Intervals

Once nucleated (by whatever means), polymer particles adsorb more monomer to be able to maintain a monomer volume fraction ϕ_m, approximated by the equilibrium swelling of the monomer in the polymer. Polymer particles thus grow with time, and this growth requires more surfactant for stabilization, as does further nucleation of new particles. Micelles are in this manner progressively depleted. When they have vanished completely, the system comprises only polymer particles and monomer droplets. The disappearance of micelles signals the end of what is called *interval I* and the beginning of *interval II*. In interval I, micelles, polymer particles, and monomer droplets coexist (as in Figure 7.1), the number of polymer particles and their size increasing at the expense of the micelles. The conversion at which interval I ends decreases with increasing water solubility of the monomer and decreasing amount of surfactant, among other things.

In interval II, the concentration of polymer particles, N, stays constant while the polymerization proceeds. Polymerization occurs within the polymer particles, which maintain a nearly constant volume fraction of monomer, ϕ_m. The incoming monomer is of course supplied by the monomer droplets via the aqueous solution. There will come a point, short of complete conversion, at which the monomer droplets are depleted and the only remaining monomer not in the polymer particles is in the aqueous phase. This marks the transition to *interval III*, which will occur at lower conversions the larger ϕ_m is. This last interval comprises consumption of the remaining monomer.

This division of an ideal emulsion polymerization into three intervals is a helpful generalization implicit in the early theories, but made more explicit later.[51] Figure 7.2 shows the demarcation between the intervals in terms of the concentration of particles N and the concentration of monomer in the form of monomer droplets, $[M]_{droplets}$. Interval I, the extent of which varies

Figure 7.2.
Three intervals of emulsion polymerization, represented in terms of the number of particles and the monomer concentration in droplets.

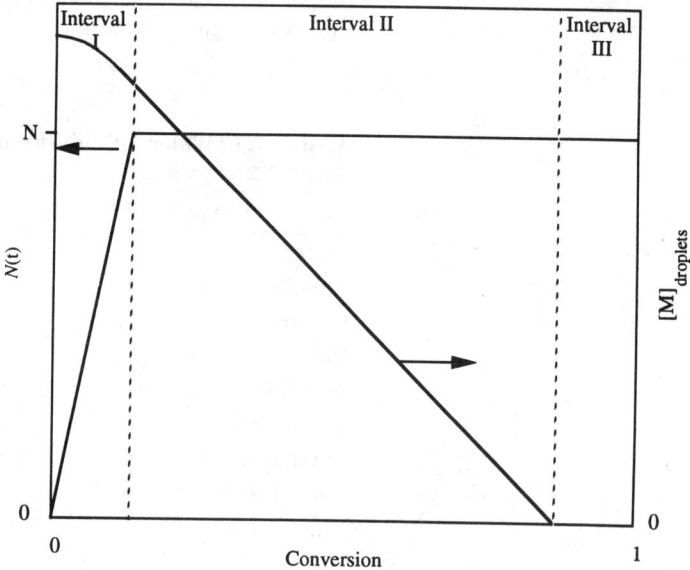

from a few percent to as high as 15% conversion,[23] is the particle formation stage; hence the final N is determined in this stage. Thus while it is appropriate to discuss the evolution of the particle concentration N in the context of interval I, intervals II and III comprise the majority of the polymerization, hence the more familiar details of rate of polymerization and molecular weights are discussed in that context.

It should be borne in mind that division of the polymerization into these intervals is an idealization, one from which many deviations may occur. It is sufficient at this point to observe that it strictly applies to batch polymerizations only, but even in nonbatch cases (as in other departures from the ideal picture), the intervals provide a useful conceptual framework.

7.3.2 Interval I

The first interval is that in which the polymer particles grow in concentration (or number) N,* which is set at the end of this interval, barring any subsequent aggregation or nucleation. Thus interval I determines the concentration and ultimate average size of the polymer particles, and so the evolution of the number of particles, $N(t)$, is of prime importance in this interval. When one considers the phenomena involved, the competition between micellar and homogeneous nucleation, and the possibility of aggregation, it is clear that an exact solution is difficult. We can begin to appreciate the complexities, though, by examining the traditional case in which micellar nucleation occurs exclusively.

The progress of micellar nucleation can be bracketed between two limiting cases.[29] The first is that in which a constant rate of radical production gives a constant rate of particle generation:

$$\frac{dN}{dt} = \rho_i \qquad (7.3.1)$$

where ρ_i is the rate of production of radicals that initiate particles. This was originally identified with the total rate of radical production R_i. Studies have shown, however, that initiator efficiency f may be rather low, and of course if oligomer formation must precede micellar nucleation, there are opportunities for termination before entry into the micelle.[47] Thus ρ_i is not equal to R_i, or even fR_i, but should be related to these. Since half-lives are generally quite long, it is a reasonable approximation to consider R_i, and thus ρ_i, to be constant.[51] The concentration of particles will thus grow linearly with time:

$$N(t) = \rho_i t \qquad (7.3.2)$$

Throughout this section, although we use the terminology of Smith and Ewart, which is commonly used with respect to emulsion polymerization, we do not use the common units for the various terms which necessitate the appearance of terms like 10^3 and N_A. A consistent set of units has been assumed in this discussion.

This linear increase continues until the total surface area of the polymer particles equals that which would be taken up by the surfactant molecules. The specific surface area of a surfactant molecule is written as a_s, and the surfactant concentration as S. The specific surface area a_s should not be thought of as a property of the surfactant only. It depends on the polarity of the surface on which it adsorbs and the interfacial tension between that surface and the aqueous medium,[52] as well as on the nature of any other kinds of surfactant present (especially those that can screen electrostatic interactions)[53] and the ionic strength of the aqueous medium.[54-56]

If it is assumed that particles, once nucleated, grow in volume with a constant rate μ; then the total surface area A_p is given as follows:

$$A_p = \rho_i \, [6\pi^{1/2}\mu]^{2/3} \int_0^t (t - t')^{2/3} dt' = \rho_i \frac{3}{5} [6\pi^{1/2}\mu]^{2/3} \, t^{5/3} \tag{7.3.3}$$

where index of integration t' is the birth time of the polymer particle, and spherical particles have been assumed. The rate of volume expansion of a particle, μ, is related to the rate of polymerization and the volume fraction of monomer ϕ_m maintained by the following equation[51]:

$$\mu = k_p \frac{\rho_m}{\rho_p} \, \phi_m (1 - \phi_m) \tag{7.3.4}$$

where ρ_m and ρ_p are the densities of the monomer and polymer, respectively. Equating A_p to $a_s S$ gives the time at which interval I ends:

$$t_1 = \left(\frac{5}{3[6\pi^{1/2}]^{2/3}} \right)^{3/5} \rho_i^{-3/5} \, \mu^{-2/5} \, (a_s S)^{3/5} = 0.420 \, \rho_i^{-3/5} \, \mu^{-2/5} \, (a_s S)^{3/5} \tag{7.3.5}$$

Through equation (7.3.2), the number of particles at this time is found to be:

$$N = N(t_1) = \left(\frac{5}{3[6\pi^{1/2}]^{2/3}} \right)^{3/5} \left(\frac{\rho_i}{\mu} \right)^{2/5} (a_s S)^{3/5} = 0.53 \left(\frac{\rho_i}{\mu} \right)^{2/5} (a_s S)^{3.5} \tag{7.3.6}$$

The dependences, or scalings, of the final particle number on initiation rate ($\rho_i^{2/5}$) and surfactant concentration ($S^{3/5}$) are worthy of note, since these can be manipulated to change the ultimate particle number and thus average particle size.

One problem with the preceding model is the absence in reality of any physical prohibition against radicals entering existing polymer particles rather than nucleating a new particle. Such entry will decrease the total number of particles obtained from that just calculated (which constitutes an upper bound), if μ is indeed constant. The most reasonable approach is to assume that the rate at which radicals enter an existing particle is proportional to its projected area, or equivalently its radius. This is rather difficult, especially when accounting for entry into existing particles. A lower bound on the number of particles, accounting for this loss of radicals, is estimated by assuming that the rate at which an individual micelle or polymer particle is attacked by an initiator is proportional to its surface area. The rate of particle formation is thus diminished from the rate given in equation (7.3.1) by the fraction of surfactant present in polymer particles:

$$\frac{dN}{dt} = \rho_i \left(1 - \frac{A_p}{a_s S} \right) \tag{7.3.7}$$

Equation (7.3.3) can then be rewritten in the following slightly more general form:

$$A_p = [6\pi^{1/2}\mu]^{2/3} \int_0^t (t - t')^{2/3} \left(\frac{dN}{dt} \right) dt' \tag{7.3.8}$$

which gives an integral equation for dN/dt. Solution yields:

$$t_1 = 0.650 \, \rho_i^{-3/5} \, \mu^{-2/5} \, (a_s S)^{3/5} \tag{7.3.9}$$

$$N = N(t_1) = 0.37 \left(\frac{\rho_i}{\mu} \right)^{2/5} (a_s S)^{3/5} \tag{7.3.10}$$

The end of interval I occurs later in time and yields a smaller particle number; the scalings of both t_1 and N, however, remain unaffected. These equations should give good bounds for the number of particles if nucleation is exclusively micellar. The lack of any explicit reference to the mechanism of nucleation in the foregoing derivation suggests that the scalings survive as long as the rate of particle nucleation (e.g., by homogeneous nucleation) is constant [or decreases in a fashion like that expressed in equation (7.3.7)] and that each particle is stabilized by the same number of surfactant molecules.[33,49]

Although our derivation does point out the variables that allow control of particle number and size (ρ_i and S, in particular), the scalings are generally violated for emulsion polymerizations of monomers that have appreciable solubility in water[32,33,45] or for cases in which the concentration of surfactant is well below the cmc.[57] These are both violations of conditions which encourage the dominance of micellar nucleation. From Table 7.1, we see that styrene should be exemplary in obeying the preceding models, which is so.[32,45,58] Homogeneous nucleation, on the other hand, should dominate emulsion polymerizations of monomers with greater water solubility, and aggregation should play a role as well.[59] Vinyl acetate is exemplary here.[46] For such monomers (and these are the majority), the foregoing scalings are not obeyed; the dependence of N on S may be stronger or weaker, and the dependence on ρ_i may vanish.[23] Needless to say, if the surfactant in any way reacts with the free radicals (especially if the surfactant acts as in inhibitor or retarder), there may be a dependence of μ on S, which would also change the apparent scaling with S. Models have been proposed to account for homogeneous nucleation, aggregation, and entry of radicals into already existing particles.[50,60] We do not review these more complicated cases here.

7.3.3 Interval II

The two intervals that follow the establishment of the particles comprise the majority of the polymerization. In these two intervals we return to issues of the sort we treated in the earlier chapters of this book—the rate of polymerization and the molecular weights of the polymer formed, as well as the particle size distribution. Let us first deal with the rate of polymerization. The overall rate of polymerization comprises the sum of the rates in all particles and is thus proportional to the concentration of particles N, the monomer concentration within those particles M, and the average number of radicals within a polymer particle $\bar{n}$:

$$R_p = k_p M \bar{n} N \qquad (7.3.11)$$

The monomer concentration M corresponds not to the overall monomer concentration but to that in the locus of polymerization, the polymer particles. It is clearly related to ϕ_m, but this statement only defers the question of what determines M. Likewise, what determines the average numbers of radicals $\bar{n}$? These two factors need further elucidation before we can evaluate equation (7.3.11) rationally.

7.3.3.1 Monomer Concentration

In interval II, as in interval I, a common assumption is that ϕ_m, the volume fraction of monomer within the polymer particles, is constant and determined by the equilibrium swelling of the growing spheres because diffusion of monomer from the aqueous phase is sufficiently rapid.[30] This equilibrium represents a balance between the free energy of mixing and the interfacial free energy. The volume fraction of monomer in a swollen sphere of radius r is given by the zero of the following equation[33,34,61,62]:

$$\frac{1}{1 - \phi_m} + \frac{\ln \phi_m}{(1 - \phi_m)^2} + \chi + \frac{2 v_m \gamma}{RT} \left[\frac{1}{(1 - \phi_m)^2 \, r} \right] = 0 \qquad (7.3.12)$$

where v_m is the partial molar volume of the monomer, γ the interfacial tension, and χ the Flory–Huggins parameter for the monomer–polymer system. The equation comprises two bulk entropic mixing terms, the bulk enthalpic term, and the interfacial term, respectively. The interfacial term vanishes as $1/r$, as it should, and also decreases with decreasing interfacial tension. For sufficiently large r and small γ, this should be equivalent to the equilibrium swelling in a bulk system. Table 7.2 lists equilibrium monomer volume fractions for different common monomers. The monomer concentration M and the volume fraction ϕ_m are of course related by a factor of the molar volume.

TABLE 7.2 / Equilibrium Monomer Volume Fractions in Latex Particles for Emulsion Homopolymerization[34]

Monomer	ϕ_m
Vinylidene chloride	0.2
Ethylene	0.2
Vinyl chloride	0.3
Butadiene	0.5
Styrene	0.6
Chloroprene	0.7
Methyl methacrylate	0.71
Ethyl acrylate	0.85
Vinyl acetate	0.85

The assumption that ϕ_m is the same for all particles regardless of size, on which the assumption of constant growth μ in interval I is based, as well, contradicts equation (7.3.12), since the volume fraction of monomer should increase as r increases. This is true, but the majority of the increase in latex radius r occurs in interval II, after the depletion of micelles and so under conditions in which the latex surfaces are no longer saturated with surfactant. Thus the interfacial tension γ increases and so compensates for the increase in r. It thus appears a reasonable approximation to use a constant monomer volume fraction of monomer in intervals I and II.[62]

It should also be noted that upon the disappearance of the monomer droplets, the majority of the monomer should be present in the polymer particles. This implies that $1 - \phi_m$ should correspond fairly well with conversion at the end of interval II.[62] Thus we find that interval II concludes at ~20% conversion for vinyl acetate, but at ~70% for vinyl chloride.[33] The discrepancy for vinyl acetate is probably due to the high solubility of vinyl acetate in water (> 3 times that of vinyl chloride: see Table 7.1), so that even at the end of interval II there is still a significant outside source for monomer.

A number of situations modify the approach embodied in equation (7.3.12). The entropic terms will of course differ for branched or crosslinked polymers made by emulsion polymerization; increasing the fraction of crosslinker will decrease rate per particle by decreasing ϕ_m, which has implications for the particle number N as well.[63,64] It has been assumed thus far that the latex spheres are themselves homogeneous. Heterogeneity may be expected in the polymerization of monomers such as vinyl chloride and vinylidene chloride for which the polymers are insoluble in the monomer but had been assumed to be negligible for polymers such as polystyrene. This assumption has been questioned since the proposal of the core–shell theory,[65] which proposes a core of polymer surrounded by a shell of unreacted monomer. The support for this model was threefold, coming from kinetic, theoretical, and morphological arguments. The original kinetic arguments rest on the interesting observation that a constant rate is maintained well past the end of interval II in experiments with polystyrene; with the decline of monomer concentration in the polymer particles in interval III, a falling rate is expected if the Trommsdorff effect is neglected.[65] The theoretical argument rests on surface depletion arising from the reduction in conformations available to a chain near a surface, ignoring the possible tethering of chains to the surface by the initiator fragment. The most convincing evidence, though, is the morphological evidence from electron microscopy and β-particle emission that in these polymerizations a core–shell structure was actually obtained.[66] Whether the core–shell morphology is an artifact of the particular experiments performed is not clear; it does seem that only the morphological evidence remains undisputed.[33]

7.3.3.2 Average Number of Radicals per Particle

The average number of radicals per particle, $\bar{n}$, is the ratio of the first to zeroth moments of the distribution N_n, the concentration of particles containing n radicals. If we proceed as we have in the past with chain length distributions, we write an evolution equation for each member of the distribution. We can use the following form[29,67,68]:

$$\frac{dN_n}{dt} = \frac{\rho_a}{N}(N_{n-1} - N_n) + k_{de}[(n + 1)N_{n+1} - nN_n]$$

$$+ \frac{k_t/2}{V_{part}}[(n + 2)(n + 1)N_{n+2} - n(n - 1)N_n] \qquad (7.3.13)$$

A particle may gain a radical by adsorption, or the particle may lose a radical by desorption or pairs of radicals by termination; the three terms on the right-hand side of the equation correspond to these three processes. The total rate of radical adsorption ρ_a is divided up among the total N particles. This will not equal the initiation rate ρ_i, however, if radical desorption can occur, because a desorbed radical may adsorb again rather than terminate in the aqueous phase. The rate constant for desorption is k_{de}; in the original Smith–Ewart formulation this rate was inversely proportional to the radius of the particle. Termination occurs much as it would in bulk polymerizations at the same ϕ_m and thus would be characterized by the same k_t.* Hence k_t at low overall conversion in emulsion polymerization is often much lower than the low conversion value in bulk polymerizations because the environment in a polymer particle corresponds to that at higher conversion. The assumption that the rates of radical entry and desorption are the same for all particles presupposes a monodisperse particle size distribution (i.e., all particles have

---

*In most papers this appears without the dividing factor of 2; however, since this term in the equation corresponds to the rate at which two radicals are lost, it is clear that this corresponds to the $2k_t$ convention not used in this book. [This is clear when one looks at case 3 kinetics, e.g., equation (17) of Smith and Ewart.[29]]

the same volume v_{part}), which is certainly not the case; we will speak more of this, and of radical desorption, shortly.

Equation (7.3.13) cannot be solved sequentially; as a consequence, the moment equations are not closed. Desorption and termination provide routes by which the number of radicals in a particle may decrease, just as in reversible step polymerization smaller chains may form from the larger. However, this is an example of a case in which the formalism of the generating function allows solution unavailable by other routes because of closure problems. In general, one obtains a partial differential equation for the generating function $G(s)$, but under quasi–steady state conditions one obtains a second-order ordinary differential equation[67,68]:

$$(1 + s) \frac{\partial G^2(s)}{\partial s^2} + m \frac{\partial G(s)}{\partial s} + \alpha\, G(s) = 0 \qquad (7.3.14)$$

Where $G(s)$ is the generating function for the radical number distribution N_n (defined here, of course, from $n = 0$ to ∞), and

$$\alpha = \frac{\rho_a v_{\text{part}}}{N k_t/2} \qquad (7.3.15)$$

$$m = \frac{k_{\text{de}} v_{\text{part}}}{k_t/2} \qquad (7.3.16)$$

The parameter α represents the importance of adsorption relative to termination, the parameter m the importance of desorption relative to termination.

Equation (7.3.14) can be solved analytically for $G(s)$[67,68]:

$$G(s) = N \frac{2^{(m-1)/2}}{I_{m-1}((8\alpha)^{1/2})} (1 + s)^{(m-1)/2} I_{m-1} \left(2\alpha^{1/2}(1 + s)^{1/2}\right) \qquad (7.3.17)$$

where $I_{m-1}(x)$ are modified or hyperbolic Bessel functions.[69] This gives the distribution:

$$N_n = N \frac{2^{(m-1)/2}}{I_{m-1}((8\alpha)^{1/2})} \frac{\alpha^{n/2}}{n!} I_{m-1+n}(2\alpha^{1/2}) \qquad (7.3.18)$$

This also allows us to calculate the average number of radicals per particle, related to the first derivative of $G(s)$:

$$\overline{n} = \left(\frac{\alpha}{2}\right)^{1/2} \frac{I_m((8\alpha)^{1/2})}{I_{m-1}((8\alpha)^{1/2})} \qquad (7.3.19)$$

Equation (7.3.19) is less than transparent, however, in telling us exactly what values $\overline{n}$ takes on. To be more quantitative, let us look at the three traditional limiting cases.[29]

Case 1 is that in which a high rate of desorption gives a small value of $\overline{n}$:

$$\overline{n} = \frac{\rho_{\text{adsorb}}}{k_{\text{desorb}} N} = \frac{\alpha}{m} \qquad \text{(Case 1)} \qquad (7.3.20)$$

This can be found by either the small α limit (if $m > 0$) of equation (7.3.19) or from the original equation (7.3.13), assuming that $N_1 \ll N_0$, and likewise that $N_n \ll N_1$ for $n \geq 2$.

Case 2 is the best known and most cited situation. It arises if desorption is negligible ($m = 0$) and termination occurs instantaneously upon the entry of a second free radical into a polymer particle already containing a radical (α is small). The appropriate limit of equation (7.3.19) gives:

$$\bar{n} = \frac{1}{2} \quad \text{(Case 2)} \tag{7.3.21}$$

Radicals arrive at a particle at a constant rate ρ_{adsorb}/N. Because termination is instantaneous, the arrival of a radical to a particle with $n = 1$ induces a transition not from $n = 1$ to $n = 2$, but rather to $n = 0$. The arrival of a radical at a particle with $n = 0$ induces a transition to $n = 1$, of course. It is easy to show that $N_0 = N_1$ and so, on the average, a given particle half the time contains a radical, and half the time no radical. Equivalently, at any given time half the particles possess one radical.

Case 3 is that in which termination is not instantaneous, and so the particle may contain several radicals simultaneously. As the particles grow in size and decrease in ϕ_m, the assumption of instantaneous termination becomes progressively less reasonable. In the extreme, the kinetics within the particles may be considered to be equivalent to those in bulk systems. In this case,

$$\bar{n} = \left(\frac{\alpha}{2} \right)^{1/2} \tag{7.3.22}$$

This can be found by assuming each particle to have the average number of radicals, $\bar{n}$, and deriving $\bar{n}$ by a macroscopic balance:

$$\frac{\rho_{adsorb}}{N} = \frac{k_t \bar{n}^2}{v} \tag{7.3.23}$$

Emulsion polymerization kinetics generally is discussed in terms of these three cases: case 1 ($\bar{n} \ll 0.5$), case 2 ($\bar{n} = 0.5$), and case 3 ($\bar{n} \gg 0.5$). Case 2 is most commonly assumed. Case 3 should be applicable for larger particles, and in interval III, when the increasing volume fraction of polymer further impedes termination. Case 1 is applicable to monomers of high water solubility and in which chain transfer is common; vinyl acetate is a good example. Chain transfer to monomer, or to agent, assists in desorption by creating a small molecule that can desorb into the aqueous phase. The importance of desorption has been realized in recent years[37,70]; it is even significant for highly hydrophobic monomers such as styrene.[71] The importance of chain transfer in effecting desorption has one interesting practical effect. In bulk, solution, or even suspension polymerizations, an ideal transfer agent decreases the polymer chain length but has no effect on the rate of polymerization not attributable to the change in chain length. In an emulsion polymerization, however, an ideal transfer agent may decrease the rate by increasing the rate of desorption and thus decreasing $\bar{n}$.

Figure 7.3 plots $\bar{n}$ as a function of α and m. From this plot it would seem that the applicability of case 2 kinetics is suspect, since desorption is required to be very low for the range of α over which $\bar{n} = 0.5$ to be sufficiently large. Figure 7.3 is somewhat misleading because it implies that α and m are independent variables. At high desorption rates, however, ρ_a is not determined by the entry of "new" radicals, ρ_i, but by the entry of desorbed radicals. Thus, α and m are not independent parameters.[33] A proper treatment demands the redefinition of the problem in terms of the independent variable ρ_i, as well as the introduction of a new parameter that indicates the relative importance of termination in the aqueous phase.[72] For negligible termination in the aqueous phase, the conclusions are much the same; but as aqueous phase termination increases, there are progressively longer ranges of α over which, at a given m, $\bar{n}$ is fairly constant and equal to 0.5.

Considering the mechanisms involved in the initiation and termination processes, a quasi–steady state solution may seem suspect. A number of treatments maintaining the time derivative have appeared, including treatments wherein radical desorption is negligible,[73] and more general cases restricted to $\bar{n} < 0.6$.[74] Results such as these generally show that quasi–steady state conditions are maintained in usual emulsion polymerization during interval II,[33] although this will not be the case if termination within the polymer particle is sufficiently slow. Approaches to quasi–steady state values may be slow, though, and so a non–steady state treatment is helpful in certain cases, such as those in which the system is subjected to some "shock" (e.g., an increase in ρ_i) or for seed polymerizations for which interval I is avoided and all particles initially have no radicals.

Figure 7.3.
Average number of radicals per particle as a function of α and m. (From reference 33, used by permission of the American Chemical Society.)

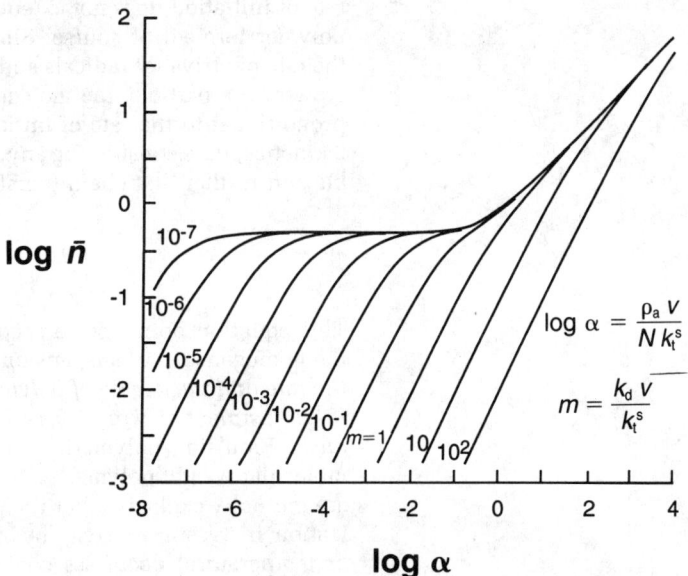

$$\log \alpha = \frac{\rho_a\, V}{N\, k_t^s}$$

$$m = \frac{k_d\, V}{k_t^s}$$

7.3.3.3 Rate of Polymerization

It is now clearer on what the rate of polymerization, given in equation (7.3.11), depends for case 2 kinetics, the rate of polymerization is:

$$R_p = k_p M \frac{N}{2} \qquad (7.3.24)$$

For homogeneous or suspension polymerizations, the initiator concentration at the time of reaction strongly affects the rate, but because the reaction is compartmentalized in polymer particles, *the rate of emulsion polymerization under case 2 kinetics is independent of the initiator concentration.* Instead, it is proportional to the number of particles, which depends on the initiator concentration of interval I. The initiator concentration serves to determine the number of particles, but thereafter it determines only the frequency with which particles become activated or deactivated, not the overall rate. Thus, while doubling the initiator concentration in interval I should increase the rate by a factor of $2^{2/5}$, doubling it in semibatch fashion during interval II should have no effect on the rate.

It should be stressed that this remarkable independence from the rate of initiation holds only for case 2 kinetics, and case 2 kinetics of course does not always hold. For highly water-soluble monomers such as vinyl acetate, case 1 kinetics should hold, and a strong dependence on initiator concentration during interval II should be seen. As the particles grow in size and case III kinetics is observed, the rate should follow the dependence on initiator concentration that is valid for homogeneous polymerizations (i.e., a square root dependence).

7.3.3.4 Molecular Weight

The independence of the rate of emulsion polymerization on the rate of initiation does not extend to the molecular weight of the polymer formed, of course. Since the rate of initiation affects the rate of arrival of radicals and thus the average time for chain growth in a particle, the molecular weight should be inversely proportional to the rate of initiation. If we consider again case 2 kinetics, thus considering any radical desorption to be negligible and further that chain transfer does not occur, we have:

$$DP_n \approx \frac{k_p M N}{\rho_a} \qquad (7.3.25)$$

This equation points out a second remarkable difference from the homogeneous or suspension polymerizations. *Because both the rate and the degree of polymerization are proportional to N, the adjustment of N provides a route to increase both simultaneously.* Emulsion polymerization allows for production of high molecular weight polymers at high rates, conditions that usually are mutually exclusive. For the other routes of radical polymerization, increasing rate (e.g., by increasing initiator concentration or temperature) decreases chain length, the only obvious exceptions being increasing temperature in photopolymerization (which is limited practically to surface applications) and increasing monomer concentration in solution polymerization.

The production of polymer with higher molecular weight than is found in bulk systems implies that chain transfer to monomer may be important. In that case, equation (7.3.25) corresponds not to DP_n but to the kinetic chain length ν. Although DP_n cannot be increased indefinitely by increasing N, it eventually reaches a limit of $1/C_m$. Similarly, equation (7.3.25) depends on the applicability of case 2 kinetics. Case 3 kinetics of course yields the result of homogeneous free-radical polymerizations, while case 1 kinetics yields a very complicated situation, since termination can occur either in the polymer particles or in the aqueous phase.

Rather than rigorously derive the molecular weight distribution of the products formed in emulsion polymerization, we shall instead make fairly qualitative arguments about the effects of this process on statistical and drift dispersion. More complete treatments have been presented in the literature.[32,73,75-79] We will restrict ourselves to linear polymerizations; both network-forming and branching polymerizations in emulsion have, however, been examined recently.[63,64,80-82]

We take as our starting point case 2 kinetics and assume that transfer is negligible. In the immediate termination of two radicals that occurs in this case, the two radicals are not equivalent; rather, one was a resident in the particle for some time and the other a new arrival. The resident chains will be relatively long (and will have a geometric distribution), whereas the newly arrived chains will be short. The resulting dead chain distribution will depend on the termination mechanism, as usual. If combination is dominant, each of the resident chains upon dying is increased by some small number of units, and the geometric distribution is roughly preserved ($Q \approx 2$). If disproportionation is dominant, there result two populations of chains of equal molar concentration, the first arising from the resident chains and having a geometric distribution with some parameter q, and the second arising from the newly arrived chains. By treating the latter as polymer, one obtains a bimodal distribution of long and short chains, which will have a polydispersity of approximately 4 (cf. Example 3.3).

As expected, the inequivalence between the two terminating chains increases statistical dispersion. The narrowing of the distribution by combination between similarly distributed chains does not occur, and a possible bimodality can arise from disproportionation. Termination is dominated by the action of short chains, even when termination rate constants are assumed to be independent of chain length. This effect of segregation is thus superimposed on the dominance of short chains in termination due to diffusional control (see Section 3.6.1). For case 3 kinetics (larger $\bar{n}$), the polydispersity will change to give bulk results, in which the inequivalence arises only from diffusion, not from the segregated nature of emulsion polymerization. Thus a shift from case 2 to case 3 kinetics may result in a decrease in statistical dispersion if the inequivalence of the terminating chains diminishes. If polymer chains are mainly ended by transfer reactions, as would be so for case 1 kinetics, a geometric distribution results.

The preceding arguments apply only for what corresponds to the instantaneous dead chain distribution in emulsion polymerization. The effect of the emulsion process on drift dispersion is even more remarkable than that on statistical dispersion. We may consider the polymer particle to be a small semibatch reactor, one that obtains a supply of radicals and monomers from its surroundings (ignoring here radical desorption). Because we have argued that ϕ_m is approximately constant, the polymer particle is a semibatch reactor that at least through intervals I and II maintains fairly constant reactor conditions. Thus in interval II, when the radical arrival rate should be fairly constant, *drift dispersion is eliminated.* Thus, if interval II covers a sufficient amount of the conversion, emulsion polymerization can result in a more narrowly distributed polymer than a corresponding bulk or suspension polymerization.

7.3.3.5 Particle Size Distribution

Many discussions of the particle size distribution have been given over the years.[51,76,77] As in the case of molecular weight distribution, we here confine ourselves to a few qualitative arguments. That the particle size distribution is not monodisperse is implicit in the derivation of N in Section 7.3.2. Particles born at different times, but all growing at the same volumetric rate μ, gives a broad size distribution. This is a typical approach to the derivation of the particle size distribution: to account for the "drift dispersion" due to different birth times. We ignore the statistical dispersion, granted by the failure of all particles born at the same time to grow at a rate equal to their average rate.* The final particle size distribution should correspond fairly closely to that at the end of interval II, disregarding volume contraction due to polymerization, intake of monomer remaining in the aqueous phase during interval III, and aggregation. If during interval II all chains grow at the same constant rate μ, the distribution narrows. Thus the shorter interval I is relative to interval II, the narrower the size distribution.

Several things can happen to counteract the self-narrowing of the particle size distribution. Case 3 kinetics will tend to broaden the particle size distribution, because the larger particles will tend to have a greater number of radicals, leading to an increased growth rate for these particles (as $\mu\bar{n}$). Bimodal (or multimodal) distributions can result if nucleation can begin anew after the end of interval I; this may happen if new micelles are introduced in semibatchwise fashion, or if surfactant desorbs during interval III as a result of the volume contraction due to polymerization.

It should be noted that any breadth of the particle size distribution allows the rate of radical entry to different particles to be different. This will in turn affect the molecular weight

---

This is easily justified by comparison with ideal anionic polymerization. The drift dispersion due to different birth times can be likened to the noninstantaneous initiation that led to the Gold distribution (see Section 3.3.4.1); it is thus clear that in cases such as this, what one ascribes to "drift" and to "statistical" dispersion is somewhat arbitrary.

distribution between the different particles, and so the final product will have a broader distribution, essentially because of a "drift" mechanism.

7.3.4 Interval III

The advent of interval III is marked by the disappearance of the separate monomer phase, the conversion at which this occurs corresponding fairly well with $1 - \phi_m$. The length of interval III thus varies quite a bit (see Table 7.2) and may comprise the majority of the polymerization, especially for monomers that greatly swell their polymer, such as methyl methacrylate, ethyl acrylate, and vinyl acetate. Despite this, and even though a rough qualitative understanding of interval III is straightforward,[33] interval III is often ignored.

The difficulties understanding interval III come from the difficulty in understanding high conversion free-radical kinetics in general. The reaction continues, with the polymeric fraction increasing and the monomer concentration decreasing. For monomers with high ϕ_m this increase in polymer fraction can be great, and so all the dramatic changes expected at corresponding conversions in a bulk polymerization occur during this interval. Thus, the same modeling concerns arise (with the added physics contributed the segregated nature of emulsion polymerization),[83] and advances in the understanding of bulk systems at high conversions help to elucidate interval III (and vice versa). Because physical conditions within the particles change with time during interval III, the mechanism of drift dispersion familiar from homogeneous polymerization is operative within the particles.

Despite these correspondences, the polymer particles in interval III cannot be treated as separate batch reactors because of radical desorption, the presence of monomer in the aqueous phase, and so forth, which lend a semibatch character to the particles. Thus is eliminated one of the physical correspondences; that the production of free radicals occurs in the aqueous phase removes the great decrease in initiator efficiency noted in bulk polymerizations.[16] This also means that the rate of polymerization is still determined by $\bar{n}$ [see equation (7.3.11)]. For example, an increasing $\bar{n}$ due to plummeting k_t (case 3 kinetics) may be sufficient to offset the decrease in rate due to monomer depletion; a maximum in rate during interval III may then be noted. Therefore, the overall rate profile will generally not be the same as that in a bulk polymerization.

7.3.5 Emulsion Copolymerization

Copolymerizations account for a good share of commercial emulsion polymerizations. The question of copolymer composition is quite difficult, however, because it entails not only the issues common to bulk free-radical copolymerization (e.g., the mode of copolymerization and the reactivity ratios) but all the factors that determine the concentrations of the different monomers at the loci of polymerization as well. The extension of equation (7.3.13) to a copolymerization is difficult at the least,[78] and that is a difficulty which is merely mathematical. The determination of the parameters is a separate, difficult issue in itself. We thus here describe only some of the differences that may be associated with a copolymerization in an emulsion.

Although there is no reason for mode of reaction or rate constants to vary simply because the reaction is in an emulsion, the apparent reactivity ratios, for example, should vary quite a bit because of the thermodynamic factors that determine the concentration of each monomer in the swollen polymer particles. Drift in copolymer composition will be difficult to counter. This phenomenon is related to the problem in suspension polymerization of nonzero solubility of monomer in the aqueous phase (see Section 7.2.1) but is much more severe here and need not be linked with differences in water solubility. However, greatly differing water solubilities do lead to a fraction of homopolymer soluble in the aqueous phase and determine the composition of newly nucleated particles or of the oligomers that enter the micelles or polymer particles. Moreover, it should be kept in mind that the water solubility of a monomer in a copolymerization system may differ from that in a homopolymerization because the comonomer may act as an extraction agent for the monomer, for example.

7.3.6 Emulsion Polymerization in Semibatch and Continuous Reactors

While the physical mechanisms and kinetics of emulsion polymerization are best described for a batch polymerization process, such is not necessarily the most convenient commercial process. The batch process may be used for specialty latices, but the semibatch process is the most popular process because of the flexibility and control it affords. Continuous processes are currently used for the production of many rubber products, such as styrene–butadiene rubber, nitrile rubber, and neoprene rubber, and for poly(vinyl chloride).[84] We outline the broad features of emulsion polymerizations in these reactor configurations; the reader should also consult the more extensive reviews available.[84–89] The importance of the process in determining the characteristics of the product should be noted.

7.3.6.1 Semibatch Processes

As discussed in Chapter 6, there is no single implementation of a semibatch process. Rather, the most common practice in emulsion polymerization is to feed monomer (usually solubilized with surfactant) to the reactor over time, and perhaps the initiator as well in a separate stream. The most visible result of this policy is direct control over the rate of polymerization (and thus the rate of heat evolution) by the rate of addition. Addition, if not too rapid, can govern polymerization, giving a relatively constant conversion in the reactor.[90,91] If this conversion is high (90–100%), as is usually the case, the reactor is said to be operated in "starved-feed" mode. This means that no separate monomer phase exists in the reactor, and thus the reactor is essentially maintained in interval III throughout the reaction. Composition drift can be virtually eliminated if the constant conversion is maintained over a large fraction of the semibatch reaction. The same would be true of molecular weights, although the increase in particle size with reaction should move the kinetics toward case 3 behavior, and so drift may occur as a result of an increase in $\bar{n}$ (as may happen in interval II of a batch polymerization). Models of such semibatch operation have been proposed which elucidate many of these issues.[92-94]

The popularity of this method is easily understood; it grants safe operation and good control of both reaction exotherm and characteristics of the product. This kind of semibatch operation is also favored because it provides great flexibility. Structured latices can be made by staging feeds of different monomers at different times. Bimodal (or multimodal) particle size distributions can be intentionally made by introducing a large amount of surfactant during the feeding. (Unintentional bimodality is also possible, and care must be taken to avoid the use of inappropriately large amounts of surfactant.)

7.3.6.2 Continuous Emulsion Processes

Most continuous processes for emulsion polymerization involve a number of continuous stirred tank reactors in series, although tubular reactors may also be incorporated into the train. The value of a series of CSTRs lies in the narrowing of the residence time distribution from the broad (exponential) distribution that would result from a single CSTR, the effects of which we briefly review here. The use of tubular reactors alone for continuous emulsion polymerization has also been considered.[95]

Consider a single CSTR into which are fed the (unreacted) ingredients of an emulsion polymerization. Steady-state operation eliminates the idea of separate intervals; regardless of the overall conversion, particle nucleation must be occurring because particles are exiting the reactor. The concentration of particles differs from that in a batch polymerization, as we can see from even the simplest of arguments. The steady-state analogue of equation (7.3.7) is:

$$\frac{N}{\theta} = \rho_i \left(1 - \frac{A_p}{a_s S}\right) \tag{7.3.26}$$

The surface area of surfactant on polymer particles (assumed to be saturated) is given as an average over the residence time distribution:

$$\frac{A_p}{N} = (36\pi\mu^2)^{1/3} \frac{1}{\theta} \int_0^\infty t^{2/3} e^{-t/\theta} \, dt = \Gamma(5/3) \, (36\pi\mu^2)^{1/3} \, \theta^{2/3} \tag{7.3.27}$$

given that $A_p \leq a_s S$ and that all particles grow at the same average rate (case 2 kinetics). For long residence times with micelles still present, the number of particles is thus given by:

$$N = \frac{a_s S}{\Gamma(5/3) \, (36\pi\mu^2)^{1/3} \, \theta^{2/3}} \tag{7.3.28}$$

This equation for N shows an independence from the initiation rate (rather than $\rho_i^{2/5}$ as in batch polymerization), a stronger dependence on the amount of surfactant (S^1 rather than $S^{3/5}$), and a dependence on the average residence time, θ.[85] Deviations from these exponents are seen just as from those for batch polymerization; styrene seems to be the only monomer that conforms to these scalings,[86] monomers more soluble in water exhibiting quite different exponents.

More important than the altered dependences is the nature of the particle size distribution: it is much broader than that in a batch reactor, just as an anionic polymerization in an HCSTR gives a broader chain length distribution than is obtained in a batch reactor. Commercial operation with a series of CSTRs (2–15) counters this trend, since the overall residence time distribution narrows as the number of reactors increases. Nucleation in a CSTR also has the disadvantage of giving rise to oscillations in conversion,[96,97] which can be eliminated by performing nucleation in a tubular reactor placed before the train of CSTRs.[98] This process refinement should also revert the dependences of N to those of a batch reactor, if the tubular reactor behaves as a PFR.

On the basis of both segregation and the broad particle size distribution, the molecular weight distribution from a single CSTR is expected to be broad. The elimination of drift dispersion during interval II in a batch reactor was conditional upon a narrow size distribution (which was assured if interval I was short) and case 2 kinetics. For a single CSTR, though, the particle size distribution is broad, and so the average arrival rates will not be consistent for all particles. Thus, there will be an effective "drift" owing to the mixing together of particles composed of polymer of different molecular weights. When a copolymerization is performed in a single CSTR, the copolymer composition will not suffer from drift, but a series of CSTRs will produce polymers that differ in composition. To counteract this effect, one may introduce intermediate feeds of the more rapidly consumed monomer(s).

7.3.7 Variations on Emulsion Polymerization

One variation of emulsion polymerization already mentioned is accomplished through semibatch means, that is, preparation of heterogeneous or structured particles that exhibit improved mechanical properties (e.g., modulus, impact strength, gas permeability) over the corresponding copolymers. Often the goal is to prepare *core–shell* particles—a shell of one polymer (often rubbery) surrounding a core of another (often glassy). The simplest process to envision is a two-stage emulsion polymerization process wherein the latex "seed" from the first polymerization offers the only locus for polymerization of the second monomer; an example would be ABS resins made by the emulsion polymerization route.[99] Core–shell morphology, however, is not always obtained. It depends on the miscibility of the second-stage monomer with the seed polymer, the miscibility of the two polymers, and the hydrophilicity of the two monomers,[100] as well as other variables such as molecular weight and viscosity at the locus of polymerization.[101] A number of morphologies are attainable:

core–shell, inverted core–shell (if the first polymer is more hydrophilic), particles with ramified, raspberrylike surfaces, and so on.[102] Grafting is expected to assist in stabilizing the interfaces between the polymer layers within the latex particles.

A second, and obvious, variation is to simply invert the phases, as in inverse suspension polymerization, so that the organic phase is continuous and the aqueous phase, which contains the monomer, is dispersed. The initiator is soluble in oil, and again a surfactant is present which forms micelles in the oil phase. Polymerization and copolymerization of acrylamide by *inverse emulsion polymerization* is used for the production of flocculants.[35,103] In practice, however, many of these production routes are more like inverse suspension polymerizations, with all polymerization occurring in the original droplets of water and monomer.[5] Such a radical departure from emulsion polymerization behavior is not merely the result of putting the initiator in the same phase as the monomer; for normal emulsion polymerizations, the use of oil-soluble initiators will not give kinetics characteristic of suspension polymerization, because of radical desorption, for example.[104]

Polymerization in the monomer droplets does not necessarily mean that the situation is reverting to a suspension polymerization. Negligible polymerization in the monomer droplets was due simply to their large size and accompanying small surface-to-volume ratio in comparison to the micelles. If the monomer droplets can be dispersed more finely, they may more effectively compete with the micelles for free radicals. *Mini–emulsion polymerization* was first accomplished by an anionic surfactant and, as cosurfactant, a long-chain fatty alcohol,[105,106] a combination known since at least 1940 to result in more stable emulsions. Mixtures with long alkanes such as hexadecane have also been found to be effective.[107,108] These recipes are able to reduce the size of the monomer phase by an order of magnitude (to 0.2 μm). The standard anionic surfactant alone, without the added long-chain molecule, can be used to prepare more finely dispersed monomer if preemulsification by extreme agitation or ultrasound is used.[109] In all these cases where there remains a competition between micelles and monomer droplets, a bimodal particle size distribution is observed, the larger particles deriving from the monomer droplets. In the extreme limit (no remaining polymerization in the micelles) the entire polymerization process would seem to lie in interval III, although the monomer concentration is initially the bulk value rather than the value determined by equilibrium swelling. The situation still differs from suspension polymerization, because radical entry and desorption occur.

One of the disadvantages of the emulsion polymerization process compared to suspension polymerization is the greater contamination of the water phase.[9] The main contaminant is the surfactant, which may migrate after the emulsion polymer (or latex) is cast as a film and dried, leading to "blooming" or "blushing." Such problems can be made less severe if the surfactant is bound to the latex, rather than simply adsorbed onto the particle surface. Thus there is an interest in using surface active monomers,[110] polymerizable surfactants,[111] or surface active initiators.[112] Another solution is to eliminate the surfactant from the process, relying instead on ionic groups from initiator fragments.[113] The sulfate ions deriving from the potassium persulfate initiator then serve to stabilize the particles, for example. These so-called *surfactant* or *emulsifier-free emulsion polymerizations* differ in many respects from the usual emulsion polymerizations. The smaller concentration of initiator compared to surfactant results in lower particle concentrations (by one to two orders of magnitude) and a corresponding increase in the size of those particles. Coagulation is also expected to be a significant mechanism for final particle formation, given the sparse surface coverage. Because of the size of the latex particles, case 3 kinetics is generally obeyed.[114]

7.3.8 Industrial Applications

Since the development of emulsion polymerization was largely influenced by the attempt to imitate the state of natural rubber, it is not surprising that the synthetic rubber industry has traditionally used and continues to use the emulsion polymerization process for the majority of its polymers.[20] Emulsion polymerization here represents a route to the desired polymer; the final product may not be in latex form. First and foremost is polybutadiene, or butadiene rubber (BR), with low crystallinity resulting from the mixed *cis* and *trans* configurations from the 1,4-addition. In addition to its direct uses as synthetic rubber, this material is the latex ingredient for high impact polystyrene and ABS resins. Other synthetic rubber materials include copolymers of butadiene and styrene for SBR, poly(butadiene-*co*-acrylonitrile) or nitrile rubber (NR), neoprene rubber (made from chloroprene), ABR, which is a copolymer of ethyl acrylate and chloroethyl vinyl ether, and "Heveaplus," which is natural rubber with styrene or methyl methacrylate grafted on. The exceptions to the rule of using emulsion polymerization for the production of rubber materials are polyisoprene and poly(ethylene-*co*-propylene) (EPM rubber), both made by the coordination catalysis route, and polyisobutylene and its copolymers made by cationic polymerization. In these exceptional cases, for the most part, the desired monomer necessitates the use of a different route.

The latex form of the resulting polymer implies use "as is" on coatings applications such as paints and adhesives.[35] Poly(butadiene-*co*-styrene), at higher levels of styrene than are present in SBR, is used for paints as well as paper coatings and nonwoven fabrics. Emulsion polymerization is used for poly(vinyl acetate) for paints or adhesives. Other polymers used in paints and adhesives are polyethylene and its copolymers, and in paints and fabrics poly(ethyl acrylate) (PEA) and the copolymer of that monomer and methyl methacrylate.

Other common coatings are made by the emulsion polymerization route, including those made from precipitant monomers. These include poly(vinyl chloride) used for plastisols and coatings, and its relative poly(vinylidene chloride). Plastisols are mixtures of PVC and a high-boiling plasticizer (such as di-2-ethylhexylphthalate) in which the polymer is insoluble at room temperatures but soluble at higher temperatures (e.g., 175°C). When heated to this temperature the resin dissolves in the plasticizer, and upon cooling the solution is virtually stable. Applications include overshoes, gloves, and PVC foams.[20] Copolymers of vinyl chloride and vinylidene chloride form the common Saran material. Other copolymers of vinyl chloride, such as that with vinyl acetate, are also produced by emulsion polymerization. The fluoroethylenes, such as tetrafluoroethylene, from which Teflon is made, and chlorotrifluoroethylene, are polymerized by this route as well.

7.4 Heterogeneous Coordination (Ziegler–Natta) Polymerization

Control or regularization of structural characteristics such as tacticity, cis–trans isomerism, and short-chain branching is as important as control of the molecular weight distribution. From Section 1.3.3, we know that regularity in configuration greatly increases the degree of crystallinity, which is often desirable. On the other hand, regularization of molecular weight—that is, narrowing of the molecular weight distribution—is more often undesirable (e.g., processing may be made more difficult because shear-thinning would begin at a higher shear rate $\dot{\gamma}$). Thus, regularity in configuration is often more important in practice than that of molecular weight. How can configurational control be attained?

The issue of configurational isomerism can arise in polymers made by both stepwise and chainwise mechanisms, but more commonly in the latter. When discussing the configurational characteristics of chains in Section 1.3.3 we limited ourselves to the chainwise polymerization of unsaturated monomers; we shall do the same here. As noted in Section 3.4.2, during a free-radical polymerization the stereochemical configuration of a monomer is determined upon the addition of the subsequent monomer, and hence is largely determined by the free-energetics of the transition state. The entropic differences involved are small, however, and so the configuration will largely be determined by the energetic differences between a transition state that will lead to a meso or a racemic penultimate dyad. In meso placement, like substituents will be on the same side of the chain, so that if one substituent is bulky and the other not (as in the case of α-singly substituted ethylene) this placement will not be sterically favored and the chain will be predominantly syndiotactic, as in the following scheme:

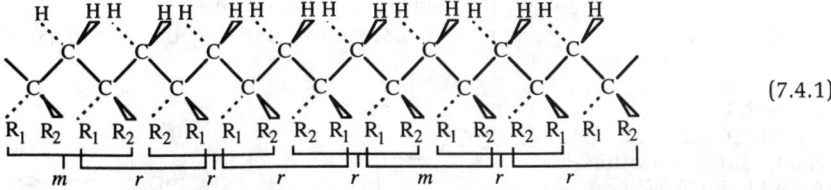

$$(7.4.1)$$

The ratio of the two dyads is determined by the particular monomer involved; free-radical polymerization of vinyl chloride results in a fairly atactic polymer with a random placement of the two types of dyad in nearly equal proportion, whereas methyl methacrylate forms a fairly syndiotactic polymer at the same temperature. The degree of syndiotacticity increases with decreasing temperature, and of course depends on the mode of polymerization, whether free radical, anionic, or cationic, and for the last two a dependence on the identity of a counterion is expected. Syndiotacticity is thus favored by a low-temperature polymerization, but the temperature required may be too low to be useful (as would be the case for vinyl chloride). In general, control with temperature alone does not seem like the best approach. Moreover, for the production of isotactic polymer, changing temperature does not at all help. What is needed is a mechanism by which monomers are directed into the desired configuration.

Control by the chain end will in general lead to a favoring of syndiotacticity as well as allowing statistical mixture of the two dyads, and so a different and more certain route is needed. If the chain end is not controlling, then something else in the near environment must be. One possibility is the insertion mechanism for anionic polymerization shown in Figure 7.4. If the substituent R is considered to be a point, the monomer is a planar molecule, the plane of which is perpendicular to the plane of the four-centered transition state. At first glance the situation appears to be symmetric, if we disregard the chain end, which will tend to encourage syndiotactic placement. The apparent symmetry relies on assuming that the counter- or gegenion G is achiral or has no specific interactions with the substituent on the now-penultimate monomer. If G is chiral, and the coordination between gegenion, chain, and incoming monomer is strong, then R will always tend to be on the same side of the plane of the polymer (in the case above, in front of the page). Thus, by this simple argument coordination of this sort will yield an isotactic product. (One might possibly worry about G general being a racemic mixture, but this is of no concern unless the gegenion is swapped between different chains, which would imply poor coordination anyhow.)

Coordination must clearly be strong, and for nonpolar monomers such as α-olefins, the catalyst environment must provide that coordination because the monomer will not tend to coordinate strongly. The history of how the present coordination catalysts were discovered is hardly as direct as this line of reasoning.[23,26,115-117] In the early 1950s at the Max Planck Institute for Coal Research in Müllheim, Germany, Ziegler and Gellert were working on the reaction of alkyl lithiums with ethylene.[118] The products of this reaction generally were only oligomeric α-olefins (formed from polymerization of ethylene terminated by a displacement reaction) and the other product of displacement, LiH,

Figure 7.4.
Insertion mechanism for anionic polymerization. (After Odian,[23] used by permission of John Wiley & Sons.)

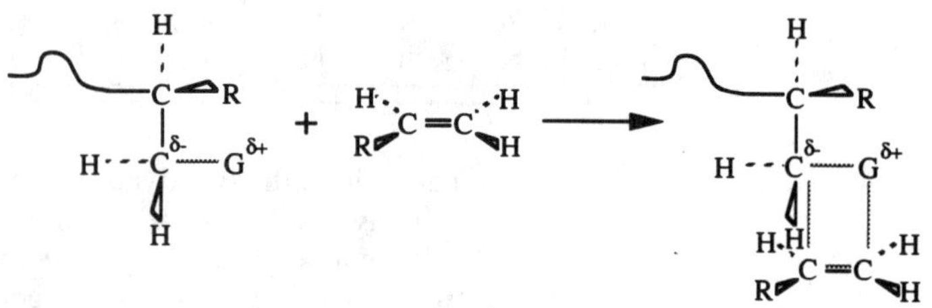

which because of its poor solubility was virtually useless for subsequent reaction. LiAlH$_4$, transformed into LiAl(C$_2$H$_5$)$_4$, was then tried because of its greater solubility, and this compound was found to polymerize ethylene more proficiently, but predominantly at the Al—C bond rather than the Li—C bond as had been expected.[119] Thus the focus was shifted from alkyl-lithiums to trialkylaluminums. Triethylaluminum and tripropylaluminum were found to polymerize ethylene fairly well, but on one occasion the latter yielded only propylene and 1-butene.[120] An impurity, a nickel salt that had been reduced by the trialkylaluminum to colloidal nickel, was implicated, and so Breil began a study of the periodic table to see which metal salts would preferentially give 1-butene. At this stage, thus, the emphasis was on finding what would promote the displacement reaction, but ironically this search led to the opposite discovery of combinations that polymerized ethylene to much higher molecular weights.[121] The first combination discovered was with zirconium acetylacetonate as the metal salt (this had also been tried by Holzkamp), but transition metal salts in general, especially TiCl$_4$, were the most active.[122]

The significant result was that high molecular-weight polyethylene could be formed at room temperature and at nearly atmospheric pressures. Moreover, the absence of methyl groups, as indicated by IR spectroscopy, served to demonstrate that the new polymer largely lacked the short-chain branches formed in the high-pressure process.[122] The possibilities of stereochemical control were not present here, since polyethylene does not contain an asymmetric carbon. The realization of the possibilities of configurational control was largely due to Giulio Natta and his co-workers at the Institute of Industrial Chemistry at the Polytechnic of Milan, supported by the Montecatini company. In 1954 Natta's group attempted synthesis of polypropylene with the Ziegler catalyst [Al(C$_2$H$_5$)$_3$ + TiCl$_4$] and found that a significant, though low, fraction of the polymer was crystalline (20–40%). The original process formed the catalyst with a particular crystal structure, β-TiCl$_3$, but various methods of preforming the catalyst before polymerization gave the α, δ, and γ forms of TiCl$_3$. These were found to dramatically increase the degree of crystallinity (80–95%). By the end of the year, Natta was able to report that isotactic polypropylene, poly(1-butene), and polystyrene had been synthesized.[123] Only half a year later, syndiotactic cis- or trans-1,2-polybutadiene and -1,4-polybutadiene were reported.[124] Later on the homogeneous vanadium-based catalysts were found to produce syndiotactic polypropylene. The details of the catalyst had also been revealed to Goodrich in the United States, and "synthetic natural rubber," 1,4-cis-polyisoprene, was developed.[125]

This brief history is worth recounting in itself,* and the discovery earned Ziegler and Natta a shared Nobel Prize in chemistry in 1963. For our purposes, the story also points out the

---

*More detailed accounts of the history are available,[26,116,117] as are the personal accounts by Ziegler et al.[122] and the Nobel Prize addresses by both Ziegler and Natta.[126] A collection of the articles of the Natta group exists.[127]

salient features of the catalysis. The nature of the nickel impurity points to the kind of catalyst that is active: a reduced transition metal salt. In general, the Ziegler–Natta catalyst systems[†] comprise the following:

1. An organometallic compound or hydride of a group I–III metal, most commonly a metal alkyl [e.g., $Al(C_2H_5)_3$, $Al(C_2H_4)_2Cl$, $Zn(C_2H_5)$].

2. A halide, hydroxide, or alkoxide of a group IV–VIII transition metal, most commonly a transition metal salt (e.g., $TiCl_4$, $TiCl_3$, VCl_3).

Nickel is a group VIII transition element, and so the colloidal nickel falls under this definition. Thus the second point is clear: not all combinations are effective for any given monomer; some may completely eliminate polymerization. Within the brief time frame of the discovery the major uses of Ziegler–Natta catalysts were established: producing more nearly linear polyethylene, polymerizing propylene (especially with stereochemical control to yield highly isotactic polypropylene), and allowing control of *cis–trans* isomerism in 1,4-polymerization of dienes. That these catalysts offered control of the structure on the backbone level, which has such a large effect on crystallinity, was thus observed early on.

It should not be thought that only Ziegler–Natta catalysts allow highly stereospecific polymerizations to proceed. In what is often referred to as the first stereospecific polymerization, and prior to the early work of Ziegler and Natta, isotactic poly(isobutyl vinyl ether) was produced by cationic polymerization initiated by BF_3 etherate in propane at $-80°C$.[128,129] Similarly, isotactic poly(methyl methacrylate) can be produced by anionic polymerization in toluene at $-78°C$ initiated by *n*-butyllithium.[130] These monomers are more polar than the α-olefins, and the greater the polarity of the monomer, the more it can participate in the coordination process, with the result that both the monomer and the gegenion play roles in the coordination process.[23] Despite the polarity of these two monomers, it is still required that the counterion be tightly bound. For example, if one repeats the same polymerization of methyl methacrylate but replaces toluene for the more polar solvent tetrahydrofuran, one obtains a syndiotactic polymer.[130] The production of isotactic polymers requires some slight degree of heterogeneity at least, in the form of a coordinated, structured environment around the growing chain end. Furthermore, for the nonpolar monomers, such as α-olefins, with which we are concerned in this section, even more coordinated conditions are needed. Thus in general we may often expect a heterogeneous catalyst to be necessary (though not all catalyst systems covered in this section will be heterogeneous, in particular those for producing linear polyethylene).

[†] *We refer to these as "catalysts" rather than "initiators" partially because of standard usage and partly because it is often more accurate: a catalyst need not be incorporated into the polymer product, it is often lodged on a support, and so on.*

It should be noted in passing that in some cases either of the components may initiate polymerization on its own.[23] For example, alkyllithium compounds may initiate anionic polymerization, while BF_3 may initiate cationic polymerization. The transition metal compound may also initiate polymerization, as $TiCl_4$ may cationically. In addition, during the reduction process by the organometallic compound, free radicals are formed which may themselves initiate polymerization. Because of this, in many cases what is claimed to be a Ziegler–Natta polymerization may actually be an uncoordinated polymerization by free radicals or ions, lacking thus in stereospecificity. Therefore care is necessary to ensure that polymerization is occurring by a coordinated route. Likewise, coordination catalysts can be used for different kinds of polymerization, most notably the polymerization of cycloalkenes (such as cyclopentene or norbornene) to double-bond-containing polymers by methathesis polymerization.[21,131]

Section 7.4.1 describes some of the features of the most common catalyst, that based on $TiCl_3$ crystals.

7.4.1 Ziegler–Natta and Related Catalysts

In the common system $TiCl_4$ + AlR_3, the oxidation state of the titanium is reduced from +4 to +3, and a crystal of $TiCl_3$ is obtained. The interaction between the two components of a Ziegler–Natta system is certainly complex, but it appears that the organometallic agent effects this reduction and alkylates the transition metal compound as well, in a way that serves to increase the stereospecificity of the catalytic site. The organometallic compound may form a complex with the active site and stabilize it or make it more stereospecific in some way. The sensitivity of the catalyst is determined by a number of factors, foremost among these being the structure of the metal alkyl and the transition metal salt (i.e., the identity of the two metals; the crystal structure and valence of the latter). The prevailing opinion—that the reaction occurs at the transition metal–carbon bond—is supported not by some crucial experiment but rather by an accumulation of circumstantial evidence. It should be made clear that the mechanisms suggested in this section are not proven but do represent those most widely believed.

There are several forms of the $TiCl_3$ crystal, the α, β, γ, and δ lattices. The β form was the form used by Ziegler, and although catalytic it does not yield a stereoregular product as Natta found. We will look at the crystal lattice of the α form. The α lattice is a regular array of chlorine and titanium layers with every third titanium absent (see Figure 7.5).[23,132] The active sites are on the edges of the catalyst. If the titanium site has not been alkylated, two configurations are possible, as shown in Figure 7.6a, where we are looking down the face of the lattice. The square is the symbol for the empty site at which polymerization can occur. The remaining bonds of the titanium atoms away from the catalytic site are not shown. The two configurations cannot be superimposed on each other, and thus the active site is chiral, as we require. If the site is alkylated, then the chirality of the active site is even more obvious, as shown in Figure 7.6b.

Figure 7.5.
Crystal structure of α-TiCl₃:
●, Ti atoms represented, ○,
Cl atoms; □, octahedral
vacancies. [From reference
23 (after reference 132),
used by permission of John
Wiley & Sons.]

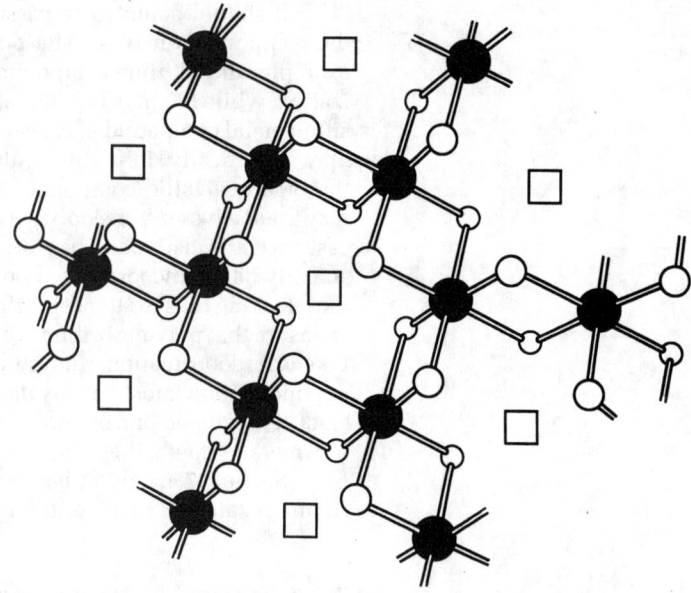

Polymerization occurs much as suggested in the introductory section, by insertion through a four-membered transition state, as shown in Figure 7.7 for propylene.[23,133,134] Note that insertion is proceeding with the opposite placement, as we have argued before for free-radical or ionic polymerizations; head-to-tail placement still results, but the apparent attack is on the substituted end of the double bond.

Figure 7.6.
Active sites of α-TiCl₃: (a)
titanium site not alkylated
and (b) titanium site
alkylated. (After Odian,[23]
used by permission of John
Wiley & Sons.)

(a)

(b)

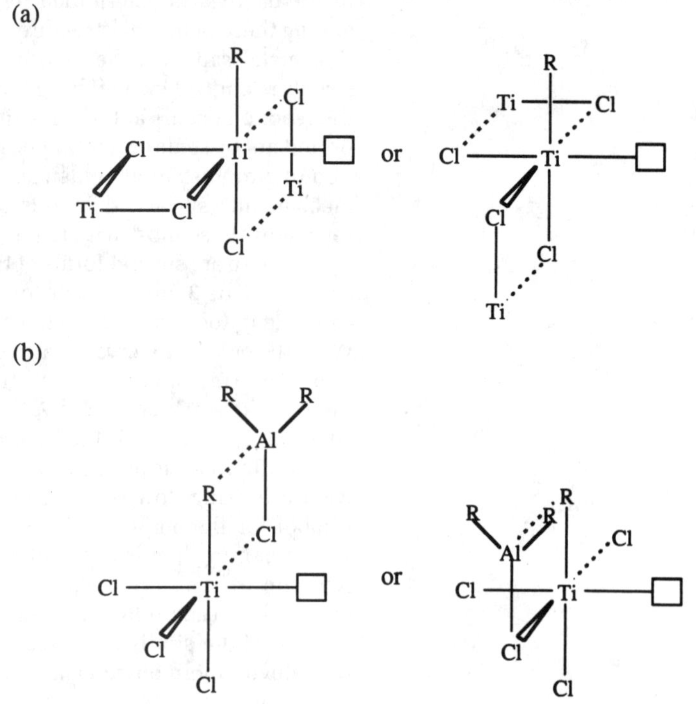

Figure 7.7.
Monometallic mechanism
for Ziegler–Natta
polymerization of
propylene yielding an
isotactic product; bond
lengths not shown to scale:
(a) adsorption, (b) formation
of four-membered
transition state, (c)
propagation, and (d)
migration. (After Cossee,[133a]
used by permission of
Academic Press.)

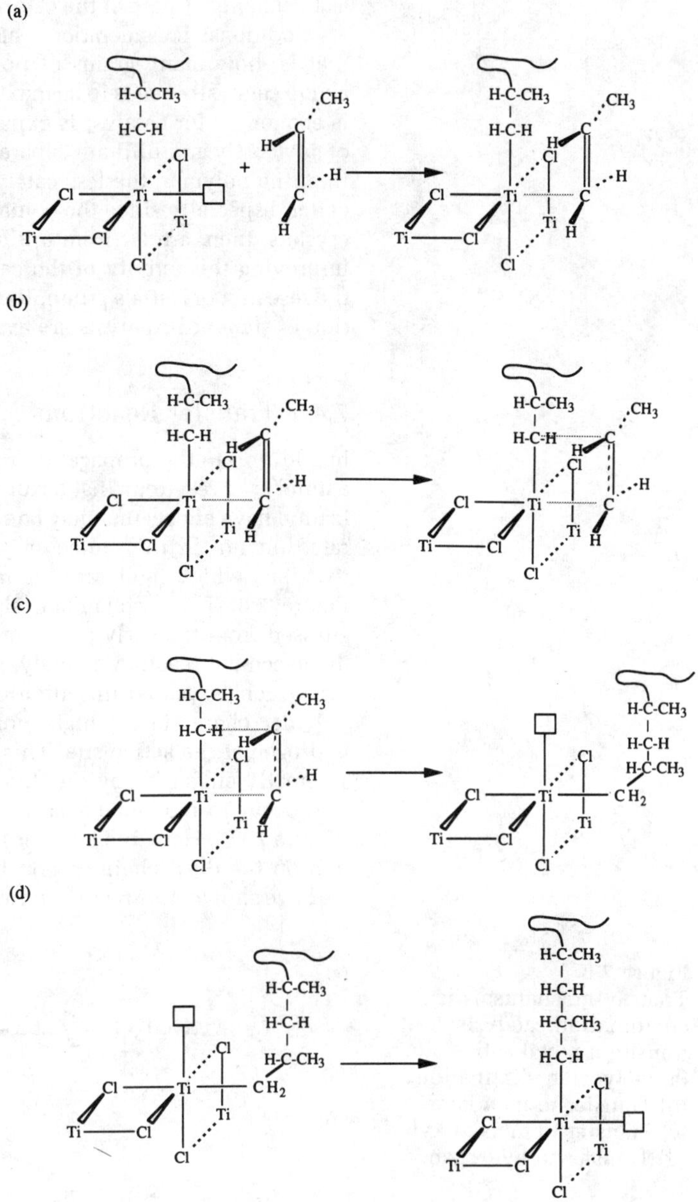

That polymerization, even with stereospecificity, can occur in the absence of the metal alkyl is considered to be one of the best pieces of evidence in favor of the thesis that reaction occurs at the transition metal–carbon bond.[116] There exists the class of related catalysts (Phillips-type catalysts) based on transition metal oxides, in particular those of chromium and, to a lesser extent, molybdenum (CrO_3 and MoO_3). These catalysts, however, are not stereospecific; thus, they are used only for the polymerization of ethylene, for which the elimination of the large share of short-chain branching still occurs.[23]

It is clear from the mention of the different stereospecificities of the α and β lattices of $TiCl_3$, that the physical state as well as the chemical state of the catalyst is of great importance. There are additional considerations, also. The activity of a catalyst—that is, how many grams of polymer it produces per gram of active catalyst—needs to be maximized for two reasons. The first is economic, for catalyst is expensive. The second is that short of any costly and difficult separation step to remove the catalyst from the polymer, the less catalyst impurity in the polymer the better, especially since the contaminant is not inert. Large TiCl₃ crystals, then, are far from optimal, and milling techniques for improving the activity of the catalyst were developed early in the research of Natta's group. Reviews that describe the preparation of standard catalysts are available.[134–136]

7.4.2 Transfer Reactions

In addition to the propagation reaction described in Figure 7.7, a number of reactions that occur during a coordination polymerization have effects that may be relatively invisible in the overall rate, but not in the molecular weight. These various transfer reactions which moderate the molecular weight are shown in Figure 7.8. The mechanism, shown in Figure 7.8a, is easily guessed from the early experiences in Ziegler's group with the displacement reaction, namely, intramolecular β-hydride transfer, which leaves an unsaturation on the end of the separated polymer chain (i.e., a high molecular weight α-olefin) and a hydrogen at the active site. This is the sort of reaction that was so greatly enhanced by the "nickel effect." Chain transfer can also occur to monomer just as in free-radical polymerizations (Figure 7.8b). Here, the transfer reaction may leave an unsaturation on the dead chain or not. For the polypropylene reaction we have shown, this results in two different kinds of unsaturated

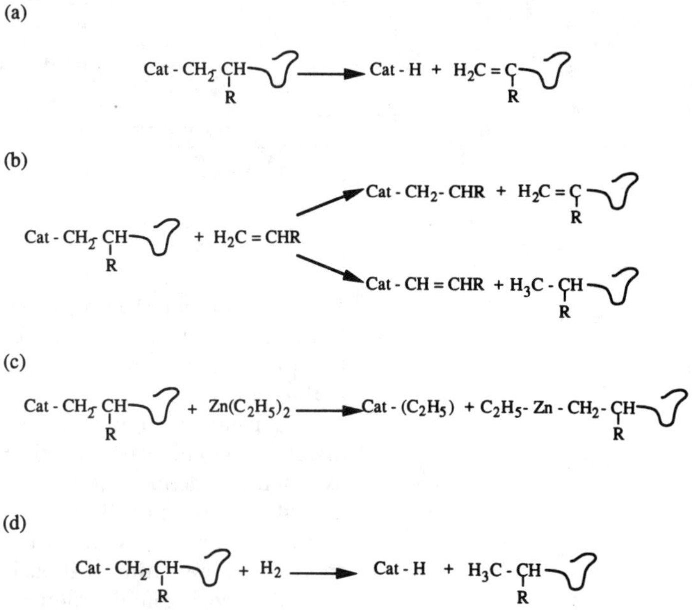

Figure 7.8.
Transfer mechanisms in coordination catalysis: Cat, transition metal active site.
(a) β-Hydride elimination.
(b) Transfer to monomer.
(c) Transfer to metal alkyl.
(d) Transfer to hydrogen.

end group, vinyl and vinylidene. In practice it appears that the situation is even more complicated, for vinylidene end groups are even found in polymerizations of ethylene, which the simple scheme of Figure 7.8b would not predict.[137]

Significantly, chain transfer can also occur to the metal alkyl, and under many circumstances these act as true transfer agents that do not poison the catalyst (Figure 7.8c). This effect was first noted early on for polymerizations of propylene with α-$TiCl_3$ and $Al(C_2H_5)_3$.[138] A more efficient chain transfer agent, though, is $Zn(C_2H_5)_2$.[139–141] Because the metal alkyl plays a significant role in the isospecificity of the polymerization, this reaction may establish a coupling between the chain length and the tacticity.

One of the problems encountered with Ziegler–Natta catalysts is the tendency to yield polymers of such high molecular weight that processing is difficult. From both Montecatini and Hercules in 1955, it was found that molecular hydrogen (H_2) acts as an effective chain transfer agent (Figure 7.8d). H_2 can have an effect on the rate and isospecificity of the reaction (decreasing the latter and either decreasing or increasing the former), but these effects seem to be reversible, in that they vanish when the H_2 is removed. For diene polymerizations by cobalt initiators, it appears that the α-olefins provide good transfer agents. One should also keep in mind that $Zn(C_2H_5)_2$ may be added as a chain transfer agent rather than as the cocatalyst. Overall, though, H_2 is the most popular chain transfer agent.

Many compounds act more like inhibitors than true transfer agents in that they remove active sites from the medium as well as decreasing the molecular weight. These poisons include alcohols, thiols (including H_2S), alkyl halides, $SnCl_4$, $SiCl_4$, CCl_4, and salts such as NaCl, NaBr, and BCl_3. The deactivating reaction is often with the metal alkyl rather than the transition metal itself.

7.4.3 Kinetics of Polymerization by Coordination Catalysts

The usual kinetic behavior of Ziegler–Natta polymerizations was reviewed early on[138] and shows among other things the importance of proper catalyst preparation (especially milling or grinding). The rate of polymerization at constant pressure (monomer concentration) is often characterized by an initial maximum at very short times, followed by a constant rate period.[5,136,138] The initial maximum may be absent when unground catalyst is used, and grinding assists in the attainment of a steady state more rapidly. In some cases the constant rate period does not exist but rather the rate monotonically decreases.

All this behavior suggests complexities in the catalyst activity, in the mass transfer of the monomer, or in both. The concentration of active sites is not necessarily a constant through the course of the polymerization, and moreover sites may not have the same activity. During the process of polymerization, the catalyst particles are fractured by polymer growing within, which exposes new catalytic sites. Likewise, grinding may create very active but short-lived sites. This suggests that active sites may disappear with time as a result of aging of the catalyst or poisoning, which could explain the lack of a constant rate period for some polymerizations.

Other indications of complexity are seen in the molecular weight distribution. As could be guessed from the preceding mechanistic description, the polymerization should give a geometric chain length distribution, so that the polydispersity Q should be 2. However, polydispersities between 4 and 30 are more common. The great regularity attained in stereochemistry apparently comes at a price of broad molecular weight distribution.[137,142] The broad molecular weight distributions have generally been attributed to multiple catalytic sites or diffusional control. Experimentally, it is difficult to determine which factor gives rise to the broad distributions, especially since both are coupled to other complications, such as catalyst fragmentation, heat transfer, and phase equilibria. Extensive mathematical modeling has been aimed at answering these questions.

Polymerization from catalytic sites on a supported catalyst leads to growth of the polymeric particles about the catalyst, with the result that the final product is obtained as a semicrystalline powder. Models for Ziegler–Natta polymerization differ chiefly in what is assumed to happen to the catalyst during polymerization or in what is assumed about the heterogeneity of the particle. In the simplest model, the "solid core model," the catalyst particle maintains its integrity at the center of the growing polymeric particle. This model cannot explain high polydispersities early in the polymerization, nor does it account for the occurrence in practice of fragmentation of the catalyst particle and dispersal throughout the polymer particle. The "polymeric flow model" assumes that after catalyst fragmentation the catalytic sites are molecularly dispersed and are conveyed with the polymer as the polymeric particle grows from the fragments.[143–147] The "multigrain model," based on the observation that separate fragments survive fragmentation, assumes something akin to a solid core model for each subparticle, thus granting heterogeneity to the particle as a whole.[148–155] The multigrain model appears to be favored in recent years despite the greater amount of computation required.

Evidence supporting both the hypothesis of diffusional limitations and the hypothesis of multiple sites can be found. Experimental support for the role of multiple sites, for example, is found in the early results that showed great variations in polydispersity between different kinds of catalysts,[156] as well as the more recent results with metallocene catalysts,[157] which have only one site and give fairly narrow molecular weight distributions.[158] Results from mathematical models over the years have been mixed; the importance of diffusion has been variously denied[159] and asserted.[143,144,148,160] However, over time it has become clearer that multiple catalytic sites definitely exist and furthermore must be invoked to obtain broad molecular weight distributions for reasonable values of the diffusional and reaction parameters.[147] Thus much recent modeling work assumes the presence of multiple sites.[161–163]

7.4.4 Processes for Coordination Polymerization

Just as free-radical polymerization can be carried out in different states (e.g., bulk, solution, suspension, emulsion), polymerization by coordination catalysis may be performed in different processes of types. There are three of these: slurry, solution, and gas phase.[134,135] The *slurry* polymerization corresponds to a precipitation polymerization, like those of vinyl chloride or acrylonitrile in bulk, wherein the polymer is not soluble in the solvent, whether it be monomer or a mixture of monomer and inert diluent. The slurry route is thus inherently heterogeneous, unlike the *solution* process, in which the polymer is soluble in the medium. *Gas-phase* polymerization is clearly heterogeneous, the monomer being gaseous and the polymer existing as a solid.

The *slurry* process is the most prevalent of the three. It is clearly a heterogeneous polymerization with respect to the polymer, but the catalyst itself may be soluble, colloidal, or heterogeneous. There is an inert diluent present, such as a low alkane (branched or linear) or cycloalkane, in which the monomer, but not the polymer, is soluble. Practically, this means that the operating temperature is kept below the melting temperature of the polymer, T_m, since the polymer often is soluble in the solvent above the crystalline melting point.

The *solution* process suggests a homogeneous system as far as the polymer is concerned, but again the catalyst may be soluble or not. The system may thus be heterogeneous with respect to the catalyst. The term "solution" here does not necessarily imply the presence of an inert diluent, as it does in free-radical polymerization, but rather that the polymer is soluble in the mixture. It thus includes bulk or mass polymerization of the monomer. Solution processes have fallen out of favor, largely because of the higher operating temperatures and the cost of solvent separation, which is higher in a system in which the polymer is soluble. Often, the solvent used for solution polymerizations will be the same as that for slurry polymerizations, only at a higher temperature ($> T_m$). Because the molecular weight of the polymer formed decreases with temperature, solution processes are still favored for the production of low molecular-weight polyethylene and also some particular linear low density polyethylenes.

The *gas-phase* process, developed in the late 1960s, presents both advantages and challenges. The obvious advantage is the absence of solvent, which eliminates separation altogether, since high activity catalysts are used which are thus contained in the polymer in such small amounts that their presence in the polymeric product is tolerated. Temperature control is generally more difficult in gas phase processes, and if local hot spots develop the polymer fluff may exceed T_m and begin to aggregate as in slurry processes.

These polymerizations can also be performed in batch or continuous modes. It should be kept in mind that for continuous processes in heterogeneous polymerizations (slurry or gas-phase), the existence of the polymer in a separate phase allows preferential removal.

7.4.5 Industrial Applications

The main applications of Ziegler–Natta catalysis are in the polymerization of ethylene, α-olefins, and dienes. Of these, by far the two most important operations are the polymerization of ethylene and propylene. Even though lower in volume, polypropylene is in one sense the more important, since it did not exist before the advent of Ziegler–Natta catalysts.

7.4.5.1 Polyethylene[134]

As we have already noted, the main distinction of polyethylene produced by the Ziegler–Natta or Phillips catalysts is the great reduction in the number of branches, often by a factor of 10. The reduction in branches increases the degree of crystallinity from nearly 50% to 90%, making this material denser and so earning it the name high-density polyethylene (HDPE). Because the issue of stereospecificity of the reaction does not arise with ethylene, both Ziegler–Natta and Phillips-type catalysts can be used for HDPE. Slurry, gas-phase, and solution processes are all used for polyethylene, but the slurry and gas-phase processes are the most important. They have in turn replaced the early solution processes because of greater ease in separation of polymer from solvent and because of lower operating temperatures. The solution processes are still used, however, for the production of some low molecular-weight polyethylenes.

With the introduction of the gas-phase processes for ethylene in the late 1960s, the economics of the low and high pressure processes became competitive, and so it was reasonable that the low-density polyethylene (LDPE) produced by high-pressure free-radical processes would be replaced by a polymer made at low pressures. To obtain similar properties, similar structure is needed, and since the branching characteristic of the high pressure process is mostly eliminated by the catalytic route, it must be engineered in. This is done by intentionally copolymerizing ethylene with α-olefins such as 1-butene, 1-hexene, and 1-octene. These agents give ethyl, n-butyl, and n-hexyl branches, which are numerous in LDPE (see Section 5.1.1). Linear low-density polyethylene (LLDPE), the polymer produced, is less crystalline and has lower T_m. Hence, in slurry and gas-phase processes, one must be more worried about hot spots and consequent aggregation.

Polyethylene is used in blow-molded, injection-molded, and extruded items such as tubing. Familiar from everyday use are milk bottles made of HDPE, bags for foods such as bread and frozen foods, and stretch-wrap films.

7.4.5.2 Polypropylene[135]

Because the properties of polypropylene that are so attractive are directly linked with high crystallinity, practically all this material is made from Ziegler–Natta catalysts, except the small amount for use as additives. Two features of polypropylene that represent improvements over polyethylene are its low density (0.90–0.91 g/mL) and its high melting temperature (170°C). These properties make it the only olefin used in the fiber industry, which is dominated by heterochain polymers or acrylics, which gain their properties from the polarity of the monomers. The low moisture absorption of polypropylene, granted by the nonpolarity of the chains, is of great advantage as is the light weight granted by its density. Fibers account for approximately a quarter of the polypropylene made in the United States. Another common application is in injection-molded items.

A class of elastomers made by coordination catalysis are those formed by a copolymerization of ethylene and propylene. Ethylene–propylene rubber (EPM) consists of a copolymer of the two, while ethylene–propylene–diene rubber (EPDM) is a terpolymer of the two with a small amount of a nonconjugated diene such as dicyclopentadiene, 1,4-hexadiene, 5-ethylidene–2-norbornene, or 5-methylene-2-norbornene.

7.4.5.3 Polymers of Other α-Olefins

The other main α-olefins that have been polymerized commercially are 1-butene and 4-methyl-1-pentene. The former has a melting temperature between that of polyethylene and polypropylene and is used primarily in piping, either for hot water or for abrasive materials. The latter has a much higher melting temperature (250°C) and so has found use in applications that demand performance at moderately high temperatures, such as laboratory and medical wares, ovens, and electronics.

7.4.5.4 Polymers of Dienes

Of the conjugated 1,3-dienes, both butadiene and isoprene are polymerized by coordination catalysis, as well as the other routes described earlier in this chapter. The polymerization of conjugated dienes by Ziegler–Natta catalysis is attractive because of the control it grants over *cis–trans* isomerism in 1,4-polymers, and thus Ziegler–Natta-type catalysts are used. Both highly *cis*- and highly *trans*-butadiene rubbers are produced by solution coordination processes. As we have mentioned, natural rubber (1,4-*cis*-polyisoprene) was synthetically produced, as was the *trans*- version of the polymer (which corresponds to balata or gutta-percha rubber). Styrene, because it is not terribly polar, can also be polymerized by Ziegler–Natta catalysts; the only industrial use of this property is in the production of the stereospecific analog of SBR rubber.

Problems

7.1. What sort of heterogeneities will arise in the UV curing of a surface coating by a photoinitiator such as benzophenone? Benzophenone exhibits absorption maxima at 250 and 350 nm, with extinction coefficients of 15,000 and 100, respectively. Why would this constitute an attractive feature for this application?

7.2. In the polymerization of vinyl acetate, chain transfer to polymer is quite extensive. Qualitatively compare the amount of branching you expect to find in a bulk (or suspension) polymerization of vinyl acetate to that in the emulsion polymerization system. If there are differences, what produces them?

7.3. What are the advantages of a continuous emulsion polymerization process? What would likely reasons be for the failure of a continuous emulsion polymerization to become a viable industrial application?

7.4. Consider an emulsion polymerization obeying case 2 kinetics. Show that in the disproportionation reaction between the resident growing chain and newly entered radical, the product will have a polydispersity near 4.

7.5. We have a number of different systems composed of a water-insoluble monomer, water, an initiator, and another component (a surfactant or a stabilizer). A number of different tests have been performed on these systems in an attempt to determine whether the polymerization occurs by emulsion or suspension. Based on the information given, decide if possible whether the polymerization system is suspension or emulsion, and give a brief reason for your opinion.
(a) The initiator was found to be soluble in the organic phase but not in the aqueous monomer phase.
(b) The amount of surfactant or stabilizer is doubled, and the rate of polymerization is found to increase by approximately 50%.
(c) The rate of polymerization was found to vary with the initiator concentration as a power law, with an exponent 0.45 ± 0.10 (there was much scatter in the data).
(d) Analysis of the monomer droplets showed a considerable content of polymer, and comparison by means of thermal analysis indicated that practically all polymerization had occurred within these droplets.
(e) Increasing the amount of surfactant or stabilizer is found to increase the molecular weight of the polymer formed.

7.6. In an emulsion polymerization, all the ingredients are charged at time $t = 0$. The times to convert various amounts of monomer to polymer are tabulated as follows.

Time (hr)	Fraction of Original Charge of Monomer Converted to Polymer
1.0	0.01
2.2	0.05
3.1	0.10
4.0	0.157
4.9	0.215

The polymer is not water soluble, nor does precipitation occur. Predict the time to reach 30% conversion.

References

1. A. Guyot, Precipitation polymerization, in *Comprehensive Polymer Science,* Vol. 4, G. C. Eastmond, A. Ledwith, S. Russo, and P. Sigwalt, Eds., p. 261. Pergamon Press, Oxford (1989).

2. K. W. Doak, Low density polyethylene (high pressure), in Ethylene Polymers, in *Encyclopedia of Polymer Science and Technology,* Vol. 6, H. F. Mark, N. M. Bikales, C. G. Overberger, and G. Menges, Eds., p. 386. Wiley, New York (1986).

3. C. H. Bamford, A. D. Jenkins, D. J. E. Ingram, and M. C. R. Symons, *Nature,* **175**, 894 (1955).

4. D. J. Walbridge, Polymerization in non-aqueous dispersions, in *Comprehensive Polymer Science,* Vol. 4, G. C. Eastmond, A. Ledwith, S. Russo, and P. Sigwalt, Eds., p. 243. Pergamon Press, Oxford (1989).

5. A. E. Hamielec and H. Tobita, Polymerization processes, in *Ullmann's Encyclopedia of Industrial Chemistry,* Vol. A21, p. 305. VCH Publishers, New York (1992).

6. E. Trommsdorff and C. E. Schildknecht, Polymerizations in suspension, in *Polymerization Processes,* C. E. Schildknecht, Ed., p. 69, Wiley-Interscience, New York (1956).

7. M. Munzer and E. Trommsdorff, Polymerizations in suspension, in *Polymerization Processes,* C. E. Schildknecht and I. Skeist, Eds., p. 106. Wiley-Interscience, New York (1977).

8. B. W. Brooks, *Makromol. Chem., Macromol. Symp.,* **35/36**, 121 (1990).

9. H. G. Yuan, G. Kalfas, and W. H. Ray, *J. Macromol. Sci., Rev. Macromol. Chem. Phys.,* **C31**, 215 (1991).

10. W. P. Hohenstein and H. Mark, *J. Polym. Sci.,* **1**, 127 (1946).

11. A. Crosato-Arnaldi, P. Gasparini, and G. Talamini, *Makromol. Chem.,* **117**, 140 (1968).

12. Z. Izumi and H. Kitagawa, *J. Polym. Sci., A-1,* **5**, 1967 (1967).

13. A. V. Ryabov, L. A. Smirnova, G. D. Panova, and L. V. Tsareva, *Tr. Khim. Khim. Tekhnol.,* **2**, 221 (1967).

14. R. Arshady, G. W. Kenner, and A. Ledwith, *J. Polym. Sci., Polym. Chem.,* **12**, 2017 (1974).

15. M. Langsam and J. T. Cheng, *J. Appl. Polym. Sci.,* **30**, 1285 (1985).

16. M. E. Adams, B. S. Casey, M. F. Mills, G. T. Russell, D. H. Napper, and R. G. Gilbert, *Makromol. Chem., Macromol. Symp.,* **35/36**, 1 (1990).

17. G. S. Bhargava, H. U. Khan, and K. K. Bhattacharyya, *J. Appl. Polym. Sci.,* **23**, 1181 (1979).

18. P. Rempp and E. W. Merrill, *Polymer Synthesis.* Hüthig & Wepf, Basel (1986).

19. L. F. Albright, *Processes for Major Addition-Type Plastics and Their Monomers 2nd ed.* Krieger, Malabar, FL (1985).

20. F. Rodriguez, *Principles of Polymer Systems,* 2nd ed. McGraw-Hill, New York (1982).

21. B. Törnell, *Polym.-Plast. Technol. Eng,* **27**, 1 (1988).

22. (a) T. Y. Xie, A. E. Hamielec, P. E. Wood, and D. R. Woods, *Polymer,* **32**, 537 (1991). (b) T. Y. Xie, A. E. Hemielec, P. E. Wood, and D. R. Woods, *Polymer,* **32**, 1098 (1991). (c) T. Y. Xie, A. E. Hamielec, P. E. Wood, D. R. Woods, and O. Chiantore, *Polymer,* **32**, 1696 (1991). (d) T. Y. Xie, A. E. Hamielec, P. E. Wood, and D. R. Woods, *Polymer,* **32**, 2087 (1991). (e) T. Y. Xie, A. E. Hamielec, P. E. Wood, and D. R. Woods, *J. Appl. Polym. Sci.,* **43**, 1259 (1991).

23. G. Odian, *Principles of Polymerization,* 3rd ed. Wiley, New York (1991).

24. I. Williams, Polymerization in the rubber industry, in *Polymerization and Its Applications in the Fields of Rubber, Synthetic Resins, and Petroleum,* R. E. Burk, H. E. Thompson, A. J. Weith, and I. Williams, Eds. Reinhold, New York (1937).

25. D. C. Blackley, Natural and synthetic rubber latices, in *Polymer Colloids,* R. Buscall, T. Corner, and J. F. Stageman, Eds., p. 247. Elsevier Applied Science, London (1985).

26. H. Morawetz, *Polymers: The Origins and Growth of a Science.* Wiley, New York (1985).

27. P. Fournier and T.-C. Cuong, *Rubber Chem. Technol.,* **34**, 1229 (1961).

28. W. D. Harkins, *J. Am. Chem. Soc.,* **69**, 1428 (1947).

29. W. V. Smith and R. H. Ewart, *J. Chem. Phys.,* **16**, 592 (1948).

30. P. J. Flory, *Principles of Polymer Chemistry.* Cornell University Press, Ithaca, NY (1953).

31. H. L. Williams, Polymerizations in emulsion, in *Polymerization Processes,* C. E. Schildknecht, Ed., p. 111. Wiley-Interscience, New York (1956).

32. K. W. Min and W. H. Ray, *J. Macromol. Sci., Rev. Macromol. Chem.,* **C11**, 177 (1974).

33. J. Ugelstad and F. K. Hansen, *Rubber Chem. Technol.,* **49**, 536 (1976).

34. J. L. Gardon, Emulsion polymerization, in *Polymerization Processes,* C. E. Schildknecht and I. Skeist, Eds., p. 143. Wiley-Interscience, New York (1977).

35. J. W. Vanderhoff, *J. Polym. Sci., Polym. Symp.,* **72**, 161 (1985).

36. G. W. Poehlein, Emulsion polymerization, in *Encyclopedia of Polymer Science and Technology,* Vol. 6, H. F. Mark, N. M. Bikales, C. G. Overberger, and G. Menges, Eds., p. 1. Wiley-Interscience, New York (1986).

37. B. S. Casey, B. R. Morrison, and R. G. Gilbert, *Prog. Polym. Sci.,* **18**, 1041 (1993).

38. R. Buscall and R. H. Ottewill, The stability of polymer latices, in *Polymer Colloids,* R. Buscall, T. Corner, and J. F. Stageman, Eds., p. 141. Elsevier Applied Science, London (1985).

39. R. H. Ottewill, The stability and instability of polymer latices, in *Emulsion Polymerization,* I. Piirma, Ed., p. 1. Academic Press, New York (1982).

40. D. H. Napper, *Polymeric Stabilization of Colloidal Dispersions.* Academic Press, London (1983).

41. D. H. Napper, *J. Colloid Interface Sci.*, **32**, 106 (1970).

42. R. Evans, J. B. Davison, and D. H. Napper, *J. Polym. Sci., Polym. Lett.*, **10**, 449 (1972).

43. J. V. Dawkins and G. Taylor, *Colloid Polym. Sci.*, **258**, 79 (1980).

44. See, for example, (a) P.-G. de Gennes, *Macromolecules*, **13**, 1069 (1980). (b) A. Halperin, M. Tirrell, and T. P. Lodge, *Adv. Polym. Sci.*, **100**, 31 (1992).

45. F. K. Hansen and J. Ugelstad, Particle formation mechanisms, in *Emulsion Polymerization*, I. Piirma, Ed., p. 51. Academic Press, New York (1982).

46. W. J. Priest, *J. Phys. Chem.*, **56**, 1077 (1952).

47. I. A. Maxwell, B. R. Morrison, D. H. Napper, and R. G. Gilbert, *Macromolecules*, **24**, 1629 (1991),

48. R. Patsiga, M. Litt, and V. Stannett, *J. Phys. Chem.*, **64**, 801 (1960).

49. C. P. Roe, *Ind. Eng. Chem.*, **60(9)**, 20 (1968).

50. R. M. Fitch and C. H. Tsai, in *Polymer Colloids*, R. M. Fitch, Ed., p. 73. Plenum Press, New York (1971).

51. J. L. Gardon, *J. Polym. Sci. A-1*, **6**, 623 (1968).

52. B. R. Vijayendran, *J. Appl. Polym. Sci.*, **23**, 733 (1979).

53. R. J. Orr and L. Breitman, *Can. J. Chem.*, **38**, 668 (1960).

54. T. R. Paxton, *J. Colloid Interface Sci.*, **31**, 19 (1969).

55. P. Connor and R. H. Ottewill, *J. Colloid Interface Sci.*, **37**, 642 (1971).

56. I. Piirma and S.-R. Chen, *J. Colloid Interface Sci.*, **74**, 90 (1980).

57. N. Sütterlin, Influence of monomer polarity on particle formation in emulsion polymerization, in *Polymer Colloids II.*, R. M. Fitch, Ed., p. 583. Plenum Press, New York (1980).

58. J. L. Gardon, *J. Polym. Sci. A-1*, **6**, 643 (1968).

59. D. H. Napper and R. G. Gilbert, *Makromol. Chem., Macromol. Symp.*, **10/11**, 503 (1987).

60. Z. Song and G. W. Poehlein, *J. Macromol. Sci.-Chem.*, **A25**, 403, 1587 (1988).

61. M. Morton, S. Kaizerman, and M. W. Altier, *J. Colloid Sci.*, **9**, 300 (1954).

62. J. L. Gardon, *J. Polym. Sci. A-1*, **6**, 2859 (1968).

63. M. Nomura and K. Fujita, *Polym. Int.*, **30**, 483 (1993).

64. H. Tobita, K. Kimura, K. Fujita, and M. Nomura, *Polymer*, **34**, 2569 (1993).

65. M. R. Grancio and D. J. Williams, *J. Polym. Sci. A-1*, **8**, 2617 (1970).

66. P. Keusch and D. J. Williams, *J. Polym. Sci., Polym. Chem.*, **11**, 143 (1973).

67. W. H. Stockmayer, *J. Polym. Sci.*, **24**, 314 (1957).

68. J. T. O'Toole, *J. Appl. Polym. Sci.*, **9**, 1291 (1965).

69. M. Abramowitz and I. A. Stegun, Eds., *Handbook of Mathematical Functions.* Dover, New York (1972).

70. (a) B. S. Casey, B. R. Morrison, I. A. Maxwell, R. G. Gilbert, and D. H. Napper, *J. Polym. Sci., Polym. Chem.,* **32**, 605 (1994). (b) B. R. Morrison, B. S. Casey, I. Lacik, G. L. Leslie, D. F. Sangster, R. G. Gilbert, and D. H. Napper, *J. Polym. Sci., Polym. Chem.,* **32**, 631 (1994).

71. I. Lacík, B. S. Casey, D. F. Sangster, R. G. Gilbert, and D. H. Napper, *Macromolecules,* **25**, 4065 (1992).

72. J. Ugelstad, P. C. Mørk, and J. O. Aasen, *J. Polym. Sci. A-1,* **5**, 2281 (1967).

73. J. L. Gardon, *J. Polym. Sci. A-1,* **6**, 665 (1968).

74. B. S. Hawkett, D. H. Napper, and R. G. Gilbert, *J. Chem. Soc., Faraday Trans. I,* **73**, 690 (1977).

75. S. Katz, R. Shinnar, and G. M. Saidel, *Adv. Chem. Ser.,* **91**, 145 (1969).

76. D. C. Sundberg and J. D. Eliassen, The prediction of particle size and molecular weight distributions in emulsion polymerization, in *Polymer Colloids,* R. M. Fitch, Ed., p. 153. Plenum Press, New York (1971).

77. (a) G. Lichti, R. G. Gilbert, and D. H. Napper, *J. Polym. Sci., Polym. Chem.,* **18**, 1297 (1980). (b) G. Lichti, R. G. Gilbert, and D. H. Napper, Theoretical predictions of the particle size and molecular weight distributions in emulsion polymerizations, in *Emulsion Polymerization,* I. Piirma, Ed., p. 93. Academic Press, New York (1982).

78. E. Giannetti, G. Storti, and M. Morbidelli, *J. Polym. Sci., Polym. Chem.,* **26**, 1835 (1988).

79. (a) G. Storti, G. Polotti, M. Cociani, and M. Morbidelli, *J. Polym. Sci., Polym. Chem.,* **30**, 731 (1992). (b) G. Storti, G. Polotti, P. Canu, and M. Morbidelli, *J. Polym. Sci., Polym. Chem.,* **30**, 751 (1992).

80. H. Tobita, *Macromolecules,* **25**, 2671 (1992).

81. H. Tobita and A. E. Hamielec, *Polym. Int.,* **30**, 177, 195 (1993).

82. (a) H. Tobita, *Polym. React. Eng.,* **1**, 357, 379 (1992/1993). (b) H. Tobita, *J. Polym. Sci., Polym. Phys.,* **31**, 1363 (1993).

83. See, for example, G. T. Russell, R. G. Gilbert, and D. H. Napper, *Macromolecules,* **26**, 3538 (1993).

84. G. W. Poehlein and D. J. Dougherty, *Rubber Chem. Technol.,* **50**, 601 (1977).

85. G. Poehlein, *ACS Symp. Ser.,* **104**, 1 (1979).

86. G. W. Poehlein, Emulsion polymerization in continuous reactors, in *Emulsion Polymerization,* I. Piirma, Ed., p. 357. Academic Press, New York (1982).

87. A. E. Hamielec and J. F. MacGregor, Latex reactor principles: Design, operation, and control, in *Emulsion Polymerization,* I. Piirma, Ed., p. 319. Academic Press, New York (1982).

88. A. Penlidis, J. F. MacGregor, and A. E. Hamielec, *AIChE J.,* **31**, 881 (1985).

89. G. W. Poehlein and J. Schork, *Trends Polym. Sci.,* **1**, 298 (1993).

90. H. Gerrens, *Kolloid-Z. Z. Polym.,* **227**, 92 (1968).

91. R. A. Wessling and D. S. Gibbs, *J. Macromol. Sci.-Chem.*, **A7**, 647 (1973).

92. R. A. Wessling, *J. Appl. Polym. Sci.*, **12**, 309 (1968).

93. J. Dimitratos, M. S. El-Aasser, C. Georgakis, and A. Klein, *J. Appl. Polym. Sci.*, **40**, 1005 (1990).

94. (a) B. Li and B. W. Brooks, *Polym. Int.*, **29**, 41 (1992). (b) B. Li and B. W. Brooks, *J. Appl. Polym. Sci.*, **48**, 1811 (1993).

95. D. A. Paquet, Jr., and W. H. Ray, *AIChE J.*, **40**, 73, 88 (1994).

96. G. Ley and H. Gerrens, *Makromol. Chem.*, **175**, 563 (1974).

97. A. R. Berens, *J. Appl. Polym. Sci.*, **18**, 2379 (1974).

98. G. W. Poehlein, *Polym. Int.*, **30**, 243 (1993).

99. M. E. Adams, D. J. Buckley, R. E. Colborn, W. P. England, and D. N. Schissel, *RAPRA Rev. Rep.*, **6(10)**, report no. 70 (1993).

100. D. I. Lee and T. Ishikawa, *J. Polym. Sci., Polym. Chem.*, **21**, 147 (1983).

101. (a) M. Okubo, Y. Katsuka, and T. Matsumoto, *J. Polym. Sci., Polym. Lett.*, **18**, 481 (1980). (b) M. Okubo, A. Yamada, and T. Matsumoto, *J. Polym. Sci., Polym. Chem.*, **18**, 3219 (1980).

102. D. R. Stutman, A. Klein, M. S. El-Aasser, and J. W. Vanderhoff, *Ind. Eng. Chem. Prod. Res. Dev.*, **24**, 404 (1985).

103. F. Candau, *Makromol. Chem., Macromol. Symp.*, **31**, 27 (1990).

104. B. W. Brooks, *J. Polym. Sci., Polym. Chem.*, **29**, 1661 (1991).

105. J. Ugelstad, M. S. El-Aasser, and J. W. Vanderhoff, *J. Polym. Sci., Polym. Lett.*, **11**, 503 (1973).

106. J. Ugelstad, F. K. Hansen, and S. Lange, *Makromol. Chem.*, **175**, 507 (1974).

107. J. Ugelstad, *Makromol. Chem.*, **179**, 815 (1978).

108. J. Ugelstad and P. C. Mørk, *Adv. Colloid Interface Sci.*, **13**, 101 (1980).

109. D. P. Durbin, M. S. El-Aasser, G. W. Poehlein, and J. W. Vanderhoff, *J. Appl. Polym. Sci.*, **24**, 703 (1979).

110. See, for example, J. H. Kim, M. Chainey, M. S. El-Aasser, and J. W. Vanderhoff, *J. Polym. Sci., Polym. Chem.*, **30**, 171 (1992).

111. See, for example, M. B. Urquiola, V. L. Dimonie, E. D. Sudol, and M. S. El-Aasser, *J. Polym. Sci., Polym. Chem.*, **30**, 2619, 2631 (1992).

112. See, for example, J. M. H. Kusters, D. H. Napper, R. G. Gilbert, and A. L. German, *Macromolecules*, **25**, 7043 (1992).

113. J. Hearn, M. C. Wilkinson, and A. R. Goodall, *Adv. Colloid Interface Sci.*, **14**, 173 (1981).

114. J. Hearn, M. C. Wilkinson, A. R. Goodall, and M. Chainey, *J. Polym. Sci., Polym. Chem.*, **23**, 1869 (1985).

115. D. O. Jordan, Ziegler–Natta polymerization: Catalysts, monomers, and polymerization procedures, in *The Stereochemistry of Macromolecules*, Vol. 1, A. D. Ketley, Ed., p. 1. Dekker, New York (1967).

116. J. Boor, Jr., *Ziegler–Natta Catalysts and Polymerizations*. Academic Press, New York (1979).

117. F. M. McMillan, *The Chain Straighteners*. Macmillan, London (1979).

118. K. Ziegler and H.-G. Gellert, *Justus Liebigs Ann. Chem.*, **567**, 179, 185, 195 (1950).

119. K. Ziegler, H.-G. Gellert, H. Kühlhorn, H. Martin, K. Meyer, K. Nagel, H. Sauer, and K. Zosel, *Angew. Chem.*, **64**, 323 (1952).

120. E. Holzkamp, *Gelenkte Polymerisation des Äthylen*, Ph.D. thesis, University of Aachen (1953).

121. H. Breil, *Über metallorganische Mischkatalysatoren*, Ph.D. thesis, University of Aachen (1955).

122. K. Ziegler, E. Holzkamp, H. Breil, and H. Martin, *Angew. Chem.*, **67**, 541 (1955).

123. G. Natta, P. Pino, P. Corradini, F. Danusso, E. Mantica, G. Mazzanti, and G. Moraglio, *J. Am. Chem. Soc.*, **77**, 1708 (1955).

124. G. Natta, *Makromol. Chem.*, **16**, 213 (1955).

125. S. E. Horne, Jr., J. P. Kiehl, J. J. Shipman, V. L. Folt, C. F. Gibbs, E. A. Willson, E. B. Newton, and M. A. Reinhart, *Ind. Eng. Chem.*, **48**, 784 (1956).

126. (a) K. Ziegler, *Angew. Chem.*, **76**, 545 (1964). (b) G. Natta, *Angew. Chem.*, **76**, 553 (1964).

127. G. Natta and F. Danusso, Eds., *Stereoregular Polymers and Stereospecific Polymerizations*. Pergamon Press, Oxford (1967).

128. C. E. Schildknecht, S. T. Gross, H. R. Davidson, J. M. Lambert, and A. O. Zoss, *Ind. Eng. Chem.*, **40**, 2104 (1948).

129. G. Natta, I. Bassi, and P. Corradini, *Makromol. Chem.*, **18–19**, 455 (1956).

130. H. Yuki, K. Hatada, K. Ohta, and Y. Okamoto, *J. Macromol. Sci.-Chem.*, **A9**, 983 (1975).

131. D. S. Breslow, *Prog. Polym. Sci.*, **18**, 1141 (1993).

132. P. Corradini, V. Busico, and G. Guerra, Monoalkene polymerization: Stereochemistry, in *Comprehensive Polymer Science*, Vol. 4, G. C. Eastmond, A. Ledwith, S. Russo, and P. Sigwalt, Eds., p. 29. Pergamon Press, Oxford (1989).

133. (a) P. Cossee, *J. Catal.*, **3**, 80 (1964). (b) P. Cossee, The mechanism of Ziegler–Natta polymerization. II. Quantum-chemical and crystal-chemical aspects, in *The Stereochemistry of Macromolecules*, Vol. 1, A. D. Ketley, Ed., p. 145. Dekker, New York (1967).

134. K.-Y. Choi and W. H. Ray, *J. Macromol. Sci.-Rev. Macromol. Chem. Phys.*, **C25**, 1 (1985).

135. K.-Y. Choi and W. H. Ray, *J. Macromol. Sci.-Rev. Macromol. Chem. Phys.*, **C25**, 57 (1985).

136. J. J. A. Dusseault and C. C. Hsu, *J. Macromol. Sci.-Rev. Macromol. Chem. Phys.*, **C33**, 103 (1993).

137. D. L. Beach and Y. V. Kissin, High density polyethylene, in *Encyclopedia of Polymer Science and Engineering*, Vol. 6, 2nd ed., H. F.

Mark, N. M. Bikales, C. G. Overberger, and G. Menges, Eds., p. 454. Wiley-Interscience, New York (1986).

138. G. Natta and I. Pasquon, *Adv. Catal.*, **11**, 1 (1959).

139. G. Natta, E. Giachetti, I. Pasquon, and G. Pajaro, *Chim. Ind. (Milan)*, **42(10)**, 1091 (1960).

140. G. Natta, I. Pasquon, and L. Giuffre, *Chim. Ind. (Milan)*, **43(8)**, 871 (1961).

141. J. Boor, Jr., *J. Polym. Sci. C*, **1**, 237 (1963).

142. U. Zucchini and G. Cecchin, *Adv. Polym. Sci.*, **51**, 101 (1983).

143. D. Singh and R. P. Merrill, *Macromolecules*, **4**, 599 (1971),

144. W. R. Schmeal and J. R. Street, *AIChE J.*, **17**, 1188 (1971).

145. W. R. Schmeal and J. R. Street, *J. Polym. Sci., Polym. Phys.*, **10**, 2173 (1972).

146. R. Galván and M. Tirrell, *Comp. Chem. Eng.*, **10**, 77 (1986).

147. R. Galván and M. Tirrell, *Chem. Eng. Sci.*, **41**, 2385 (1986).

148. J. R. Crabtree, F. N. Grimsby, A. J. Nummelin, and J. M. Sketchley, *J. Appl. Polym. Sci.*, **17**, 959 (1973).

149. E. J. Nagel, V. A. Kirillov, and W. H. Ray, *Ind. Eng. Chem. Prod. Res. Dev.*, **19**, 372 (1980).

150. S. Floyd, K. Y. Choi, T. W. Taylor, and W. H. Ray, *J. Appl. Polym. Sci.*, **31**, 2231 (1986).

151. S. Floyd, K. Y. Choi, T. W. Taylor, and W. H. Ray, *J. Appl. Polym. Sci.*, **32**, 2935 (1986).

152. S. Floyd, T. Heiskanen, T. W. Taylor, G. E. Mann, and W. H. Ray, *J. Appl. Polym. Sci.*, **33**, 1021 (1987).

153. R. A. Hutchinson, C. M. Chen, and W. H. Ray, *J. Appl. Polym. Sci.*, **44**, 1389 (1992).

154. P. Sarkar and S. K. Gupta, *Polymer*, **33**, 1477 (1992).

155. M. Sau and S. K. Gupta, *Polymer*, **34**, 4417 (1993).

156. H. Wesslau, *Makromol. Chem.*, **26**, 102 (1958).

157. H. Sinn, W. Kaminsky, H.-J. Vollmer, and R. Woldt, *Angew. Chem., Int. Ed. Engl.*, **19**, 390 (1980).

158. E. L. Hoel, C. Cozewith, and G. D. Byrne, *AIChE J.*, **40**, 1669 (1994).

159. J. W. Begley, *J. Polym. Sci., A-1*, **4**, 319 (1966).

160. V. W. Buls and T. L. Higgins, *J. Polym. Sci., A-1*, **8**, 1025, 1037 (1970).

161. K. B. McAuley, J. F. MacGregor, and A. E. Hemielec, *AIChE J.*, **36**, 837 (1990).

162. A. B. de Carvalho, P. E. Gloor, and A. E. Hamielec, *Polymer*, **30**, 280 (1989).

163. A. B. M. de Carvalho, P. E. Gloor, and A. E. Hamielec, *Polymer*, **31**, 1294 (1990).

INDEX

Degree of polymerization
(*cont.*)

free-radical crosslinking
copolymerization, 211,
222–224, 240
in HCSTR, 267–68
long-chain branching,
241–250
gelation, 247–250
probability of a finite
structure, 225–227
Normal stress, 22–24
Nuclear magnetic resonance
(NMR) spectroscopy,
14, 15, 177, 196, 198,
199, 205
Nucleation. *See* Particle
nucleation
Nylon. *See* Polyamide.
Nylon 6, 107, 119, 209
Nylon 6/6, 56, 62, 209, 292
Nylon 6/10, 291, 57
Nylon 11, 40, 58

Olefins, α-, 2, 16, 106,
118–119, 338, 349
Oligomer, 5
Osmometry, 26–27, 198
Oxidative coupling, 60

Parameter estimation, 75–76,
154–55, 195–198
Particle identity point, 310
Particle morphology, 323–24,
334–35
Particle nucleation, 317–322,
333–34
Particle number, 318–321,
333–34
Particle size, 33
in emulsion polymerization,
330–31, 333–34
in suspension
polymerization, 309–
10
Pendant material, 227–28
Penultimate model, 167–68,
180–185
Percolation, 239–40
Phase separation, 6, 15, 32,
163–165, 240, 305–06,
311
Phenolic resins, 2, 211–12
Photoinitiation, 126–27
Plasticizer, 11, 14, 34, 336
Plastisol, 336
Plug flow reactor (PFR),
280–282
definition, 280
with recycle, 280–282
anionic polymerization
(ideal) in, 280–282

Poisson distribution
in chain length, 110,
113–115, 124, 270, 274
as limit of binomial
distribution, 115
Polyacetal. *See*
Polyoxymethylene.
Polyacrylamide, 127, 312, 335
Polyacrylonitrile, 2, 106, 305
Polyamide, 2, 56–58, 75, 83,
119, 209
equilibrium constant, 83
Polybutadiene, 3, 13, 106, 339
Polybutylene, 12
Polycarbonate, 59, 293–294
Polychloroprene, 3, 106, 332,
336
Polycyanate, 252
Polydimethylsiloxane, 11, 62,
107, 119
Polydispersity, 20
gelation, 216
for geometric distribution,
48
convoluted, 140
for Gold distribution, 122
as inadequate measure of
breadth of distribution,
268, 285
for Poisson distribution, 113
Polyester, 2, 58–60, 75–76, 83,
119, 210
equilibrium constant, 83
Polyester resins, unsaturated,
2, 59–60, 211
Polyether, 60, 107, 212
Poly(ethyl acrylate), 106, 128,
139, 157, 331, 336
Polyethylene, 2, 11, 106,
205–06, 209
high-density (HDPE), 2, 30,
339, 347, 348
linear low-density (LLDPE),
2, 347, 348
low-density (LDPE), 2, 30,
31, 286, 348
branching, 205–06
Poly(ethylene oxide), 107
Poly(ethylene terephthalate),
40, 58–59, 62, 76,
291–92, 305
Poly(hydroxyethyl
methacrylate), 106
Polyimide, 57–58
Polyisobutylene, 119
Polyisoprene, 18, 106, 313,
336, 339, 349. *See also*
Balata (gutta-percha)
rubber; Natural rubber;
Rubber